Xenobiotics in Fish

Xenobiotics in Fish

Edited by

David J. Smith

United States Department of Agriculture
Fargo, North Dakota

William H. Gingerich

United States Geological Survey
LaCrosse, Wisconsin

and

Maria G. Beconi-Barker

G. D. Searle, Co.
Skokie, Illinois

Kluwer Academic / Plenum Publishers
New York, Boston, Dordrecht, London, Moscow

Proceedings of an American Chemical Society symposium on Xenobiotic Metabolism in Fish, held April 1–2, 1998, in Dallas, Texas

ISBN 0-306-46189-7

233 Spring Street, New York, N.Y. 10013

10 9 8 7 6 5 4 3 2 1

A C.I.P. record for this book is available from the Library of Congress.

Printed in the United States of America

PREFACE

Fish are becoming an increasingly important and preferred protein source for humans. The potential for xenobiotic chemicals to be present in the tissues of cultured and wild fish has also increased dramatically in the last two decades due to the widespread incorporation of chemicals in the environment. To protect the consumer of fish flesh from exposure to harmful levels of xenobiotics, regulation of residues in fish is an important component of efforts by regulatory agencies throughout the world. Two types of xenobiotic agents can generally be identified. Chemicals posing perhaps the greatest concern to regulatory agencies are those that have the potential to be taken up by the fish as incidental contaminants, either through diet or waterborne sources. Such chemicals are generally classed as environmental contaminants from sources as diverse as industrial pollutants and agricultural operations. A second smaller and newer source of xenobiotic agents are those introduced intentionally to fish for management or culture purposes. This class of chemicals consists of mainly waterborne and oral medicinal drugs or anesthetic agents.

Regardless of the source of the chemicals, agencies charged with protection of either the environment or human food resources are faced with ensuring that the xenobiotics do not pose significant health or safety problems either for the fish exposed to them or for humans who consume fish as part of their diet. Because fish have been less intensively studied than other food producing animals, much initial information concerning the animal safety and residue chemistry guidelines that has been used to drive regulatory decisions have been based on studies made in mammalian models.

Regulatory agencies are now benefitting from recent research with fish models. These models have helped establish a more realistic understanding of fish as test systems. This knowledge has subsequently given regulators more insight into fish as research subjects which has allowed the agencies to set more realistic guidelines by which to address issues of contaminant and veterinary drug residues in fish.

In April of 1998 an international symposium on xenobiotic pharmacokinetics and metabolism by fish was held at the American Chemical Society annual meeting in Dallas, Texas. Papers presented at that symposium have been consolidated into this book. The intent of the symposium was to provide a forum to present current knowledge of the disposition of xenobiotics by fish. Moreover, the book presents in selected chapters how this knowledge is being used by those who regulate environmental chemicals and veterinary drugs and how the models are used to build a data base of scientifically defensible data on which regulatory agencies may make informed decisions concerning environmental and human food safety issues.

The book is organized into three broad segments. The first segment is devoted to issues of fish xenobiotic residue chemistry as viewed by regulatory agencies in the U.S. and Europe. An overview of the depth and breadth of use of fishery management and fish culture chemicals used in the U.S. precedes chapters that present perspectives of regulatory agencies in controlling the use of veterinary drugs in both the United States and Europe. A final chapter deals with requirements, experimental design and data analysis topics considered important for the registration of waterborne pesticides by the U.S. Environmental Protection Agency.

The second segment of the book is devoted to presentations that review topics for development of compartmental and physiologically based pharmacokinetic models for xenobiotics in fish. The information required from compartmental models generally drives the scope and complexity of the experiments from which the information is derived. Data derived from experiments with small fish generally are limited to whole body time/concentration profiles and usually do not include residue characterization. Experiments with intermediate and large size fish, while generally more costly to implement, allow for estimates of bioavailability, clearance constants and volumes of distribution for both the parent and metabolites. Parameters derived from such experiments allow for the calculation of residue half lives, bioconcentration, and bioaccumulation factors. These parameters are useful to regulatory agencies that can then reference the parameters to mammalian models. Physiologically based pharmacokinetic models, once developed, promise wide utility for regulatory agencies to conduct risk assessments. Such models, developed for mice and rats for organochlorine solvents have been used by regulatory agencies to assess risk and set residue tolerances to protect humans. Physiologically based pharmacokinetic models have been developed and validated for two fish species and have been used successfully to predict the uptake and disposition of waterborne organic chemicals in fish. The utility of the models to predict similar kinetics for oral chemicals is currently constrained by adequate models for dietary uptake and metabolism. Despite these current constraints, recent advances in an understanding of these processes in fish promise to make these powerful models more available as tools to assess risk from a broad range of xenobiotics.

The final segment of the book is devoted to case studies of how these models have been applied to address issues of fish residue chemistry studies with fish. Types of chemicals evaluated include waterborne medicinals including the anesthetic benzocaine and medicinal triarylmethane dyes including malachite green. Chapters are also devoted to waterborne pesticides TFM, Bayluscide, and Trifluralin. Two chapters provide examples of the use of compartmental models to describe the disposition of the oral drugs sulfamonodimethoxine and sarafloxacin in fish. Finally, a chapter is presented that describes a method to predict withdrawal times from oxytetracycline therapy in cultured chinook salmon using a physiologically based pharmacokinetic model.

The intent of this book is to provide investigators with a snapshot of current regulatory needs to control xenobiotic residues in fish tissues and how the needs are being addressed by current research efforts. It is significant that the presentation of the physiologically based pharmacokinetic models presented here may also represent a new tool that, when validated and accepted by the regulatory agencies, will become a tool of choice for risk assessments by regulatory agencies.

D. J. Smith
Agricultural Research Service, United States Department of Agriculture

W. H. Gingerich
Biological Resources Division, United States Geological Survey

M. G. Beconi-Barker
G. D. Searle, Co.

CONTENTS

1

USE OF CHEMICALS IN FISH MANAGEMENT AND FISH CULTURE

Past and Future

Rosalie A. Schnick

Michigan State University
3039 Edgewater Lane
La Crosse, Wisconsin 54603-1088

1. ABSTRACT

Chemicals have been used in fish management since the 1930's and in the culture of fish since the late 1800's. Intensive efforts were made from the 1950's to the late 1970's to find, develop, and register general and selective fish toxicants and aquatic herbicides, but only two selective toxicants, two general fish toxicants, two algicides, and seven aquatic herbicides were registered or are being reregistered. The expense of registration or reregistration is prohibitive for most potential products; only federal sponsorship has maintained the registration of fish toxicants. Aquatic herbicide registrations have been maintained by their registrants because there are other uses of aquatic herbicides beyond that of fishery management and culture.

The first efforts to gain aquaculture drug approvals were through a partnership of federal and state agencies during the mid-1960's. These endeavors resulted in the approvals of five therapeutants, one anesthetic, and registration of one pesticide as a therapeutant (i.e., trichlorfon) from 1970 to 1986. In 1990, regulatory scrutiny increased for drugs and chemicals used in aquaculture. As a result, partnerships between federal, state and private organizations have been formed with the goal of increasing the availability of drugs for fish culture. For example, the federal government and 37 state natural resource agencies formed a unique partnership in 1994 with the goal of gaining approval for seven therapeutants and one anesthetic. Collectively, the efforts of a variety of public and private partnerships could result in approximately 20 drugs gaining initial approvals, expansions, or extensions by the year 2002.

Xenobiotics in Fish, edited by *Smith et al.*
Kluwer Academic / Plenum Publishers, New York, 1999.

2. INTRODUCTION

Chemicals have been used in fish management in the United States since the 1930's and in the culture of fish since the late 1800's. The use of these chemicals is coincidental to, or made necessary by, human manipulation of natural waters, the effects of pollution on fish populations, a demand for recreational fisheries, the introduction of exotic species into surface waters, over harvesting, the decline of commercial fisheries, an increase in seafood demand, and the intensification of fish culture.

Regulation of aquatic chemicals has intensified during the last 35 years; drugs are approved through U.S. Food and Drug Administration's Center for Veterinary Medicine (CVM) and pesticides are registered by the U.S. Environmental Protection Agency (EPA). Increased scrutiny of chemicals by CVM and EPA has decreased the availability of fish management chemicals, but the potential for a major increase in the number of aquaculture drugs available exists through federal, state, and private partnerships. Indeed, all current fish toxicants and drugs have been registered or approved as a result of partnerships.

3. REGISTRATION AND USE OF FISH MANAGEMENT CHEMICALS: PAST AND FUTURE

3.1. Background

Reasons for the use of fishery management chemicals or fish toxicants have changed over time. Originally, fish toxicants were used mainly to eradicate "undesirable" fish so that "desirable fish" could be stocked and managed without competition, predation, or other interference by the "undesirable fish" (Lennon *et al.*, 1970). Today, fish toxicants also are used to: (1) restore threatened or endangered species; (2) eradicate vectors of disease; (3) treat drainages prior to reservoir impoundment; (4) aid in quantifying populations of aquatic organisms; and (5) eliminate competing species in hatcheries (Finlayson and Schnick, unpublished, 1996). The careful and professional use of fish toxicants was encouraged by the Dingell-Johnson Act in 1950, establishment of the Cooperative Fishery Unit program in 1960, the development of training and manuals in the 1960's and 1970's, and the increased availability of sophisticated delivery equipment (Lennon *et al.*, 1970).

Thousands of chemicals were screened by U.S. Fish and Wildlife Service (FWS) scientists during the 1950's at the Hammond Bay Biological Station (Millersburg, MI) and the Eastern Fish Disease Laboratory (Kearneysville, WV) in efforts directed at identifying selective toxicants to control the sea lamprey (*Petromyzon marinus*). In addition, general and selective toxicants targeted at undesirable fish were also sought (Hollis and Lennon, unpublished, 1954; Wood, unpublished, 1953; Applegate *et al.*, 1957). These two screening programs were successful in identifying classes of active compounds that were good candidates for further development. In 1962, many of the same chemicals were transferred to the Forest, Wildlife, and Range Experimental Station at the University of Idaho to identify selective toxicants to control the northern squawfish (*Ptychocheilus oregonensis*) (MacPhee and Cheng, unpublished, 1974). Squoxin (1,1′-methylenedi-2-naphthol), a selective toxicant to the northern squawfish was identified but never developed. During the early 1960's and into the late 1970's, intense efforts were made to find, develop, and register fish toxicants by FWS at the Fish Control Laboratory (FCL) at La Crosse, WI, a laboratory created by FWS to identify chemical controls for undesirable fish, especially the common carp (*Cyprinus carpio*) (Lennon, 1961; Marking, 1970, 1972, 1974).

Several promising fish toxicants were identified from the screening programs of the 1950's and early 1960's; these were developed for registration. The lampricides (3-trifluoromethyl-4-nitrophenol; TFM and Bayluscide™; niclosamide, Bayer 73), and a general fish toxicant (antimycin) are examples of compounds registered in the 1960's as a result of the early screening efforts. In 1984, the U.S. Department of Energy and the Bonneville Power Administration funded an investigation into the registration requirements for squoxin. It was determined that it would cost between $436,600 and $2,070,000 to register squoxin (1984 dollars) with the EPA. As a result of the high development costs, no company or organization funded the research necessary for gaining its registration (Rulifson, unpublished, 1984).

In 1998, only two selective toxicants (lampricides = TFM and Bayluscide) and two general fish toxicants (rotenone and antimycin) were registered or are being reregistered (Joint Subcommittee on Aquaculture, personal communication, 1994). Maintenance of the registrations for these compounds have been possible only through funding from FWS, U.S. Department of State through the Great Lakes Fishery Commission (GLFC), or the Biological Resources Division of the U.S. Geological Survey.

A number of aquatic herbicides and algicides were also registered from the 1950's to the 1970's through the efforts of FWS laboratories. These pesticides include copper chelates, copper sulfate, acids blue and yellow, dichlorbenil, diquat dibromide, endothall, fluridone, glyphosate, and 2,4-D (Joint Subcommittee on Aquaculture, personal communication, 1994). These products were needed in aquaculture facilities and natural waters where aquatic vegetation and algae had become too dense to sustain healthy fish populations.

Congress passed amendments to the Federal Insecticide, Fungicide, and Rodenticide Act (FIFRA) in 1972 and 1988 that had dramatic impacts on the development and registration of minor use pesticides. The 1972 amendment required that all pesticides be registered and reclassified for each use and that EPA develop regulations for certification of pesticide applicators. In addition, federal agencies were no longer exempt from requirements of an experimental use permit. All of the aquatic pesticides in use in 1972 were affected by this amendment. Economic return dictated the amount of commitment that registrants of aquatic pesticides had in maintaining registrations. In many instances, registrants withdrew registrations and refused to authorize the federal government to proceed with the registration of products that were effective for aquatic use. A case in point was the lack of interest expressed by the registrants of copper sulfate to consider extension of the registration to include therapeutic uses in addition to the algicidal uses. Without this extension, the therapeutic uses of copper sulfate are illegal.

Reclassification of pesticides occurred too slowly for Congress so they passed the 1988 amendment to FIFRA. This amendment required that pesticides registered before December 1, 1984 be reregistered to prove safety to humans and the environment. As a result, registrations of several aquatic pesticides were withdrawn because of the cost to maintain the registration. Examples of registrations that were withdrawn included amitrole, dichlone, potassium permanganate, simazine, trichlorfon, and xylene (Schnick *et al.*, personal communication, 1989; Joint Subcommittee on Aquaculture, personal communication, 1994).

3.2. Current Registrations

3.2.1. Rotenone. The roots of rotenone-bearing plants have been used for centuries to stun and kill fish by native peoples. In the United States, rotenone use began in the 1930's when it was applied mainly to ponds and lakes. In the early 1960's rotenone application to streams for either complete or partial reclamation was initiated. By 1949, 34 states were using rotenone routinely for managing fish populations. In addition to its use as a fish toxicant, rotenone is also utilized in fisheries as a reference toxicant and sampler of fish populations (Schnick, 1974b).

Rotenone was originally registered in 1947 by S. B. Penick & Company (now AgrEvo Environmental Health, Montvale, NJ). A challenge to the registration occurred in 1976 when EPA considered listing rotenone as a possible tumorigenic agent. Claims of tumorigenicity were proven false in 1981 but the classification set in motion regulatory requirements that resulted in a federal-state cooperative effort to reregister rotenone. The U.S. Fish and Wildlife Service joined with the International Association of Fish and Wildlife Agencies (IAFWA) to fund the reregistration effort at the National Fisheries Research Laboratory at La Crosse, WI. Data from the reregistration package indicate that rotenone is safe to the environment and humans (Sousa *et al.*, 1987). In 1989 FWS submitted a 17-volume response to the Registration Standard on rotenone that had been developed by EPA. Since 1989, AgrEvo has led the way to complete the re-registration of rotenone with some assistance from the National Fisheries Research Laboratory.

The American Fisheries Society (AFS) recognized in 1993 that guidelines for rotenone use were needed to address safety issues raised by environmental and animal-rights groups (Finlayson and Schnick, unpublished, 1996). This recognition led to the submission of a proposal by AFS to the FWS Division of Federal Aid suggesting the use of administrative funds for the development of a manual for fishery managers. The proposal, entitled "Rotenone Stewardship Program", was funded by FWS in December 1997.

3.2.2. Antimycin. Antimycin was isolated from a culture of *Streptomyces* sp. at the University of Wisconsin in 1945. It was recognized as a potential fish toxicant in 1962. The FCL, Wisconsin Alumni Research Association, and Ayerst Laboratories (New York, NY) collaborated to develop the data needed for antimycin's registration. Three antimycin formulations were registered beginning in 1966 (Lennon, 1966). The FCL continued to maintain antimycin's registration and to demonstrate its efficacy for several general and selective purposes. Antimycin is particularly effective in selectively removing scaled fish from catfish ponds (Schnick, 1974a). In 1975, Aquabiotics Corporation (Bainbridge Island, WA) took over the registration of one of Ayerst's formulations and continues to produce antimycin. Recently, the company has increased production by contracting the initial fermentation process to another company (Romeo, personal communication, 1998).

The 1988 amendment of FIFRA included antimycin even though EPA considered antimycin to be a very minor use pesticide. Aquabiotics Corporation is in the process of re-writing the label based on discussions with EPA. Antimycin is being re-classified as a restricted use pesticide because of its current specialization in controlling scaled fish in catfish ponds, fish in shrimp ponds, and fish that compete with endangered species (e.g., Gila trout, Apache trout, and salamanders) (Romeo, personal communication, 1998). Two of three regulatory committees within EPA have recommended the re-registration of antimycin, but the Safety Committee within that agency has until 2002 to make a recommendation.

3.2.3. Lampricides.

3.2.3.1. 3-Trifluoromethyl-4-nitrophenol (TFM). The sea lamprey invaded the upper Great Lakes and destroyed the lake trout fishery during the 1940's and 1950's. In 1946, the Congress directed FWS to develop measures to control the sea lamprey and appropriated funds for that purpose in 1949 (Schnick, 1972). Screening of over 6,000 chemicals was initiated in 1951 at the Hammond Bay Biological Station. A doctoral candidate, Philip J. Sawyer (University of Michigan), investigated 179 compounds of which six were selectively toxic to sea lamprey. After a seven year effort, TFM, was determined to be toxic to, and selective for, the sea lamprey; its toxicity profile to non-target organisms was appropriate for widespread use of TFM.

The FWS originally registered TFM in 1964. In May 1970, the U.S. Department of Agriculture, Pesticides Regulation Division (replaced by EPA later in 1970), notified FWS that all the uses of TFM in water would be canceled on December 31, 1970 unless tolerances were obtained for water and fish. The FWS designated FCL as the facility responsible for maintaining the registration of TFM; the U.S. Department of State provided funding for the project through GLFC. In early 1971, the FCL submitted a proposal for the reregistration of TFM to EPA and received acceptance of most of the proposal. In 1972, EPA provided a list of data requirements for establishing tolerances. Based on these requirements, FCL contracted the mammalian safety studies and conducted the required residue chemistry studies. These data were submitted to EPA in 1976. The EPA accepted most of the data but found some deficiencies. Clarification of the data and additional studies were then requested by EPA (Meyer and Schnick, 1980). By mid-1983, EPA was prepared to publish a *Federal Register* notice on the establishment of tolerances and the successful reregistration of TFM. However, the Toxicology Branch of EPA was concerned with aspects of the registration package because studies were old, data reviews were brief, only one teratology study had been conducted, and some of the studies did not fall into EPA's current classification system.

From 1983 to 1986 EPA continued to review studies and to request additional data or clarifications of data; FWS responded to these requests in a timely manner. In 1987, EPA stated that FWS could respond to questions on TFM tolerances as follows: "You could say that a petition, submitted by the U.S. Fish and Wildlife Service, has been filed with the Agency proposing the establishment of these tolerances and that, although the current toxicological data base is not sufficient to support the establishment of these tolerances based on the currently available data, the Agency has found no adverse effects of human health concern with TFM." In 1989, FWS submitted nine volumes of additional data to EPA, petitioned EPA for a classification of TFM as a minor use pesticide, and requested waivers from any additional data requirements. Currently, TFM remains available for use and the FWS continues to provide EPA with data to support the re-registration efforts. Re-registration is expected to be completed by 2001.

3.2.3.2. Bayluscide™. Bayluscide™ was registered as a lampricide and lamprey population assessment tool in 1968 by Chemagro Chemical Company (Kansas City, MO). When the 1988 amendment to FIFRA was issued, the Mobay Corporation (formerly Chemagro) chose not to support Bayluscide's reregistration because of the cost. Mobay agreed to transfer the registration of Bayluscide™ to FWS. Since its transfer to FWS, FCL (currently the Upper Mississippi Science Center; UMSC) has been pursuing the reregistration of Bayluscide™ in conjunction with that for TFM. Resolution of the reregistration process on Bayluscide™ should also be completed by 2000 or 2001.

4. REGISTRATION AND USE OF FISH CULTURE CHEMICALS: PAST AND FUTURE

4.1. Background

Chemicals are used mainly to control disease, aid spawning, manipulate gender, control vegetation and algae, and disinfect waters where fish are cultured. Sodium chloride was the main drug used until 1909 when it was reported that formalin was useful to control fish disease. Since then, many drugs have been tested for their propensity to alleviate stress and facilitate rearing. Nationwide surveys were conducted in 1966, 1980, and 1982 by the U.S. Bureau of Sport Fisheries and Wildlife (USBSFW, a bureau of FWS) and in 1985, 1988, 1990, and 1995 by public and private organizations to determine priorities for chemicals needed in aquaculture. The 1966 survey identified 95 chemicals for registration or approval including 23 herbicides, 18 internal therapeutants, 16 external therapeutants and disinfectants, 13 control agents for pests other than fish, eight fish control agents, seven transport aids, seven anesthetics, and three dyes (Schroettger and Hunn, unpublished 1967). The later surveys have also indicated a desire by public and private organizations for increased drug availability.

4.2. Aquaculture Drug Approval Efforts

The Division of Veterinary Medicine (forerunner of CVM) notified the USBSFW in March 1964 that chemicals used at federal, state and private aquaculture facilities must be approved for specific applications against designated species under defined conditions (Lennon, unpublished 1964). Because the aquaculture market was too small for potential drug sponsors, the USBSFW in cooperation with the states of Minnesota, Wisconsin, and Missouri, selected several compounds to develop for approval. The target molecules were sulfamerazine, mercurials, MS-222, methyl pentynol, quinaldine, and Roccal™. Approval packages were developed for sulfamerazine and MS-222 in the mid 1960's (Lennon, personal communication, 1967, 1971, 1972). From 1970 to 1986, oxytetracycline, nifurpirinol, sulfamerazine, MS-222, Romet-30™, and formalin also gained limited approvals as a result of these efforts (Schnick *et al.*, personal communication, 1986, 1989).

The FWS consolidated its drug approval efforts in 1972 and designated the FCL as its lead facility. Existing funding and personnel were deemed adequate to accomplish this mission. Since this designation, the FCL has devoted a portion of its annual budget to the registration of fish toxicants and the approval of drugs. In 1972, the priority for approval was: formalin, Wescodyne™, nifurpirinol, potassium permanganate, Hyamine 1622™ (or Roccal™), sulfadimethoxine + ormetoprim, trichlorfon, MS-222:quinaldine sulfate, malachite green, and broad spectrum antibacterials (streptomycin, erythromycin, or chloramphenicol). Various FWS laboratories were asked to coordinate efficacy studies regarding Wescodyne™, nifurpirinol, formalin, formalin-malachite green, potassium permanganate, and Hyamine 1622™ (Schnick and Meyer, 1978).

Because the progress toward drug approvals was slow, FWS requested a series of meetings with CVM in the early 1980's to discuss problems encountered in the approval process (Meyer and Schnick, 1978). Problem areas identified were inconsistencies in regulatory guidelines, frequent changes in study requirements, poor communication among CVM divisions, and a general lack of support at several levels within FWS.

In 1983, CVM amended its animal drug regulations by codifying policy regarding data collection requirements for the approval of minor use drugs. The regulation allowed the use of data supporting the approval of a major use drug to support a NADA for a minor use of the same drug. This action was taken to encourage sponsors to submit NADAs for minor uses.

In recognition of the need for a concerted effort to increase the availability of drugs for use in minor species, the Interregional Research Project Number 4 (IR-4, now the National Research Support Program Number 7; NRSP-7) broadened its scope in 1982 to include minor use drugs. Annual meetings in the 1980's addressed the needs of producers and data requirements for new approvals (Schnick, unpublished, 1984, 1989; Schnick, 1987, 1991a). The CVM also established a liaison to interact with IR-4 and the minor use industries. In 1985, the IR-4 Aquaculture Work Group established its priorities for approval of aquaculture therapeutants as follows (in priority order): fungicides, parasiticides, oxolinic acid, chloramine-T, erythromycin, oxytetracycline, formalin, and benzocaine.

From the late 1970's to the present, the FCL continued to search for candidate drugs for use in aquaculture (Meyer and Schnick, 1989). For example, FCL assisted the sponsor in registering calcium hypochlorite as a disinfectant and worked with and National Fishery Research Center-Seattle toward the approval of nifurpirinol (aquarium fish only). Cooperation with the National Fish Health Research Laboratory led to the successful approval of sulfadimethoxine plus ormetoprim (Romet-30™). Efforts were made to convince the registrants of copper sulfate to expand their registrations to therapeutic uses on aquatic species, but no registrations were expanded. At this time, allowances (not approvals or registrations) were obtained by FCL for selective applications of acetic acid (glacial), benzalkonium chloride, benzethonium chloride, carbonic acid, lime, potassium permanganate, povidone iodine compounds, sodium bicarbonate, and sodium chloride (Schnick, unpublished, 1991).

A worldwide search to find a replacement for malachite green was begun in the early 1980's (Alderman, 1985; Bailey and Jeffrey, 1989). Malachite green has been the treatment of choice to control fungal infections since 1933, but it caused developmental anomalies in several species exposed to high concentrations and was considered to be potentially hazardous to humans (Meyer and Jorgenson, 1983; Schnick, personal communication, 1988). Hence IR-4 identified the replacement of malachite green as its number one priority need (Schnick, unpublished, 1984, 1989; Schnick, 1987). Possible replacements have been considered, but none were acceptable with the exception of hydrogen peroxide and Pyceze™. Scientists at UMSC investigated the fungicidal activities of numerous compounds and selected hydrogen peroxide as likely candidate to replace malachite green; they have since developed the data for its approval (Marking *et al.*, personal communication, 1994; Dawson *et al.*, 1994; Schreier *et al.*, 1996; Rach *et al.*, personal communication, 1997).

At the same time that efforts were being expended to find a replacement for malachite green in the United States, researchers in the United Kingdom did likewise. Pyceze™, a substance used in various applications (e.g., toothpastes, sunscreens) for over 30 years in more than 40 countries was identifies as a candidate fungicide by UK scientists (Anonymous, 1997). Grampian Pharmaceuticals Limited (Glasgow, Scotland) is currently pursuing approval in the United Kingdom and the United States (Braidwood, personal communication, 1998).

After 10 years of negotiation, the data requirements for formalin were accepted by CVM in 1983, but it took another three years to gain a sponsor to complete the approval (Schnick, 1974c; Schnick, 1987). In 1986, CVM approved the first New Animal Drug Application (NADA) on formalin. The 13-year interim period illustrates the difficulty of bringing a drug through the approval process. Since formalin's approval, no NADAs for aquaculture applications have been granted for additional drugs by CVM.

The federal Joint Subcommittee on Aquaculture (JSA), created by the National Aquaculture Act of 1980, recognized the need in 1987 to establish a national effort to gain aquaculture drug approvals that involved the private aquaculture sector in addition to FWS (JSA, unpublished, 1988). The JSA appointed a Task Force on Therapeutic Compounds representing six federal agencies. This task force identified 24 drugs that were considered to be urgently needed by the rapidly growing private aquaculture industry. Identification of these priorities was restricted to food fish because of the expense of the approval process. It was estimated that three million dollars annually would be required for five consecutive years to support the approvals of the following top priority drugs: antivirals, oxolinic acid, oxytetracycline (extensions to nine aquatic species), Romet-30 (extensions to five aquatic species), flumequine, trifluralin, formalin (extensions to 3 aquatic species), formalin (extensions to eight aquatic species), replacement for malachite green, trichlorfon for fish consumed by humans, praziquantel, diflubenzuron, and benzocaine. Drugs of lower priority included streptomycin, chloramine-T, erythromycin (extensions to four aquatic species), copper-free parasiticide, di-n-butyltin oxide, and MS-222 (extensions to two aquatic species). This report was submitted to Congress for funding but no action was taken.

In 1990, CVM increased its scrutiny of drugs and chemicals used in aquaculture and efforts were begun to intensify drug approval efforts (Schnick, 1991a, 1991b; Schnick, personal communication, 1992). The JSA responded to this increased scrutiny by establishing the Working Group on Quality Assurance in Aquaculture Production (Working Group) in 1990. The Working Group's mission was to increase understanding of, and compliance with, Federal requirements regarding drugs, chemicals, and biologics used in aquaculture production in the United States. The mission was to be achieved through education and coordination of efforts of government, industry and academia, related to the development and implementation of quality assurance programs. Initial efforts included (1) development of a comprehensive booklet summarizing Federal regulation of drugs, chemicals, and biologics used in aquaculture production; (2) development of a central databank of drug and chemical information for compounds used in aquaculture production; (3) development of a guide to the proper use of drugs, pesticides, and vaccines; (4) development of producer quality assurance programs; and (5) clarification of the regulatory status of compounds used in aquaculture production (JSA, personal communication, 1992, 1994; Schnick and Armstrong, personal communication, 1997).

The Working Group also determined that there was a need for a national coordinator of Investigational New Animal Drug (INAD) exemptions and NADAs. As a result, a National INAD Coordinator office was funded by the U.S. Department of Agriculture (USDA) on a quarter-time basis from September 1992 to May 1995. The success of this program led to the recognition of the need for a full-time effort directed toward guiding NADA packages through the approval process. The office of the National Coordinator for Aquaculture New Animal Drug Applications was established in May 1995 and is sup-

Table 1. Joint Subcommittee on Aquaculture 1995 List of priority aquaculture drugs for approval through the compassionate INAD process.

A. Therapeutants and Anesthetics
Amoxicillin—systemic antibacterial (Gram positive and Gram negative bacteria)
Benzocaine—anesthetic (replaced by Aqui-S™)
Chloramine-T—external microbicide (bacterial gill disease and flavobacteriosis)
Copper sulfate—external microbicide (fungus, external parasites and external bacteria)
Cutrine-Plus™—external microbicide (fungus, external parasites and external bacteria)
Diquat dibromide—external microbicide (fungus and external bacteria)
Erythromycin—systemic antibacterial (Gram positive and Gram negative bacteria)
Formalin—extensions as external microbicide (fungus on fish and their eggs, external parasites)
Hydrogen peroxide—external microbicide (fungus, external parasites, external bacteria, sea lice)
Neomycin sulfate—systemic antibacterial (vibriosis)
Oxytetracycline—extensions and expansions as systemic antibacterial (Gram positive and Gram negative bacteria); extensions as fish marking agent
Potassium permanganate—external microbicide (fungus, external parasites and external bacteria)
Praziquantel—anthelminthic (trematodes and cestodes)
Sea Lice Control therapeutants (acyl urea, azamethiphos, cypermethrin, hydrogen peroxide)
Trichlorfon—external parasiticide, pesticide (external parasites, predaceous insects, zooplankton, clam shrimp); Special Local Need registrations
B. Spawning and Gender Manipulation Aids
Common carp pituitary—spawning aid
Human chorionic gonadotropin—spawning aid
Luteinizing hormone-releasing hormone analogue—spawning aid (replaced by Ovaprim™ and gonadotropin-releasing hormone analogue)
17α-methyltestosterone—gender manipulation aid

ported by agencies, organizations, and companies interested in facilitating drug approvals for aquaculture.

In 1995, CVM requested that the Working Group develop a list of high priority drugs to develop for approval through the compassionate INAD process (Schnick, unpublished, 1995). Public and private aquaculture groups responded with a list of 19 drugs; large segments of the aquaculture industry are working toward obtaining NADA approvals for compounds on this list (Table 1). Public and private aquaculture organizations formed partnerships with the goal of supporting new aquaculture drug approvals. These partnerships (described below) facilitate the drug approval process through consolidation and disclosure of compassionate INADs, development of new compassionate INADs, and exploration of funding strategies for funding drug approval research.

4.2.1. Federal-State Aquaculture Drug Approval Partnership. Personnel at UMSC recognized that more funding was needed to develop the data required for approvals and pursued the use of state Sport Fish Restoration Federal Aid matching funds. Several meetings were held in August and September 1992 with representatives of the AFS, IAFWA, National Aquaculture Association, National Fisheries Institute, FWS, FDA, and USDA to determine an effective course of action. As a result of these discussions, the IAFWA Executive Committee directed the IAFWA ad Hoc Committee to (1) gather information on the aquaculture drug needs of each state and (2) investigate funding possibilities. Based on the tremendous and positive responses by states, the IAFWA Executive Committee selected as the funding mechanism a multi-state federal aid project with funding being shared with federal appropriated dollars at UMSC. The project is called the IAFWA Project or the Federal-State Aquaculture Drug Approval Partnership. This unprecedented partnership is one of the largest and most important agreements ever forged on behalf of fish management, production, and disease control (Schnick, personal communication, 1996, Schnick *et al.*, 1996).

The IAFWA Project was envisioned to last from July 1994 through June 1999. At its inception, 39 states agreed to contribute funds to the initiative for five consecutive years; to date, only 37 states are active participants. The objectives of the IAFWA Project included the development of data to (1) extend the NADA approval of formalin to additional fish species and eggs; (2) expand the existing NADA approval for oxytetracycline to include additional diseases and other species; (3) gain NADA approvals for all important fish species for use of benzocaine, chloramine-T, copper sulfate, hydrogen peroxide, potassium permanganate, and sarafloxacin; and (4) develop data required to support the concept of crop grouping by CVM. The crop grouping concept is based on the idea that selected fish species can serve as surrogates during the approval process for many of the cultured fishes in the United States.

Progress is being made on eight drugs targeted by the IAFWA Project and the crop grouping research. Highlights of the project include (1) identifying sponsors for all IAFWA Project drugs; (2) acceptance by CVM of data on formalin for control of fungus on fish eggs and parasites regardless of species; (3) acceptance by CVM of the human food safety for hydrogen peroxide and both human food and environmental safety data for copper sulfate; (4) submission to CVM of technical sections on the efficacy and target animal safety of formalin, hydrogen peroxide, and chloramine-T; (5) selection of florfenicol and Aqui-S™ as the replacement oral antibacterial and anesthetic for sarafloxacin and benzocaine, respectively; (6) requests for data for formalin, oxytetracycline, and chloramine-T; (7) development of analytical methods for chloramine-T and oxytetracycline; (8) the design of pivotal efficacy studies; and (9) development of compartmental and physiologically based pharmacokinetic models (Dawson *et al.*, 1994; Dawson and Davis, 1997; Rach *et al.*, 1997; Rach *et al.*, personal communication, 1997; Schreier *et al.*, 1996).

The IAFWA Drug Approval Oversight Subcommittee has recognized the need for IAFWA Project to be extended until 2002. Anticipated products for the extended IAFWA project include (1) extended labels for formalin; (2) expanded and extended labels for oxytetracycline; (3) approval of copper sulfate; (4) approval and extensions for chloramine-T; (5) approval, expansions and extensions for florfenicol; (6) approval for potassium permanganate; (7) approval and extensions for Aqui-S™; (8) approval, expansions, and extensions for hydrogen peroxide; and (9) completion of the crop grouping research.

4.2.2. Other Aquaculture Partnerships. In 1989, the Bonneville Power Administration provided funding for an approval project for use of erythromycin as a feed additive and injectable in salmonids. In 1992, at the request of the Pacific Northwest Fish Health Protection Committee, this drug approval effort was expanded so that field trials with erythromycin could be conducted at facilities throughout the Pacific Northwest. The University of Idaho coordinated the field trials and conducted all of the laboratory studies needed for approval (Moffitt and Haukenes, personal communication, 1995). Most of the data have been accepted by CVM and a sponsor is being sought.

In 1994, FWS established a national INAD office in Bozeman, MT that coordinates FWS-INAD related activities (e.g., protocol development, data collection and analysis). These activities involve 12 individual INAD exemptions, 90 facilities, and approximately 300 INAD units (INAD/facility x facility). The National INAD Office has coordinated, organized, and performed several pivotal efficacy studies at cooperating national fish hatcheries and has involved FWS fish health professionals. In April 1998, FWS extended its INADs to the members of the former Western Regional INAD Project and will offer this same service to all public and private facilities who desire to be attached to the FWS INADs by January 1999. This consolidation will streamline the collection and reporting of data to CVM, encourage the use of standardized protocols, and facilitate regulatory compliance (U.S. Fish and Wildlife Service, unpublished, 1998).

The natural resources departments of twenty states have 61 individual INAD exemptions for eight therapeutants, one anesthetic, and three spawning aids. Many of these INAD holding states have offered to consolidate INADs with other states, and in some cases, with the private sector. The Texas Parks and Wildlife Department, in particular, has added the private producers of tilapia, sturgeon, and hybrid striped bass to its INAD on oxytetracycline (Fries, personal communication, 1998).

Several universities have been involved in the drug approval process in recent years. Their contributions have mainly been efficacy and target animal safety studies. Southern Illinois University has established a Good Laboratory Practices program to perform target animal safety studies. Some of the other universities who have assisted in this process include Auburn University, University of Arizona, University of Florida, University of Idaho, Louisiana State University, Mississippi State University, Rutgers University, Cornell University, Ohio State University, Texas A & M, and University of Wisconsin-Madison. Much of the work performed at these universities have been funded by the NRSP-7.

The CVM has recognized the need to cooperate with the aquaculture industry to resolve the drug availability situation. Over the last few years, CVM has developed several programs that have clarified data requirements for drug approvals and have created flexibility in the approval process. The major programs include classification of drugs, pesticides, or disinfectants; formal compassionate INADs; formal low regulatory status; flexible labeling; innovative approaches to efficacy and target animal safety studies; encouragement of crop grouping research; and encouragement of disclosure and consolidation of INADs (Schnick and Armstrong, personal communication, 1997). In addition, two major animal drug regulations were recently implemented in the United States that may have an impact on the approval of aquaculture drugs: The Animal Drug Availability Act of 1996 and the recently finalized regulations for the Animal Medicinal Drug Use Clarification Act of 1994. The former codifies the efforts by CVM to facilitate minor use drug approvals and the latter clarifies extra-label use in aquaculture.

Since 1995, major efforts have been made to attract pharmaceutical and chemical companies to develop their products for the aquaculture industry. Because the size of the aquaculture market has limited drug development opportunities, partnerships have been offered to many companies. As a result, pharmaceutical firms, chemical companies, and feed manufacturers are working with public agencies, and private aquaculture producers and groups. In 1998, prospective NADAs with sponsors from existing INADs are: amoxicillin, Aqui-S™, chloramine-T, common carp pituitary, copper sulfate, EarthTec™, erythromycin, florfenicol, formalin (extensions), fumagillin, gonadotropin-releasing hormone analogue, human chorionic gonadotropin, hydrogen peroxide, 17α-methyltestosterone, Ovaprim, oxytetracycline (extensions and expansions), Pyceze™, pet fish therapeutants, potassium permanganate, and sea lice control agents. There are a number of additional potential sponsors of non-disclosable INADs pursuing other aquaculture

NADAs. In addition, several states have pursued and obtained Special Local Need registrations for trichlorfon to control predaceous insects and zooplankton.

4.2.3. International Cooperation and Harmonization Efforts. Efforts by federal agencies and other organizations have been made since the early 1990's to communicate with regulatory agencies and aquaculture organizations worldwide to maintain global awareness of developments on the use of drugs in aquaculture production and to press for harmonization of approval requirements and guidelines (Schnick, personal communication, 1991; Schnick, 1992a, 1992b). To this end, CVM and the Canadian Pest Management Regulatory Agency co-sponsored a meeting in September 1996 to consider the regulatory jurisdiction for sea lice bath treatments. Sea lice treatments are regulated as pesticides in Canada but are regulated as drugs in the United States. A joint Canadian and United States Aquaculture Working Group has been established to facilitate approvals and registrations for each country (Schnick *et al.*, personal communication, 1997).

Two International Harmonization Workshops for Aquaculture Drugs and Biologics were held in 1997 and 1998 to create an educational forum for the exchange of information and to identify issues between public and private sectors and international organizations. The goal of these workshops was to initiate follow-up strategies to advance harmonization of drug maximum residue levels, aquaculture drug approval standards, and biological licensure. Progress has been made in identifying researchers, drug and vaccine priorities, and remaining data requirements for drug approvals worldwide. A follow-up workshop and round table to the 1997 International Harmonization Workshop for Aquaculture Drugs and Biologics was held in September 1997. This meeting helped to identify currently approved aquaculture drugs throughout the world, identify drugs that are being pursued for approval, and determine where cooperative efforts may begin (Schnick *et al.*, personal communication, 1997).

4.3. Aquaculture Drug Approvals—the Future

CVM has provided the aquaculture industry a "window of opportunity" to obtain drug approvals through the compassionate INAD process. Pharmaceutical and chemical companies have become interested in the development of their products for the aquaculture industry. Partnerships with agencies, organizations and companies have helped the aquaculture industry to attract both domestic and international firms because the cost of approvals has been shared. With the limited resources available for approvals, partnerships remain a realistic avenue through which drug approvals may be gained.

The future efforts of the aquaculture industry will continue to center on forming more partnerships to support NADA approvals and taking on more responsibilities in the NADA process. Aquaculture industry experts are likely to take an increasingly significant role as investigators on behalf of pharmaceutical sponsors and as professional consultants contributing to NADA package preparation and submission.

5. SUMMARY

This paper reviewed the process and progress made toward approvals and registrations of both drugs and pesticides and what chemicals managers and producers can anticipate being able to legally use after 2002. Obviously, few, if any new pesticides will be developed for fish control and no new aquatic herbicides will be developed. Because of a variety of partnerships, it is anticipated that approximately 20 drugs will have gained approvals for use in aquaculture by 2002.

6. REFERENCES

Alderman, D.J., 1985, Malachite green: a review. J. Fish Dis. 8:289–298.

Anonymous, 1997, High hopes for fungal treatment. Fish Farming Int. 24(12):20.

Applegate, V.C., Howell, J.H., Hall, A.E. Jr., and Smith, M.A., 1957, Toxicity of 4,346 chemicals to larval lampreys and fishes. U.S. Fish Wildl. Serv. Spec. Sci. Rep. Fish. No. 207. 157 pp.

Bailey, T.A., and Jeffrey, S.M., 1989, Evaluation of 215 candidate fungicides for use in fish culture. U.S. Fish Wild. Serv. Investig. Fish Control No. 99. 9 pp.

Dawson, V.K., and Davis, R.A., 1997, Liquid chromatographic determination of chloramine-T and its primary degradation product, p-toluene sulfonamide in water. J. Assoc. Off. Anal. Chem. 80(2):316–318.

Dawson, V.K., Rach, J.J., and Schreier, T.M., 1994, Hydrogen peroxide as a fungicide for fish culture. Bull. Aquacult. Assoc. Can. 94(2):54–56.

Lennon, R.E., 1961, The Fish Control Laboratory. Wis. Conserv. Bull. 26(3):7–8.

Lennon, R.E., 1966, Antimycin—a new fishery tool. Wis. Conserv. Bull. 31(2):4–5.

Lennon, R.E., Hunn, J.B., Schnick, R.A., and Burress, R.M., 1970, Reclamation of ponds, lakes, and streams with fish toxicants: A review. F.A.O. Fish. Tech. Pap. 100. 99 pp.

Marking, L.L., 1970, Juglone (5-hydroxy-1, 4-naphthoquinone) as a fish toxicant. Trans. Am. Fish. Soc. 99(3):510–514.

Marking, L.L., 1972, Salicylanilide 1, an effective non-persistent candidate piscicide. Trans. Am. Fish. Soc. 101(3):526–533.

Marking, L.L., 1974, Toxicity of the 2-(digeranylamino)-ethanol, a candidate selective fish toxicant. Trans. Am. Fish. Soc. 103(3):736–642.

Meyer, F.P., and Jorgenson, T.A., 1983, Teratological and other effects of malachite green in development of rainbow trout and rabbits. Trans. Am. Fish. Soc. 112:818–824.

Meyer, F.P., and Schnick, R.A., 1978, The approaching crisis in the registration of fishery chemicals. Proc. Annu. Conf. Southeast. Assoc. Fish Wildl. Agencies 30:5–14.

Meyer, F.P., and Schnick, R.A., 1980, Potential problems in the registration of sea lamprey (*Petromyzon marinus*) control agents. Can. J. Fish. Aquat. Sci. 37(11):2093–2102.

Meyer, F.P., and Schnick, R.A., 1989, A review of chemicals used for the control of fish diseases. Rev. Aquat. Sci. 1(4):693–710.

Rach, J.J., Howe, G.E., and Schreier, T.M., 1997, Safety of formalin treatments on warm- and coolwater fish eggs. Aquaculture 149:183–191.

Schnick, R.A., 1972, A review of literature on TFM (3-trifluoromethyl-4-nitrophenol) as a lamprey larvicide. U.S. Fish Wildl. Serv. Investig. Fish Control No. 44. 31 pp.

Schnick, R.A., 1974a, A review on the literature of the use of antimycin in fisheries. U.S. Fish and Wildlife Service, National Fishery Research Laboratory, La Crosse, Wisconsin. National Technical Information Service (NTIS) No. PB-235 440. 85 pp.

Schnick, R.A., 1974b, A review of the literature on the use of rotenone in fisheries. U.S. Fish and Wildlife Service, National Fishery Research Laboratory, La Crosse, Wisconsin. NTIS No. PB-235 454. 130 pp.

Schnick, R.A., 1974c, Formalin as a therapeutant in fish culture. U.S. Fish and Wildlife Service, National Fishery Research Laboratory, La Crosse, Wisconsin. NTIS No. PB-237 198. 145 pp.

Schnick, R.A., 1987, Aquaculture Work Group Session Report. Vet. Hum. Toxicol. 29 (Supplement 1):28–35.

Schnick, R.A., 1991a, An overview of current programs on drug development and regulation for aquaculture: U.S. Fish and Wildlife Service perspective. Vet. Hum. Toxicol. 33(Supplement 1):4–5.

Schnick, R.A., 1991b, Crisis in chemical and drug registration. Fisheries 16(2):3.

Schnick, R.A., 1992a, An overview of the regulatory aspects of chemotherapy in aquaculture. Pages 71–79 in C. Michel and D. J. Alderman. Chemotherapy in Aquaculture: from Theory to Reality. Office International des Epizooties, Paris, France.

Schnick, R.A, 1992b, Trends in international cooperation for aquaculture drug registration. Pages 549–558 in C. Michel and D. J. Alderman. Chemotherapy in Aquaculture: from Theory to Reality. Office International des Epizooties, Paris, France.

Schnick, R.A., Gingerich, W.H., and Koltes, K.H., 1996, Federal-State Aquaculture Drug Registration Partnership: A success story in the making. Fisheries 21(5):4.

Schnick, R A., and Meyer, F.P., 1978, Registration of thirty-three fishery chemicals: Status of research and estimated costs of required contract studies. U.S. Fish Wildl. Serv. Investig. Fish Control No. 86. 19 pp.

Schreier, T.M., Rach, J.J., and Howe, G.E., 1996, Efficacy of formalin, hydrogen peroxide, and sodium chloride on fungal-infected rainbow trout eggs. Aquaculture 140:323–331.

Sousa, R.J., Meyer, F.P., and Schnick, R.A., 1987, Re-registration of rotenone: A state/federal cooperative effort. Fisheries 12(4):9–13.

2

U.S. FOOD AND DRUG ADMINISTRATION'S PERSPECTIVE ON AQUACULTURE DRUG RESIDUES

Julia A. Oriani

Food and Drug Administration
Center for Veterinary Medicine
Division of Human Food Safety
Rockville, MD 20855

1. ABSTRACT

The U.S. Food and Drug Administration's (FDA) Center for Veterinary Medicine (CVM) evaluates the safety and efficacy of each animal drug prior to its approval. Human food safety data must be provided on the safety of the drug residues remaining in the animal's edible tissue after treatment and the time that residues take to deplete to an acceptable level. Conducting human food safety studies with aquatic species is very different from similar studies in terrestrial animals. Sponsors are permitted to generate preliminary data under compassionate investigational new animal drug applications (INAD's). CVM's recent studies to bridge microbiological and chemical methods, FDA's compliance program for seafood, and the implication of recent legislation on drugs approved for minor use is discussed.

2. INTRODUCTION

The human food safety requirements for the Food and Drug Administration's (FDA) Center for Veterinary Medicine (CVM) approval of a new animal drug application (NADA) for use in fish will be described. The emphasis of the discussion will be on the nature of the residue chemistry studies and how the conduct of studies in fish differ from those in terrestrial animals. Human food safety refers to an evaluation of the toxicology of the drug (the potential to harm humans), and the residue of the drug in the fish (that component of the drug left in the edible tissue).

Xenobiotics in Fish, edited by *Smith et al.*
Kluwer Academic / Plenum Publishers, New York, 1999.

3. HUMAN FOOD SAFETY STUDIES

To support approval of a new animal drug, a series of toxicity studies and residue chemistry studies are performed in order to demonstrate the safety of food products derived from the treated animals. The human food safety studies must be conducted in compliance with good laboratory practices (GLP's) under 21 CFR 58 (U.S. OFR, 1995). In applying the risk standard that drug residues in edible tissues from treated animals can be consumed daily in the human diet for a lifetime with no adverse effects, FDA calculates an acceptable daily intake (ADI) derived from toxicological studies. FDA further derives safe concentrations for total residues in edible tissues by adjusting the ADI for the amount of tissue consumed. Additional consideration is given to the rate of residue depletion under conditions of use. The probability of a unique metabolite of toxicological concern occurring from the proposed use is evaluated through a metabolism study in the target species.

When the drug proposed for use in fish (which are considered minor species) is already approved for use in a major food species (cattle, swine, poultry), it is not generally necessary for the fish drug sponsor to generate all of the human food safety data typically required to demonstrate safety of a new animal drug approval in major species. FDA may be able to extrapolate human food safety data generated for the major food species approval to the fish drug use when the sponsor of the major use approval provides authorization to FDA to refer to the data on behalf of the minor species sponsor.

4. PROPOSED LABEL CLAIM DRIVES DESIGN OF HUMAN FOOD SAFETY STUDIES

When a sponsor seeks guidance on conducting human food safety studies, the proposed label claim drives the design of those studies. The protocols for the studies are individually tailored for each drug claim to address the unique aspects of the chemical entity, target animal species, intended route of administration, along with the drug formulation. The fish should be dosed with the intended product at the maximum proposed dose and duration. The water quality parameters such as temperature and salinity must be considered as to whether the drug will be used in warm water, cold water, all year around, fresh- or saltwater. The species that the sponsor intends to be included on the label and the maximum size at time of treatment must be considered in the study design. Is the sponsor seeking an "all fish" claim or a claim specific to certain fish species? Can the drug possibly be used in market size fish where there is no time between last treatment and slaughter or is the drug for use in fingerlings only where the fish will grow to adult size after 2–3 years? What factors will prevent the drug from being used at later life stages—a disease that occurs only in an early life stage or a drug product formulation that prevents use for larger fish?

5. TOXICITY STUDIES

The basic toxicity studies that should be conducted for assessment of human food safety include a genetic toxicology battery, 90 day oral toxicity studies, multigeneration reproduction study, and teratology study (U.S. HHS, 1994). Additional toxicity studies may be necessary to clarify the results obtained in the basic studies or to provide chronic data to support a higher ADI. The toxicity studies are performed in mice or rats, and in dogs as models of human toxicity in order to determine the biological and pharmacological effects, mutagenicity, carcinogenicity, and reproductive effects.

The results of the toxicity studies are used to establish an ADI which represents an amount of drug that can be consumed over the lifetime of the individual without harmful effect. The ADI is calculated by dividing the No Observed Effect Level (NOEL) in the most sensitive toxicological species by the safety factor reflecting the type of study in which the NOEL was set. FDA will normally use the following safety factors: 100 for a chronic study, 100 for a clear indication of maternal toxicity in a reproduction/teratology study, 1,000 for other effects observed in a reproduction/teratology study, and 1,000 for a 90-day feeding study. The ADI is used to calculate the safe concentration for residues of the drug in the human diet. The safe concentration is calculated by multiplying the ADI by 60 kg (the weight of the average human), and dividing by tissue consumption value. For fish, the consumption value is 300 grams of muscle with adhering skin/day.

It is usually unnecessary for the fish sponsor to perform the toxicity studies when the drug is approved for use in a major food species and FDA has permission to refer to that toxicity data on behalf of the fish sponsor. This extrapolation from major to minor species is particularly common for data used to support the safe concentration since the ADI is common to all edible portions of food animals.

6. RESIDUE CHEMISTRY STUDIES

The following studies typically are conducted to assess the residue chemistry of a new animal drug: total residue and metabolism study, comparative metabolism study, development of a regulatory method, and a tissue residue depletion study. These studies focus on the nature and amount of drug residue remaining in the edible tissue of the fish after the drug treatment. It may not be necessary for the fish sponsor to perform the total residue and metabolism study and the comparative metabolism study when the drug is approved for use in a major food species (cattle, swine, chickens, turkeys) and FDA has permission to refer to that data on behalf of the fish sponsor.

6.1. Edible Tissue

FDA is concerned with the safety to consumers of residues that remain in the edible tissue of the animal after drug treatment. For fish, the edible tissue is considered to be muscle with adhering skin, except for those fish such as catfish where the skin is not consumed. Even though residues may be higher in kidney, liver, and bile, these fish organs are not routinely consumed in this country. Drug residues usually do not distribute uniformly across the muscle. Some drugs may be associated more with fat so the fattier parts of the muscle have higher residue levels. Therefore, when measuring residues for a food safety study, the entire fillet with adhering skin should be ground up and then subsamples taken from the homogenized fillet. This approach reduces sample variability and helps to ensure that the drug concentration in the analyzed sample is representative of the entire edible tissue.

6.2. Total Residue and Metabolism Study

This study is performed in order to characterize the depletion, distribution, and nature of the residues in the edible tissue of the target animal. The radiolabeled drug (typically with ^{14}C) is administered to the fish by the proposed route of administration at a high enough dose and specific activity so that residues can be monitored down to at least 0.1 ppm. Total radioactivity is measured in the tissue usually at four timepoints (including zero time) after the termination of drug treatment. The edible tissue in which the radiolabeled drug residues persist for the longest period of time is called the target tissue. Since for fish, the only edible tissue is muscle with adhering skin that is considered the target tissue. Total residues for each individual fish, not composites, should be measured, so that variation between each fish is reported.

Metabolites that make up at least 10% of the total residue are characterized so that the marker residue can be identified and selected. The marker residue is the residue (metabolite or parent) that persists for the longest time in the target tissue. The marker residue also must be a compound for which a suitable analytical method may be developed. Once the marker residue is selected, the tolerance can be calculated.

6.3. Tolerance

The tolerance is the concentration of the marker residue in the fish muscle with adhering skin as measured by the regulatory method at the time when the total radiolabeled residue depletes to the safe concentration. When the depletion curve for the total residues and the depletion curve for the marker residue are plotted on the same graph (concentration vs. time), a horizontal line may be drawn from the safe concentration to the intersection of the total residue curve (see Figure 1). A vertical line may then be drawn from the intersection point down to marker residue depletion curve. The concentration at which the vertical line intersects the marker residue depletion curve is called the tolerance. The tolerance for each drug is listed in Code of Federal Regulations, 21 CFR 556 (U.S. OFR, 1998).

6.4. Comparative Metabolism Study

The purpose of the comparative metabolism study is to compare qualitatively the metabolic profile of a toxicological test species, typically rats, with the metabolic profile obtained in the fish muscle. This is to ensure that the same major metabolites are present in the edible portion of the fish and in the toxicological test species.

6.5. Regulatory Method

A determinative method must be developed for routine monitoring of the marker residue. A second method, a confirmatory one, must be developed using a different ana-

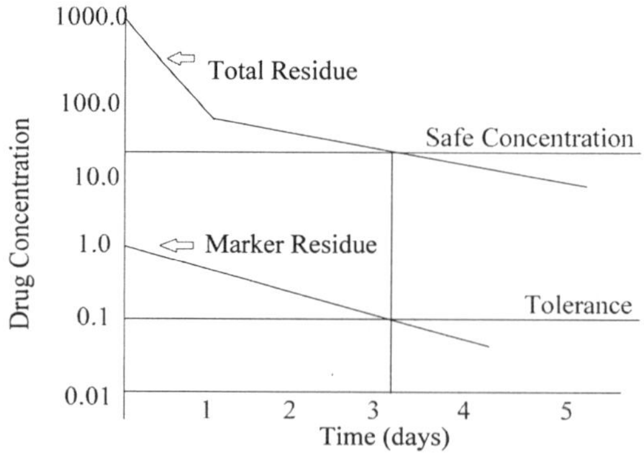

Figure 1. Total Residue and Marker Residue Depletion Curves.

lytical approach (typically mass spectrometry) that can confirm the identity of the marker residue. The determinative method must undergo a method trial during which three laboratories evaluate the method performance to determine that the method is specific, accurate, and reliable. Once the method trial is completed, FDA Center for Food Safety and Nutrition (FDA/CFSAN) uses the regulatory method to monitor for drug residues in fish. If the drug is approved in another species, a regulatory method should already exist and should be used for the food safety studies. In that case, typically all that would be required would be a validation package confirming that the method also has reliable performance in fish tissue.

6.6. Tissue Residue Depletion Study

Once the tolerance is calculated and a regulatory method is chosen, a tissue residue depletion study is conducted to measure the depletion of residues from the edible tissue over time in order to set the withdrawal time using a procedure outlined below. The withdrawal time is defined as that time after the last administration of the drug to the time at which residues deplete to the tolerance.

The unlabeled drug is administered to the fish under field conditions. The "worst case" scenario should be used so that the withdrawal time is based on the slowest depletion rate. The highest anticipated dose and duration is used at the lowest water temperature with the largest size fish in which use is anticipated (typically market size fish). Groups of fish are sampled at a minimum of three timepoints and residues are measured in the muscle with adhering skin with the regulatory method. Again, residues should be measured and reported in individual fish, not composites.

The collection timepoints should be on the linear portion of the depletion curve (see Figure 2). It may be necessary to complete a pilot study to determine the appropriate sampling intervals. Ideally, there should be at least two sampling timepoints when residue levels are above the tolerance and one sampling timepoint when residues are below the tolerance so that the point at which the depletion curve crosses below the tolerance is captured. A common error is to collect data points too early in the depletion curve which can lead to an overestimation of the withdrawal time if the data are not linear. Another common error is to collect the samples too late in the depletion curve when most of the residues are below the limit of quantitation of the method. When it is necessary to measure residues in inedible tissue (i.e., liver and kidney) for efficacy studies, residues in edible tissue should not be measured at the same timepoints since muscle/skin

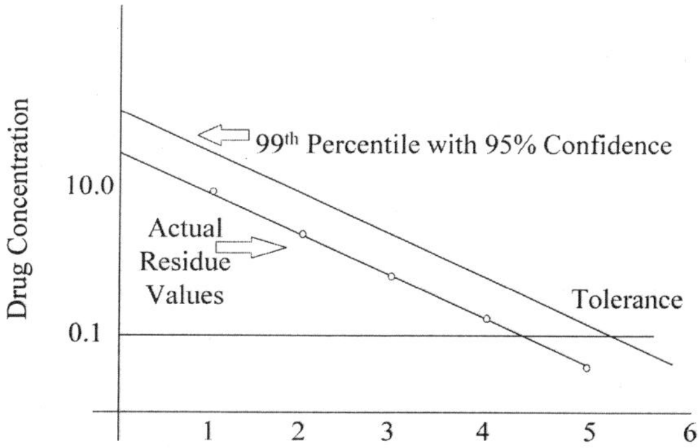

Figure 2. Tissue Residue Depletion Curve.

usually depletes much faster than other organs. A withdrawal time can not be calculated with values below the limit of quantitation of the method.

The results of the tissue residue depletion study are used to calculate the upper bound of the 99th percentile tolerance limit with a 95% confidence interval. A linear fit of the upper bound of the tolerance limit is performed and the time required for this resulting curve to deplete to below the tolerance is determined. The calculation is greatly affected by variability in the depletion data, and the use of fewer fish per time period may lead to an increased withdrawal time.

7. CHALLENGES WITH CONDUCTING RESIDUE CHEMISTRY STUDIES WITH FISH

Problems that make conducting residue chemistry studies with fish particularly challenging are:

a) large inter-fish variation—Large differences in residue levels are often observed among fish within a species. This may be true particularly in feeding studies when it is difficult to get the same amount of medicated feed to each fish. The large inter-fish variation means that the numbers of fish to be sampled at each timepoint is relatively high. It is recommended that 15 to 20 animals per timepoint are collected.
b) environmental influences—Drug absorption and metabolism in fish are known to be affected by the environment, particularly water temperature and perhaps salinity. Typically warm water fish (catfish) deplete drugs faster than cold water fish (salmon). The cold water can significantly lengthen the depletion time of the drug. Withdrawal studies must be conducted in the target species at the minimum water temperature in the type of water (fresh or salt) for which an approval is sought.
c) Numbers of species—There are at least a hundred different fish species being farmed and the data on human food safety are limited to just a few. Data collection would be expedited by grouping species together so that separate studies were not required for each species. This grouping together of different fish for data collection purposes is called crop grouping. A representative species would be chosen and a complete human food safety data package developed. Limited studies would be conducted in the other species in that group. FDA/CVM traditionally has considered salmonids to be a crop group with Atlantic salmon, Pacific salmon and rainbow trout serving as the representative species of all salmonids. This crop group may not be appropriate for all drugs. It is hoped that it will be possible to form other crop groups in the future but at this point, there is very little information to support other groups.

8. INVESTIGATIONAL USE OF UNAPPROVED DRUGS

Unapproved drugs can be used for investigational purposes in fish. A sponsor can apply for an investigational new animal drug (INAD) exemption which permits the use and interstate shipment of the unapproved drug. For food animals, an authorization is assigned to allow the slaughter of those animals treated with the drug under specified conditions of use and appropriate withdrawal time periods. Each fish farm that is using unapproved drugs operates under an INAD (has a specific number) and keeps treatment and withdrawal time records.

Since there are so few drugs approved for aquaculture and there were so many requests for investigational use of unapproved drugs, CVM started the compassionate INAD program since it was recognized that there is a need to permit the regulated use of drugs that are not yet approved. Through compassionate INAD's, the producer has access to drugs for diseased fish while data are being collected to support the approval of the drug.

9. NONFOOD VS FOOD FISH

FDA has been asked repeatedly to consider the immature life stage of fish such as eggs, sac-fry, fry, juveniles, or brood fish as nonfood. A nonfood designation would eliminate the need for human food safety data for a drug approval, potentially facilitating drug approvals. CVM has resisted doing so because the food or nonfood designation is not related to life stage. There is no precedent for calling the early life stages of terrestrial animals as nonfood. For example, chickens are considered food animals even though ten day old birds generally are not consumed by humans. There is a problem with approving a drug without a human food safety evaluation for use in an early life stage when the drug might be diverted for use in later life stages where there is significant human food safety concern. Also, defining the point when each fish species is considered a food fish would be extremely difficult. If the drug can only be used at the early life stage then the human food safety concern is lessened and typically the data requirements are reduced. FDA has relied on long inherent withdrawal times to determine that the residues of an animal drug product do not represent a human food safety concern.

Baitfish (golden shiners, fathead minnows and goldfish) and ornamental fish are currently considered to be nonfood fish. A new animal drug approved for use in nonfood fish would have to address the same information as food fish (effectiveness, target animal safety, labeling, chemistry, manufacturing, and controls, environmental assessment) except that human food safety data would not be needed.

10. COMPLIANCE PROGRAMS

FDA/Center for Food Safety and Nutrition (CFSAN) currently does not have any active compliance programs for seafood. Compliance programs previously were initiated but which are not active currently targeted domestic and imported shrimp for chloramphenicol, domestic and imported salmon for oxolinic acid, domestic catfish for malachite green, and imported shrimp for oxytetracycline.

11. BRIDGING STUDIES

Investigators are often interested in using chemical methods rather than microbiological methods for food safety studies with antibiotics because it can be difficult to find a GLP laboratory capable of running microbiological assays. When a tolerance is based on microbiological activity (as for many of the old antibiotics) and the regulatory method is a microbiological assay, a bridging study must be completed in order to adapt a new chemical method. The microbiological and chemical methods are run using aliquots of the same samples (incurred and fortified) in order to establish a relationship between the two methods. Bridging studies have been completed for oxytetracycline in shrimp, amoxicillin in catfish, and oxytetracycline in rainbow trout. In all three studies, the microbiological and chemical methods have produced essentially identical results over the particular range of values tested including the tolerance values.

12. CURRENT LEGISLATION—AMDUCA AND ADAA

Two recent legislative acts potentially change the availability of certain drugs for minor species. The Animal Medicinal Drug Availability Act (AMDUCA, Pub. L. 103–396) and the Animal Drug Availability Act (ADAA, Pub. L. 104–205) have implications for minor use drugs.

Under AMDUCA (passed in 1994), the extralabel use of approved therapeutic drugs is legal with a valid veterinarian-client relationship. The drugs cannot be administered in feeds and cannot be for non-therapeutic uses like spawning. Although the availability of drugs for aquatic species is improved under AMDUCA, legal extralabel use does not result in an increase in approvals of drugs for aquaculture. There may be a reduction in incentives for the pharmaceutical sponsor to pursue aquaculture drug applications. While AMDUCA makes many products available, it does not address the lack of good label information about dosage, target animal safety, and withdrawal periods for food-producing animals. Also, the drug needs of many minor species cannot be met by use of the approved product formulations.

Under ADAA (passed in 1996), the FD&C Act has been amended in several ways, including the modification of the definition of "substantial evidence" of effectiveness and establishing Veterinary Feed Directive Drugs. Also, FDA was required to develop a proposal to Congress by April 9, 1998 that describes options to encourage the development and approval of minor use drugs. A discussion draft of the proposal and CVM's request for comments on the draft are posted in the On-Line Library section of CVM's web site. The proposal is under review within the Agency.

13. CONCLUSION

Sponsors and investigators are strongly encouraged to meet with CVM early in the drug approval process to develop a drug approval plan. Decisions on the type and number of studies to be conducted and on the designs of those studies can be made in cooperation with CVM and will greatly facilitate the process.

Additional information on human food safety requirements for drugs used in aquaculture is provided in the 1994 guideline, "General Principles for Evaluating the Safety of Compounds Used in Food-Producing Animals" (U.S. HHS, 1994), and the 1997 document "Draft Guidance for Industry #6, FDA Approval of Animal Drugs for Minor Uses and for Minor Species" (U.S. HHS, 1997). These documents are available from the FDA/CVM home page at www.fda.gov/cvm.

14. REFERENCES

U.S. Department of Health and Human Services, 1994, General Principles for Evaluating the Safety of Compounds Used in Food-Producing Animals. Food and Drug Administration, Center for Veterinary Medicine, Rockville, MD.

U.S. Department of Health and Human Services, 1997, Guidance for Industry #6, FDA Approval of Animal Drugs for Minor Uses and for Minor Species. Food and Drug Administration, Center for Veterinary Medicine, Rockville, MD.

U.S. Office of the Federal Register, 1998, Good Laboratory Practice for Nonclinical Laboratory Studies. Code of Federal Regulations, Title 21, Part 58, U.S. Government Printing Office, Washington, D.C.

U.S. Office of the Federal Register, 1998, Tolerances for Residues of New Animal Drugs in Food. Code of Federal Regulations, Title 21, Part 556, U.S. Government Printing Office, Washington, D.C.

3

REQUIREMENTS FOR THE APPROVAL OF VETERINARY THERAPEUTICS OR GROWTH ENHANCERS USED IN FISH PRODUCTION

European Union

David J. Alderman

CEFAS Weymouth Laboratory
The Nothe, Weymouth
Dorset, U.K.

1. ABSTRACT

This paper attempts to provide a brief overview of the complexities of veterinary medicines legislation in the European Union in relation to aquaculture medicines. The harmonization of 15 different legislatorial environments into one has meant that the last 15 years have been years of continuous change. The legislation involved, the terminologies employed and the data currently required for aquaculture drug approvals in Europe are described. Implications of increased regulatory demands and resultant increased costs are discussed in relation to the ability of the aquaculture industry to support such costs.

2. INTRODUCTION

The process of harmonising the legal controls on veterinary (and human) medicines between the member states of the European Union (EU) has now been underway for over 15 years. This has meant that the legislation of 15 countries, some with radically different approaches to the control of medicines, has had to be modified and adapted to produce a single uniform legislatorial environment across all member states. The progress towards the creation of the open market within the European Community in 1993 increased the needs for the harmonization of medicines regulation across the Community. Much of the law created during this process also has wider European Economic Area application.

This harmonization process has been carried out at a time when the regulatory environment would inevitably have had to become more demanding as greater consumer safety

Xenobiotics in Fish, edited by *Smith et al.*
Kluwer Academic / Plenum Publishers, New York, 1999.

requirements were introduced during the 1990s. The net result has been the introduction of what seemed to many to be a remorselessly increasing range of new controls on veterinary medicines. It must be realised that medicines used in aquaculture form only a very small part of the animal health market, but must conform to all of these new regulatory controls. These controls place demands which are expensive and difficult to satisfy, even for areas of veterinary medicine which are much more extensive and profitable than aquaculture medicine can ever be. It is appropriate to draw attention to the costs and profitability and the effects of increased regulatory demands at the beginning of this overview of European legislation because of the impact of these on availability of aquaculture medicines. Aquaculture medicines will only become and remain available if the companies manufacturing and supplying them can cover their costs and obtain a reasonable profit from sales. Despite the growth of aquaculture in the last 25 years in Europe, the market for veterinary medicines in aquaculture remains small and the regulatory costs and demands place considerable constraints on continuing availability. Not only have regulatory demands increased, but attitudes of governments in Europe have also changed, such that the full rather than the marginal costs of the regulatory procedures themselves are now recovered from pharmaceutical companies in the form of license fees.

The first stages of the harmonization process commenced with Directive 81/851/EEC "On the approximation of the laws of the Member States (MS) relating to veterinary medicinal products". This entered into force in 1983 and was the first stage in bringing together the disparate medicines legislation of the (then) 12 Member States. A parallel Directive (81/852/EEC), made provision for harmonization of the analytical, pharmacotoxicological and clinical tests and trials of veterinary medicinal products. It requires that safety trials be conducted to Good Laboratory Practice (GLP) standards. These two Directives and their amending Directives set the initial basic standards for regulating veterinary medicines in the European Community.

As part of the harmonization process the European Medicines Evaluation Agency (EMEA) was formed in 1993 (Council Regulation 2309/93). The EMEA is responsible for both human and veterinary medicine, and the EU's Committee for Veterinary Medicinal Products (CVMP) is now a committee within its structure. The EMEA is responsible for the determination of applications submitted under Centralised Procedures within the EU; currently, in the veterinary area, only applications for biotech products and new molecules are obliged to use this route.

3. DEFINITIONS

Three definitions may be valuable to readers at this stage.

3.1. Veterinary Medicine

The definition of a veterinary medicine may be taken from that used in the British Medicines Act 1968 in which a medicine is defined as follows. A medicinal product is a product designed for administration (to animals) for a Medicinal Purpose. A Medicinal Purpose is defined as including the treatment or prevention of disease, diagnosis of disease, for contraception, anesthesia or for otherwise preventing or interfering with normal operation of a physiological function. This very wide definition, thus the use of salt (NaCl) to treat bacterial gill conditions is a medicinal use and, strictly, falls under these controls.

3.2. EU Law: Regulations and Directives

There are now more than 45 items of EU legislation which apply to the use of veterinary medicines in aquaculture. The majority of these relate to two main areas, marketing authorization (licensing) and control of residues in edible tissues. European laws are of two types, Regulations and Directives. Regulations are laws effective throughout the EU, applying simultaneously in all Member States. Directives, in contrast, are pieces of community law, which Member States must implement in their national legislation by a specified deadline. In addition to Council Regulations, subsidiary Regulations made by the European Commission exist; these modify primary Regulations, relevant examples being the series of Commission Regulations placing substances in the different Annexes of Regulation 2377/90. Such Commission Regulations comprise the majority of items of legislation relating to veterinary medicines.

3.3. Marketing Authorization

A medicinal product may not be marketed until a Marketing Authorisation has been obtained. Stages by which a product may be used on animals in order to generate the data necessary are as follows. The initial extensive laboratory stages to determine efficacy and safety are carried out under research authorization (e.g., the Animals (Scientific Procedures) Act,1986, in the UK). Once sufficient data has been generated, trials on farms may be carried out under the authorization of an Animal Test Certificate (ATC). This enables the gathering of additional efficacy data under practical farm conditions prior to the final application for Marketing Authorization. The term "Marketing Authorization" replaces the term "Product License" used previously in the UK. In the near future (when legislation is in place), before field trials of substances new to medicine can be undertaken a Maximum Residue Limit (see below) will need to be obtained.

4. BASIC CRITERIA FOR APPROVALS

Applications for veterinary medicines approvals are assessed in the EU, as in most other regulatory regimes, against three basic scientific criteria. These are quality, safety and efficacy. The requirements for satisfying each of these criteria are presented in detail in the relevant volumes of "The Rules Governing Medicinal Products in the European Union" (European Commission 1991, 1995a, 1995b).

Quality is concerned principally with the chemistry and pharmacy of all components of the medicinal product, details of the processes of manufacture, packaging and stability under storage, especially in relation to formulation and intended route of application (i.e., bath or in feed). It includes details of the chemical nature and identity of excipients and, especially for medicated premixes, details of particle size and distribution.

Data are required to support manufacturer's claims of efficacy. This can sometimes be a difficult matter to demonstrate. With bacterial and viral diseases, effective laboratory challenges will normally be available enabling the demonstration of comparative relative percent survivals (RPS), but with other agents (e.g., the protozoan disease Proliferative Kidney Disease) field challenges may be essential. For these, unless a research field site is available, problems will arise because farmers will not be happy with the presence of significant numbers of unprotected fish on site during an outbreak of disease. This is quite understandable, but without those control fish it will be impossible to demonstrate that a challenge really has occurred and that the product under test is effective. Use of controls medicated with an existing product are useful comparisons, but do not confirm efficacy.

Although many of the quality requirements are in themselves safety considerations, safety itself is a separate specific consideration and forms by far the largest part of any new drug application. Applicants are required to demonstrate that the product is:

- safe to the consumer;
- safe to the user (farm staff, during feed medication on farm or in the feed mill);
- safe to the target species (fish);
- safe to the environment.

4.1. Consumer Safety

The primary consumer safety consideration is now addressed via Maximum Residue Limits (MRL), established by Council Regulation EEC/2377/90. The MRL defines the maximum level of residues of any component of a veterinary medicine that may be present in foodstuffs of animal origin without presenting any harm to the consumer. It should be noted that, although there is no formal MRL regulation established in USA, the equivalent is the Tolerance established by the regulatory authorities. There is a conflict in terminology between the two regulatory regimes, in that within the EU the term tolerance specifically refers to target animal's tolerance of excess treatment doses.

Regulation 2377/90 provided that all pharmacologically active substances in veterinary medicines must be allocated to one of four annexes established under the Regulation. For new active ingredients, allocation to an Annexe was required beginning in January 1992 and from that date, no consideration of Marketing Authorization for such new molecules could progress until MRL allocation. For all existing veterinary medicinal products, both actives and excipients, a *post quem* date of 31 December 1996 for allocation was set. After that date, if no allocation had been made it was intended that products would have to be withdrawn for use in food species. This intention was too ambitious, and it become evident by 1994 that if this closure date were to be applied strictly, then many important and valuable veterinary medicines would be lost at least temporarily for use in food species. The EMEA was therefore required to publish a list of those medicines for which manufacturers had indicated an intention to support by submission of preliminary data. This list, referred to as the list of Defended Substances, appeared in 1997 (European Agency for the Evaluation of Medicinal Products, 1997) and contains a number of substances of interest to aquaculture (Table 1).

The MRL for any substance is determined from data submitted by manufacturers or suppliers to the Safety of Residues Working Party, a sub committee of the CVMP. This determination is ratified by the CVMP and adopted into law by a Regulatory Committee in the form of Commission Regulations. The MRL is determined by an iterative process from a range of safety data, the most important of which is the Acceptable Daily Intake (ADI). The ADI is defined as the level of a substance that may be consumed daily without presenting a hazard to the consumer. It is based on a suitable no-observed effect level (NOEL) or from observations in humans, divided by a safety factor, often 100. For a discussion of MRL determination and ADI relationships see Woodward (1996).

Table 1. List of Defended Substances (European Agency for the Evaluation of Medicinal Products (1997)).

Substance	Potential Aquacultural Use
Apramycin	Antimicrobial
Benzalkonium	Disinfectant chloride
Benzocaine	Anaesthetic
Cypermethrin	Sea lice control
Dichlorvos	Sea lice control
Fumagillin	Antiprotozoal
8-Hydroxyquinoline	Antibacterial
Oxolinic acid	Antibacterial

The above are extracted from a list of 309 defended substances, many of which CVMP does not regard as falling within the scope of Regulation 2377/90. The defence of those substances listed above as being of aquaculture interest does not mean that the sponsors defending them have any interest in aquacultural applications, nor that the defences will ultimately prove successful in attaining allocation to Annex I, II, or III.
Erythromycin, flumequine and thiamphenicol are currently are placed in Annex III, Provisional MRL.

The Annexes to Regulation 2377/90 are as follows:

- Annex I Full MRL set;
- Annex II Safe, no MRL needed to protect the consumer;
- Annex III Sufficient data to set a Provisional MRL, but additional data needed to allocate full MRL;
- Annex IV On safety grounds no MRL can be set. Substances placed in this Annex are prohibited for use in food animal species, although they may still be used in pet species.

Substances once of interest to aquaculture currently placed in Annex IV include chloramphenicol, the nitrofurans (including nitrofurazolidone and nifurpirinol) and dimetridazole.

MRL data packages are large and consist of two main parts, the safety file and the residue file. Although MRLs are species specific, much of the data in the safety file (e.g., mammalian toxicity studies and the ADI) may be reused for new species MRL applications. The unique part that is required for each species MRL is the residue depletion study. This requires the determination of the marker residue which may either be the parent drug or a metabolite. Depletion of the marker residue characterizes the depletion of the total residue (parent drug and metabolites) from the relevant tissues, such that the total residues in edible tissues will be below the safe concentration when the marker residue is below the MRL.

It should be noted that, in the EU as in the USA, for fish, edible tissues are defined as muscle with adherent skin in natural proportions (European Commission, 1991). A factor for consumption of 300 g (substituting for red meat as a dietary component) is taken to determine the fish meat contribution to the Acceptable Daily Intake (ADI) in the calculation of the MRL. Muscle and skin, in natural proportion, is therefore the target tissue for determining withdrawal periods in fish.

Initially, for fish, applicants were required to present data for each active and excipient to enable the determination of individual marker residues, but very recently the EMEA has published a Note for Guidance (Committee for Veterinary Medicinal Products, 1998). In this, the EMEA states that, although only a limited number of MRLs have been established for fish, where these have been evaluated, the marker residue determined in Salmonidae has been identical to that established in other animal species. A pragmatic approach should therefore apply to the establishment of MRLs for substances used in Salmonidae and other finfish as well. Thus an extrapolation can be considered acceptable, whereby if an MRL has been established for a substance in muscle in a major mammalian species it may be applied to Salmonidae and other fin fish as well. Also, the parent compound is normally acceptable as a valid marker residue in Salmonidae and other fin fish.

This newly adopted pragmatic approach recognises the difficulties imposed by the costs of developing data for MRL determinations against expectations of slight financial returns by potential applicants. In another relevant Note for Guidance (Committee for Veterinary Medicinal Products, 1997), the CVMP notes that, since the implementation of Regulation 2377/90, very few MRLs have been allocated for minor species. This in turn, may lead to animal welfare problems from the increasing shortage of medicines approved for use in minor species. Although not expressed, there is evident anxiety that, by imposing excessive demands on data for minor species medicines, a lack of such medicines will develop. This in turn will mean that the cascade system (see below) will be exploited in such a way that the majority of drugs used come via this exemption and may thus result in a public health problem. The EMEA Note for Guidance concerns the establishment of MRLs for minor animal species. It states that, since in the EU Salmonidae are major food producing species, where a substance is already included in Annex I, II or III of Regulation 2377/90 for Salmonidae, then that MRL may be extended to relevant other minor species (i.e., other fin fish). Provisions concerning validation of analytical methods for tissues of minor species and a description of an abbreviated data package for substances exclusively used in minor animal species are also included in this note.

There is however one problem not properly covered by Regulation 2377/90, nor so far addressed by any other legislation, EU or national. This is that, without an *application* to set an MRL a substance is effectively in "limbo". Its use in food species is not authorised; equally it is not a prohibited Annex IV substance. More than most areas of veterinary medicine, aquaculture has used a range of "traditional remedies" whose use has persisted to the present day. Malachite green is the classic case—it is most unlikely that anyone will make a formal application to establish an MRL for this product. It can hardly be expected that, even if the effort were to be limited to the cost of assembling published data and then of submitting it for consideration by the Working Group on the Safety of Residues, any application for an MRL allocation could be funded by a commercial sponsor. However, without such an application and accompanying data package a substance like malachite green will not be scrutinised by the Working Group and therefore it will not be entered into Annex IV and will not become a prohibited substance. This appears to be a gap in the existing legislation that opens an avenue for failure of consumer protection. It is however true that in any tissue residue monitoring program, substances such as malachite green, without MRL or authorization in any food species, would be regarded as illegal residues.

4.2. Withdrawal Periods

The establishment of an MRL allows the setting of a withdrawal period for the product. The applicant normally proposes the withdrawal period for the Marketing Authorization and presents data to support that proposal. The licensing authority of the

Member State concerned assesses the data against the proposal and agrees with the proposal. Alternatively, the licensing authority may discuss and agree to an alternative with the applicant or may refuse on grounds of inadequate data. In most food-animal species withdrawal periods are defined in days, but with fish the EU requires data to be presented from trials conducted at at least two water temperatures relevant to the proposed conditions of use. If depletion of residues is found to be temperature dependant then a withdrawal period in degree days will be set making the withdrawal period a function of temperature and time. If the data does not indicate a temperature effect on depletion then a day-based withdrawal can be accepted. The degree day concept in some ways parallels the temperature band withdrawal periods used in Canada and Norway. It avoids the illogicality of doubling the withdrawal period for a water temperature change of less than one degree (10°C to 9°C from 40 to 80 days, Canada) and makes provision for the likely effect of changes in water temperature during the withdrawal period.

4.3. Target Species Safety

Tolerance studies must be carried out to determine the safety of the product to the target fish species. Details are provided in the “Rules” for both therapeutants (European Commission, 1995a) and immunologicals (European Commission, 1995b). In general these rules require the application of a double dose of the product followed by a period of close observation. There are provisions for single dose and repeated dose indications.

In the UK a post authorization surveillance scheme (Suspected Adverse Reactions Reporting Scheme, SARRS) is in operation by which any suspect adverse reaction should be reported to the UK Veterinary Medicines Directorate. Any such reports are followed up and in severe cases this can result in modification or suspension of authorizations.

4.4. User Safety

As for environmental safety, this is an area with parallel provision of protective legislation. Medicines’ legislation requires that user safety be assessed in the safety package and that the product label must include advice and warnings to the user giving guidance for safe use. Areas involved exclude the pharmaceutical manufacturing plant but do include hazards associated with feed medication, whether in the feed mill, or on the farm. Finally any hazard to the final user (fish farm staff) must be considered. Such hazards include any tendency to produce respirable dust; the need for use of gloves, masks, or other protective clothing; and the risks of skin absorption and hand contamination. Advice must be provided on packaging labels and inserts and details of these labels are assessed as part of the approval process.

Apart from respirable dust hazards, sensitization, absorption and explosion hazards (e.g., hydrogen peroxide, fine powders) may need to be considered in assessing aquaculture medicines. Aquaculture vaccines present the known user hazard of self-injection, where oil based vaccines can produce vascular spasm, ischemia and tissue necrosis. Once self injected it may be necessary for the individual to be prohibited from further involvement with fish vaccination.

Health and Safety legislation protecting workers exposed to potential hazards in the workplace also apply to veterinary medicines and require employers to have in place proper assessments of hazards associated with the use of veterinary medicines in the farm environment. The user safety data included on label and package inserts should provide sufficient information for such occupational safety assessments to be made.

4.5. Environmental Safety

The two original Directives, 81/851 and 81/852 included requirements to assess environmental safety of the use of any product passing through the authorization process. A twofold approach was required, Phase I consisting of an environmental impact assessment which could be largely literature based, followed by Phase II involving studies based on the findings of Phase I. In the UK a more detailed ecotoxicological assessment guideline has been produced (Veterinary Medicines Directorate, 1996) after discussions with the relevant veterinary pharmaceutical industry members. This guideline requires a three tiered approach in which the results obtained in each tier dictate the need for and type of tests needed in the next tier. If the product is demonstrated to be free of problems in a lower tier, then there is no need to carry out further tests at the higher tier level. Specific areas of concern at each tier are, physico-chemical, environmental fate and ecotoxicity.

Acceptable results under the medicines' authorization process will mean that a Marketing Authorization (subject to all other necessary data) will be granted. However, as with user safety a further and separate area of consideration is then involved once the product is on the market and in use by farms. This is the area of environmental protection in which the authorities responsible for protection of the aquatic environment may determine how it is used, either generally or in specific localised environments. The latter control is reasonable and inevitable since, in the process of Marketing Authorization, only general environmental aspects may be taken into consideration. In the real environment, extreme cases will occur e.g., sea lochs with an abnormally high density of sea cage farms where the environmental impact of multiple sources of pharmacologically active substances may have significant environmental impact. Aside from these "extreme" cases, in the separation of regulation of environmental impact from that of veterinary medicines authorization, there is considerable opportunity for regulatory conflict. Where environmental authorities are regional rather than nationally based, there are opportunities for unequal interpretation of regulations to occur. At the inter-member state level there are also opportunities for uneven application of controls since the harmonization of environmental protection legislation across the EU is far less developed than is that for veterinary medicines.

Conflict may also occur between requirements of any environmental agency and the need to conduct preliminary field trials (e.g., ATC) on substances still under development, where field trials are a necessary part of developing the full environmental assessment package.

5. CASCADE SYSTEM (OFF LABEL USE)

Directive 90/676/EEC amended Directive 81/851 to establish the prescribing cascade. This limits veterinarians treating food-producing animals to prescribing veterinary medicines containing only substances authorised for use in food-producing species. Thus, where there is no suitable product to treat fish, a suitable product approved in other food animal species may be prescribed. Use of this prescription route requires the imposition of a standard withdrawal period by the veterinarian. For fish in the EU the standard withdrawal period has been set at 500 degree days. The standard withdrawal period is also imposed on any special emergency authorizations for use granted whilst product data packages are still insufficient.

Further significant restrictions exist on use of the cascade system. Even if used in different species, products may still only be applied by the exact route for which they are authorized. Thus cattle-pour on and pig oral-drench ivermectin formulations may not be "cascaded" for use on fish feed, only a formulation approved for use in feed for another species may be used in fish feed in this way. This particular product points to a potential hazard in that research has clearly demonstrated (Palmer *et al.*, 1997) that ivermectin in salmon can be extremely persistent, requiring withdrawals of at least 1000 degree days for full depletion to limit of detection. Thus an in-feed formulation of ivermectin approved for another species may be prescribed under the cascade provisions, but if the EU standard 500 degree day withdrawal is observed, residues will be detectable if slaughter occurs before depletion has completed. Since there is no MRL set for ivermectin in salmonids, the presence of such residues at slaughter are in breach of the requirements of Regulation 2377/90, although no breach was involved before slaughter. It is incumbent on veterinary surgeons prescribing via the cascade route to take note of any known pharmacokinetic characteristics of the product that they are prescribing and, where necessary, apply an appropriately extended withdrawal period.

The final restriction on the use of the cascade route of medication is that it is specifically restricted to use on a "small number" of animals. As Woodward (1996) commented, the interpretation of "a small number" is difficult when applied to aquaculture. With farmed fish all individuals in a cage or pond are at equal risk so that provided a proper diagnosis has been made, it might be reasonable to interpret "a small number" in aquaculture as applying to that cage or pond. In any case, proper records should be kept of such cascade use including:

- dates of examination;
- owners name and address;
- numbers treated;
- diagnosis;
- product prescribed;
- dosage administered;
- duration of treatment;
- withdrawal period specified.

Such records are, in any case, now required under the provisions of Directive 96/23 (see Residues Surveillance below) to enable efficient monitoring of tissue residues in residue monitoring programs.

6. MEDICATED FEEDINGSTUFFS

Where aquaculture medicines are applied in feed, then the provisions of Directive 90/167/EEC "laying down the conditions governing the preparation and placing on the market and use of medicated feeding stuffs in the Community" apply. In the UK final implementation of this for the "on fish farm" mixing of medicated fish feeds had been delayed, but is scheduled for implementation from 1 April 1998. This requires all fish farms that produce medicated feeds on site to register as on farm medicated feed producers and be subject to regular inspection for adherence to standards of quality and evenness of mixing of medicated product.

These requirements may prove difficult for fish farms to adhere to, but on-farm mixing of medicated feeds continues to be important in the UK. This is because of the transit times involved in transporting mill-mixed medicated feeds from mills (generally in central Scotland) to farms in remote areas of the Highlands and Islands or to southern England. Delivery of medicated feed from feed mills, where manufacture and transit times may exceed two days, is a guarantee of animal welfare problems and ineffective therapy during a rapid bacterial disease outbreak. At the time of writing, the plans for final implementation of Directive 90/167 were still in a late stage of development and any effects on the industry will not be seen for some time.

Drug formulations approved for incorporation in medicated feeds must be specifically approved for use via that route and must be in the form of a single authorized medicated premix. Provision is made that if a feedingstuff already contains medication, no further medication of the same type may be added. These requirements also apply in full to products prescribed as medicated feeds employing the provisions of the cascade system.

7. IMMUNOLOGICALS

The original Directives 81/851 and 81/852 did not apply to immunological products such as vaccines. In January 1992 these were included in the EU medicines regulatory environment by the implementation of Directive 90/677/EEC. Requirements for testing the safety and efficacy of vaccines are detailed in the Rules Governing Medicinal Products in the European Union (European Commission, 1995b) that includes not only general requirements for vaccine production, but defines test procedures for bacterial and viral fish vaccines. The European Pharmacopoeia Commission (EP) (a Council of Europe body, rather than an EU or EC body) has also published a number of monographs on vaccine test protocols for the more important bacterial diseases of salmon. In this author's opinion, these EP monographs are somewhat over restrictive, in some cases naïve in design, and certainly one recently aired proposal to restrict numbers of fish in a vaccine trial to ten fish (on welfare grounds) do not accord with best scientific practice in this specialized field. Unlike vaccines for other animal groups, batch testing solely for presence of antigen, will not guarantee potency in many fish vaccines.

8. RESIDUE SURVEILLANCE

There is, of course, little value in developing an elaborate and expensive system for ensuring that no unacceptable residues are present in food of animal origin (i.e. Regulation 2377/90) if no action is taken to confirm and enforce compliance. This next stage of the EU veterinary medicines harmonization program took the form of Directive 96/23/EC, enacted in April 1996 and brought into action in January 1998. This Directive introduces fish meat, poultry meat, milk and honey into Member States monitoring programs that had previously been limited to red meat. It requires member states to produce a monitoring programme to search for illegal or excessive drug residues in fish meat. It is also a third country Directive in that countries exporting to the EU have to demonstrate that they are able to ensure that no unacceptable residues are present in fish meat exported to the EU by means of a suitable residues control and monitoring program.

The non-statutory UK program, which had been in operation since 1993, has been replaced by a statutory program in compliance with Directive 96/23. For farmed fish, a monitoring level of one sample (= one or more fish) per 100 tons of production is required. A maximum of two-thirds of samples may be taken a wholesale level provided traceability to farm is guaranteed, the rest must be collected from the farm itself. Farms must keep full records of medications used and provide access to these during sampling inspections. Samples will be analysed for:

- for presence of residues of approved substances in excess of MRL;
- for residues of illegal substances and for substances which it is believed may be in use under the provisions of the cascade system, but for which there is no MRL in fish;
- for presence of Annex IV substances.

9. GROWTH ENHANCERS

Antimicrobial growth enhancers are unlikely to be cost effective to supplier or to a fish farmer in EU. Attitudes to their use within other areas of animal husbandry within Europe are hardening and with the high food conversion factors and rapid growth rates that are already being obtained there is unlikely to be any interest in the development of such products from any part of the industry.

Hormone based growth enhancers will not be acceptable in the EU. Problems in other areas of animal husbandry have resulted in a negative view of such uses. Even if the scientific case were to be that such products were harmless, consumer acceptance is not likely to be forthcoming. There is some interest in the use of probiotics, but for disease prevention rather than growth enhancement. Such uses would appear to fall under Feed Additives rather than Medicines legislation, although so far this has not been formally tested.

10. LICENSING AUTHORITIES

Until recently almost all veterinary medicines licensing has been carried out by the authorities of Member States. In the UK veterinary medicines regulation is the responsibility of the Veterinary Medicines Directorate an executive agency of the Ministry of Agriculture Fisheries and Food, advised by the independent Veterinary Products Committee. The Licensing Authority per se is the Ministers of Agriculture and Health. From 1 January 1998 what had been referred to as EU future systems came into operation. This means that Member States options of completely independent licensing decisions are no longer available. National procedures still exist however and the national Licensing Authority must determine applications made by this route within 210 days of receipt of an application.

Provision is made for applications for Marketing Authorizations to be made by a Centralised Procedure to the EMEA. This route is mandatory for some biotech products such as genetically modified organisms and for the moment these and applications for new molecules are the only applications which are accepted centrally. The majority of applications are still made by the Decentralised Procedure. This may be in the form of a serial application in which an application is made to the Licensing Authority of one MS which if successful, may be followed by applications to other MS. Those MS Licensing Authorities have 90 days to accept authorization of the first MS or to present cogent scientifically based reasons for any refusal. Alternately, parallel applications may be made simultaneously to several MS, once the first MS approves the others have 90 days to accept or produce cogent scientific reasons why they will not. New applications are required for changes in active ingredient, addition or change of indication or target species, change of MRL or change of withdrawal period

11. CONCLUSIONS

In 1998 we are at an intermediate stage in the development of a common legislatory framework for the control of all aspects of veterinary medicines within the EU. The initial legislation (Directives 81/851, 81/852 and Regulation 2377/90) laid what appeared to be a reasonable framework targeted especially at protection of the consumer. Subsequently, increased protection for environmental aspects and further consumer protection actions have been introduced. At the same time the recognition has dawned that the aims of the original legislation, whilst extremely well intentioned, were over ambitious both from timing and from the point of view of the costs imposed.

The impossibly tight schedule imposed by Regulation 2377/90 has had to be eased by the introduction of the Defended Substances list. The assumption sometimes made by consumer lobbies that vast profits are to be made from veterinary pharmaceuticals is far from reality, particularly in niche market areas such as that of aquaculture medicines. For niche and minor species markets, sponsors of species specific MRLs have not been forthcoming (as in N. America, see Schnick this volume) so that properly authorized medicines are likely to be less available, presenting animal welfare problems. This in its turn leads to reliance on cascade (off label) prescription, which if continued to an extreme, could reduce rather than improve, consumer safety. The welcome recent pragmatic approach by the CVMP to MRL and target residue determination in these niche market areas, reflects these pressures.

Pressures on environmental safety requirements are now increasing and in the future may present both the same and additional conflicts. Costs of environmental safety data packages (and parallel demands from environmental agencies) are likely to increase continuing restrictions where other safety demands are becoming somewhat more pragmatic. In this case the possibility must be considered that production of aquaculture products could reduce in Europe and transfer to other areas where, although the consumer safety controls on imports to Europe will still apply, environmental safety requirements may be more lax.

The aquaculture industry is fortunate in that very effective vaccines have been developed for the control of many of the more important diseases that affect European farmed stocks. The use of these vaccines have resulted in an exponential fall in the use of antibiotics in European salmon culture (Alderman, 1998) reliving pressure on the limited range of approved antimicrobials where antibiotic resistance was an increasing problem. No such vaccines however are available for sea lice control or for control of protozoan ecto- and endoparasites.

Current and foreseeable EU regulatory requirements for fish vaccines are relatively economic to comply with, but to produce vaccines for sea lice or protozoan control may well require entry into the biotechnology arena. Requirements in this area may well preclude development of effective and safe vaccines as sponsors weigh the cost of generating of the consumer and environmental safety data against income which could be generated from the relatively small aquaculture market.

The conflicts between costs which can be afforded by aquaculture and safety requirements are unlikely to disappear.

ACKNOWLEDGEMENT

The author would like to acknowledge the support and funding of the Ministry of Agriculture, Fisheries and Food in the preparation of this manuscript.

12. REFERENCES

Alderman, D.J., 1999, Chemicals in aquaculture, Proceedings, 2nd International Conference on Sustainable Aquaculture, Holmenkollen, Oslo, Nov 1997, Balkema, Netherlands. pp. 205–220.

Committee for Veterinary Medicinal Products, 1997, Note for guidance on the establishment of maximum residue limits for minor species, *EMEA/CVMP/153a/97-FINAL*, European Medicines Evaluation Agency, Veterinary Medicines Evaluation Unit, London, 6pp.

Committee for Veterinary Medicinal Products, 1998, Note for guidance on the establishment of maximum residue limits for *Salmonidae* and other fin fish, *EMEA/CVMP/153b/97-FINAL*, European Medicines Evaluation Agency, Veterinary Medicines Evaluation Unit, London, 2pp.

Council of the European Communities, 1981a, Council Directive No. 81/851/EEC on the approximation of the laws of Member States relating to veterinary medicinal products, *Official Journal of the European Communities*, No. L 317/1.

Council of the European Communities, 1981b, Council Directive No. 81/852/EEC on the approximation of the laws of Member States relating to analytical, pharmacotoxicological and clinical standards and protocols in respect of testing of veterinary medicinal products, *Official Journal of the European Communities*, No. L 317/16.

Council of the European Communities, 1987, Council Directive No. 87/20/EEC amending Directive 81/852/EEC on the approximation of the laws of Member States relating to analytical, pharmacotoxicological and clinical standards and protocols in respect of testing of veterinary medicinal products, *Official Journal of the European Communities*, No L 15/34.

Council of the European Communities, 1990a, Council Directive No. 90/167/EEC laying down the conditions governing the preparation, placing on the market and use of medicated feedingstuffs in the Community, *Official Journal of the European Communities*, No L 92/42.

Council of the European Communities, 1990b, Council Directive No. 90/676/EEC amending Directive 81/851/EECon the approximation of the laws of Member States relating to veterinary medicinal products, *Official Journal of the European Communities*, No L 373/15.

Council of the European Communities, 1990c, Council Directive No. 90/677/EEC extending the scope of Directive 81/851/EEC on the approximation of the laws of Member States relating to veterinary medicinal products and laying down additional provisions for immunological veterinary medicinal products, *Official Journal of the European Communities*, No L 373/26.

Council of the European Communities, 1990d, Council Regulation No. 2309/93/EEC laying down Community procedures for the authorisation and supervision of medicinal products for human and veterinary use and establishing a European Agency for the Evaluation of Medicinal Products, *Official Journal of the European Communities*, No L 214/1.

Council of the European Communities, 1990e, Council Regulation No. 2377/90/EEC laying down a Community procedure for the establishment of maximum residue limits of veterinary medicinal products in foodstuffs of animal origin, *Official Journal of the European Communities*, No L 224/18.

Council of the European Communities, 1996, Council Directive No. 96/23/EC on measures to monitor certain substances and residues thereof in live animals and animal products and repealing Directives 85/358/EEC and 86/469/EEC and Decisions 89/187/EEC and 91/664/EEC, *Official Journal of the European Communities*, No L 125/10.

European Commission, 1991, Establishment by the European Community of maximum residue limits MRLs for residues of veterinary medicinal products in foodstuffs of animal origin, in: *The Rules Governing Medicinal Products in the European Union*, Volume VI, Office for Official Publications of the European Communities, Luxembourg.

European Commission, 1995a. Efficacy of veterinary medicinal products for use in farmed aquatic species, in: *The Rules Governing Medicinal Products in the European Union, Guidelines for the testing of veterinary medicinal products*, Volume VII, pp 145–161, Office for Official Publications of the European Communities, Luxembourg.

European Commission, 1995b, Specific requirements for the production and control of live and inactivated vaccines intended for fish, in: *The Rules Governing Medicinal Products in the European Union, Guidelines for the testing of veterinary medicinal products* Volume VII, pp 245–250, Office for Official Publications of the European Communities, Luxembourg.

European Agency for the Evaluation of Medicinal Products, 1997, Communication of the European Agency for the Evaluation of Medicinal Products pursuant to Article 1 of Council Regulation EC No 434/97, *Official Journal of the European Communities*, No C 97/ 165/03.

Palmer, R., Coyne, R., Davey, S., and Smith, P., 1997, Case notes on adverse reactions associated with ivermectin therapy of Atlantic salmon. *Bur. Eur. Assoc. Fish Pathol.*, 17:62–6.

United Kingdom, 1986, *The Animals (Scientific Procedures) Act 1986*, The Stationary Office, London.

Veterinary Medicines Directorate, 1996, Ecotoxicity testing of medicines intended for use in fish farming. *VMD Guidance Note*, AMELIA 11, VMD, Addlestone, U.K.

Woodward, K.N., 1996, The regulation of fish medicines—U.K. and European Union aspects, *Aquacult. Res.*, 27:725–734.

4

PESTICIDE BIOACCUMULATION AND METABOLISM

Study Requirements, Experimental Design, and Data Analysis

Mace G. Barron[1] and Kent B. Woodburn[2]

[1]Hagler Bailly Services
Inc., 1881 9th St., Suite 201
Boulder, Colorado 80302
[2]Health and Environmental Research Laboratories
The Dow Chemical Company
1803 Building
Midland, Michigan 48674

1. ABSTRACT

Registration of pesticides under the U.S. Federal Insecticide Fungicide Rodenticide Act requires multiple laboratory and field studies assessing bioaccumulation, persistence, and biotransformation in fish. The specific study requirements are dependent on the intended use of the pesticide, its physical/chemical properties, and its expected environmental persistence. Typical studies include laboratory bioconcentration studies quantifying the rate and extent of pesticide uptake, biotransformation, and elimination, and environmental field studies quantifying bioaccumulation and persistence in aquatic biota under actual-use conditions. This chapter presents an overview of pesticide bioaccumulation and metabolism study requirements, experimental design and analytical chemistry approaches, and data analyses.

2. INTRODUCTION

The U.S. Federal Insecticide, Fungicide, and Rodenticide Act (FIFRA) is the predominant statute for regulating chemicals used as pesticides. FIFRA defines pesticides as any substance or mixture of substances intended to prevent, destroy, repel, or mitigate insects, rodents, fungi, weeds, and other forms of life (Touart, 1995). FIFRA is unique in that it licenses toxic materials intended for release into the environment. The

Xenobiotics in Fish, edited by *Smith et al.*
Kluwer Academic / Plenum Publishers, New York, 1999.

U.S. Environmental Protection Agency (EPA) is responsible for approving and regulating pesticides under FIFRA. The pesticide registrant (e.g., manufacturer or licensee) is responsible for meeting EPA's data requirements for pesticide registration, including demonstrating that the intended uses will not result in unreasonable adverse effects to the environment. EPA determines allowable uses of a pesticide by considering risks to human health and the environment, as well as economic and other benefits of pesticide use.

Pesticides include many specific chemical substances that can be grouped according to the type of pest they are intended to control: herbicides, insecticides, fungicides, rodenticides, and animal repellents. Pesticides may include synthetic as well as naturally-derived chemicals or chemical mixtures. Because of their dispersive applications, such compounds may enter surface water by one or more of the following routes: direct application, drift, mobilization in ephemeral stream channels, overland flow, and leaching. Properties of the chemical (e.g., vapor pressure, solubility in water, and octanol: water partition coefficient) and properties of the environment (e.g., temperature, moisture, topographic characteristics) interact to produce the particular behavior (e.g., movement, persistence, and fate) observed in the environment. This chapter summarizes the study requirements, experimental design and analytical approaches, and data analysis procedures for bioaccumulation and metabolism studies required for U.S. pesticide registration. Additionally, U.S. registration requirements are compared and contrasted with relevant Organization for Economic Cooperation and Development (OECD) testing guidelines.

3. STUDY REQUIREMENTS

Bioaccumulation and metabolism study requirements for U.S. registration of a pesticide under FIFRA are dependent on the intended use, chemical properties, and environmental fate of the chemical (Figure 1). Registration of pesticides intended for use on rotational or irrigated crops (e.g., insecticides and herbicides applied to crops such as corn) generally require a laboratory bioconcentration study. Any chemical intended under FIFRA for aquatic food crop uses, aquatic noncrop uses (i.e., aquatic weed control, antifouling paints, and other outdoor protective uses), or for any direct aquatic impact uses, must also undergo both field residue testing in an aquatic field study and residue identification and characterization in a laboratory fish metabolism study. As discussed below, the bioconcentration and fish metabolism studies quantify accumulation, biotransformation, and elimination under controlled laboratory conditions, but they can differ in study duration, species of fish, and exposure concentration. Fish metabolism studies and aquatic field studies are registration requirements for licensing aquatic uses of a pesticide, regardless of the properties of the chemical (Figure 1). In contrast, a bioconcentration study is generally only required for registration with the OECD (OECD, 1996) when the pesticide has sufficient hydrophobicity (octanol:water partition coefficient, $K_{ow} > 1000$), as noted in the 1996 Official Journal of the European Communities, Commission Directive. The Directive noted that bioconcentration studies on fish are required only on compounds that are *"likely to partition into fatty tissues (such as log $K_{ow} \geq 3$). . . . unless it can be justified that exposure leading to bioconcentration is not likely to occur."*

For U.S. registration, until finalization of new guidelines, fish bioconcentration studies under FIFRA are currently directed by two sets of U.S. EPA guidelines: (a)

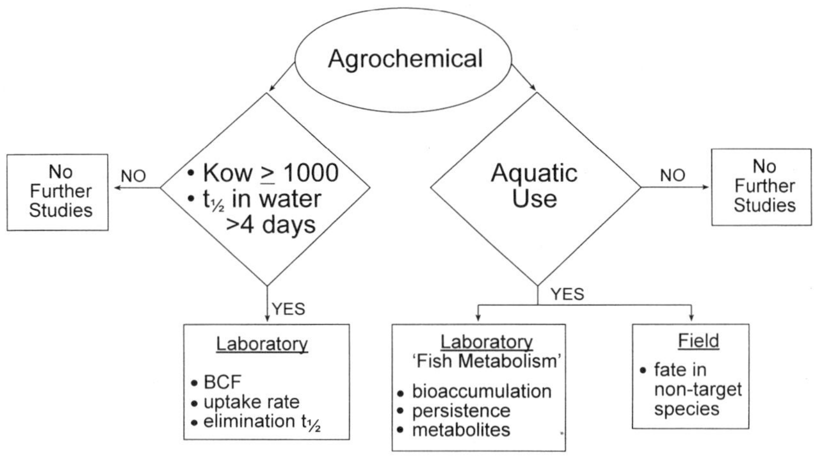

Figure 1. Bioaccumulation and metabolism study requirements for U.S. pesticide registration.

Guideline 165-4 (Subdivision N-Environmental Fate) on "*Laboratory studies of pesticide accumulation in fish*" (U.S. EPA, 1982a); and (b) Guideline 72-6 (Subdivision E-Wildlife and Aquatic Organisms) on "*Aquatic organism accumulation tests*" (U.S. EPA, 1982b). The guidelines fundamentally do not require fish accumulation data if the active ingredient will not persist in water (i.e., has a half-life of 4 days or less), has a relatively low potential for accumulation in fish as indicated by an octanol : water partition coefficient less than ~1,000, or if the active ingredient or its principal degradation product(s) does not accumulate in the organs and tissues of mammals or avian species (U.S. EPA, 1982a; 1982b). In these guidelines, the agency has concluded that bioconcentration is a time-dependent process that requires both lipophilicity and persistence in the water column for a chemical to accumulate in fish tissue; the scientific rationale for an octanol : water partition coefficient of ≥1,000 and an aqueous half-life value of ≥4 days as indicators of bioconcentration potential was detailed by Gillett (1983). Similarly, the Brazilian pesticide registration requirement for fish bioconcentration studies (Brazil, 1995) is dependent on the compound having a "$\log K_{ow} > 2$ or water solubility <1 mg/L and hydrolysis half-life > 4 days or (evidence that) the product is not readily biodegradable."

3.1. Bioconcentration Study

The objective of the laboratory bioconcentration study is to generate a "worst case" estimate of the potential for bioconcentration of a pesticide to occur in fish. This study quantifies the bioconcentration factor (ratio of chemical in fish and water; BCF), the rate of uptake (e.g., time to reach 90% of steady state; t_{90}), and the elimination half-life ($t_{1/2}$) of a pesticide in nontarget fish and other aquatic organisms. Bioconcentration studies are typically conducted under flow-through conditions with a sublethal concentration of the pesticide. Both the parent chemical and metabolites are quantified, and major metabolites are identified as described in Section 4 below. Study data are intended to be used in assessing the risks of pesticide bioaccumulation in fish and other aquatic organisms (EPA, 1982; EPA, 1996a; OECD, 1996).

3.2. Fish Metabolism Study

The objective of the fish metabolism study is to determine the metabolic fate of an aquatic-use pesticide in fish. This study is categorized as a "nature/magnitude of residue study" under FIFRA. A fish metabolism study quantifies the parent chemical and metabolites in fish at exposure concentrations greater than the maximum-proposed use rate (EPA, 1996b). Study data are intended to be used in dietary risk assessments and to establish acceptable residue levels (tolerances) for enforcement purposes.

3.3. Aquatic Field Study

The objective of the aquatic field study is to determine residues in fish and other biota under actual-use conditions. This study assesses the fate of the pesticide in the environment of intended use and quantifies the nature, magnitude, and dissipation of residues resulting from the maximum-proposed use rate (EPA, 1996c). Study data are intended to allow evaluation of the potential bioaccumulation and environmental persistence of the aquatic-use pesticide under actual-use conditions.

3.4. Additional Study Requirements

Additional study requirements for licensing a pesticide include acute and chronic toxicity studies with a variety of fish and other aquatic test species. These studies address toxicity, rather than bioaccumulation and metabolism, and have been summarized by Touart (1995). EPA has also established a tiered approach to environmental fate degradation testing to support the federal registration of pesticides (key studies are listed in Table 1). Pesticides introduced into water may undergo hydrolysis (chemical interaction with water), and the rate of hydrolysis and the identity of degradation products can be determined in laboratory studies under sterile, controlled pH conditions (Guideline 161-1 in Table 1). If the pesticide has the appropriate quantum yield and absorbs light at wavelengths greater than 290 nm, the compound may undergo photochemical degradation in water, a process that may also occur with pesticide residues on soil or in the vapor phase. Experimental data on the rate of photolysis and the decay products arising from this process are the desired endpoints of specific laboratory studies (Guidelines 161-2 and 161-3 in Table 1). The data generated from hydrolysis and photolysis studies are used to determine the cumulative rate of chemical decay of a pesticide in the environment and to identify the nature of the degraded residues.

The environmental fate guidelines listed in Table 1 with a "162" prefix describe experiments designed to assess the relative importance of aerobic and anaerobic metabolism of a pesticide in both soil and water media. The data generated by such studies are important, along with hydrolysis and photolysis studies, in determining the nature and persistence of pesticide residues in the environment. Pesticide mobility in soil is addressed by Guideline 163-1 in Table 1; the other mobility guidelines ("163" prefix; not shown in Table 1) involve vapor phase transport of volatile pesticides. Data generated from these studies are useful in assessing the potential for off-site movement of a pesticide, with the possible resulting impacts on wildlife, food products, and water resources. Nonvolatile pesticides bind to soil particles to varying extents and potentially move with the water front (i.e., leach), depending on physical and chemical properties such as affinity for adsorption surfaces, water solubility, and pK_a. These studies are designed to measure the leaching potential of pesticides and their degradation products in soil.

Table 1. FIFRA environmental fate studies.

Study Title	OPP Guideline* No.	General Description
Hydrolysis	161-1	Used to establish the significance of chemical hydrolysis as a degradation route of the pesticide and to identify hydrolytic products.
Photodegradation in Water	161-2	Data on photolysis half-life and photoproducts formed are needed to establish importance of this degradation process.
Photodegradation on Soil	161-3	Data on the rate of pesticide photolysis and photoproducts formed are needed to establish importance of this degradation process.
Aerobic Soil Metabolism	162-1	Used to determine the rate of formation and pattern of pesticide metabolites under aerobic soil conditions.
Anaerobic Soil Metabolism	162-2	Used to determine the rate of formation and pattern of soil metabolites under anaerobic soil conditions These conditions might occur due to the effects of flooding or waterlogging of soils.
Anaerobic Aquatic Metabolism	162-3	Used to assess the effects on a pesticide of exposure to anaerobic conditions in water or sediment.
Aerobic Aquatic Metabolism	162-4	Used to assess the effects on a pesticide of exposure to aerobic conditions in water or sediment.
Leaching (Adsorption/ Desorption)	163-1	To examine the probability of movement of pesticide residues by leaching through the soil profile.
Pesticide Accumulation in Fish	165-4	Used to determine the extent of pesticide bioconcentration in fish (see Table 2).

*U.S. Environmental Protection Agency, Office of Pesticide and Toxic Substances, Pesticide Assessment Guidelines Subdivision N-Chemistry: Environmental Fate. 1982. EPA 540/9-82-021.

4. EXPERIMENTAL DESIGN AND ANALYTICAL APPROACHES

4.1. Bioconcentration Study

The experimental design of a bioconcentration study is intended to allow estimation of the bioconcentration potential of a pesticide. Available guidance on experimental design includes ASTM (1996), EPA (1982a,b; 1996a), and OECD (1996). Table 2 summarizes recommended bioconcentration study conditions. Phases of a bioconcentration study include preliminary work (determination of exposure concentration, flow-rates, test duration), analytical method development, and experimental uptake and elimination phases (Figure 2). During the uptake phase, fish are exposed to a constant concentration of the pesticide, and then are allowed to depurate residues in clean water during the elimination phase. Generally ^{14}C-labeled pesticides are used to facilitate identification of total residues (parent plus metabolites) in the fish. The radiolabel should be on a portion of the molecule with anticipated stability (e.g., ring-labeled rather than on an alkyl chain, which may be cleaved from the pesticide moiety). Pesticides of greater than 98% radiochemical purity are preferred (OECD, 1996), although generally a purity of ≥95% is acceptable. Typically a stock solution of the pesticide is prepared in a polar solvent (e.g., methanol, dimethylformamide, etc.) to provide high miscibility with water and allow preparation of a stable high concentration of the pesticide for dosing; the maximum-allowable solvent concentration is 0.1 mL/L and this level should be the same in control and exposure aquaria (OECD, 1996). Analytical chemistry methods are developed to assure adequate extraction efficiency (e.g., >90% recovery of parent spiked into

Table 2. Laboratory bioaccumulation study conditions required or recommended for U.S. EPA pesticide registration.

Study Component	Study Conditions: Bioconcentration	Study Conditions: Fish Metabolism
Test species	Standard test species (e.g., trout, bluegill)	Site and application relevant
Fish size	2 to 12 cm; <3 x size variation	Species specific
Test substance	Prefer radiolabeled material (e.g., ^{14}C-labeled), with purity >95%	Same
Test concentrations	Two separate dose studies, preferably 10x apart; higher dose <1% of LC_{50} and >10x aquatic detection limit; ±20% variation during exposure	Single dose at 1 to 10x maximum application rate (typical: 2–4x)
Solvent	<0.1 mL/L	Same
Photoperiod	12:12 or 16:8 (Light/Dark)	Same
Temperature	Species specific; ±2°C	Same
Dissolved oxygen	>60% saturation	Same
pH	6 to 8.5, but within ±0.5 for a given test	Same
Flow-rate	≥5 volume replacements/day; semi-static possible if test concentration conditions met	Same
Fish loading	0.1–1.0 g fish/L/day	Not specified
Mortality	<10%	Same
Feeding	1–2% total body weight/day	Same
Uptake phase duration	28-day standard; if steady-state is not achieved, to a maximum of 60 days	Minimum 3 days (based on environmental fate)
Steady-state definition	Three successive fish samples ±20% (2-day interval)	Not applicable
Elimination phase duration	Period sufficient for a 95% reduction of steady-state body burden; if this exceeds 2x the exposure time frame (e.g., 56 days), a 90% reduction may be used	Not required; sample fish within 24 hours of exposure termination
Water samples	Uptake: ≥5; demonstrate test concentrations are ±20% of mean elimination: ≥4 samples	Same; no elimination samples required
Fish samples	Uptake: ≥5 Elimination: ≥4	Same; no elimination samples required
BCF	Whole fish; edible/non-edible desirable	Same
Rate constants	Required; non-linear parameter estimation recommended	Not required
Metabolite identification	If BCF ≥1000; identify/quantify degradates ≥10% of total residues at steady-state	Identify 90% of total radioactive residues (TRR); see Fig. 7
Lipid measurement/ normalization	Required; chloroform/methanol lipid extraction recommended	Not required

water and tissue blanks) and storage stability (e.g., >4 weeks or within the holding time of all water and tissue samples).

The dosing and exposure system mixes the pesticide with water using precisely controlled flow rates of stock solution and laboratory dilution water (Figure 3). Control vessels receive the same concentration of solvent and water flow rates as pesticide exposure vessels. Components of the test system are constructed of silicone, glass, stainless steel, and Teflon to limit adsorption of the pesticide. Dosing and exposure system performance is refined to meet required flow rates, fish loading, and water quality requirements (Table 2). Before the start of the uptake phase, fish are acclimated to the test conditions (e.g., temperature, dilution water). The exposure system consists of one or

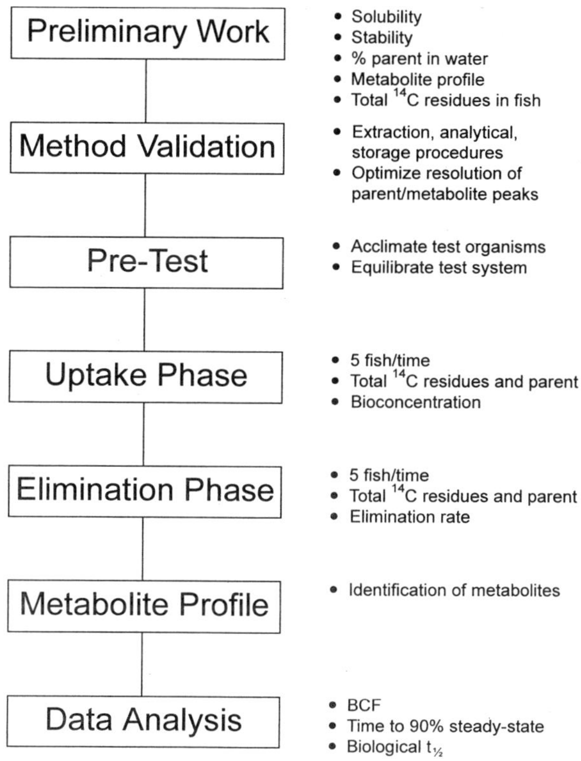

Figure 2. Bioconcentration study phases.

more exposure and control vessels (e.g., 40 L glass aquaria). For pesticide registration, the bioconcentration study requires two separate dose exposures, both at less than 1% of the 96-hr median lethal concentration (LC50) and greater than 10 times the analytical detection limit of the pesticide. The control vessel is used to evaluate mortality during

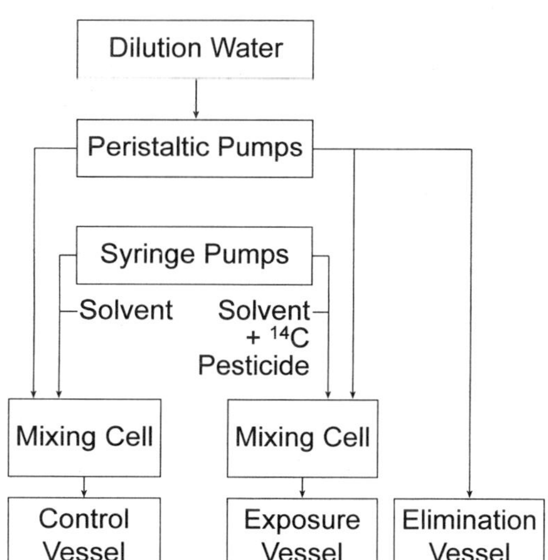

Figure 3. Schematic representation of a dosing and exposure system.

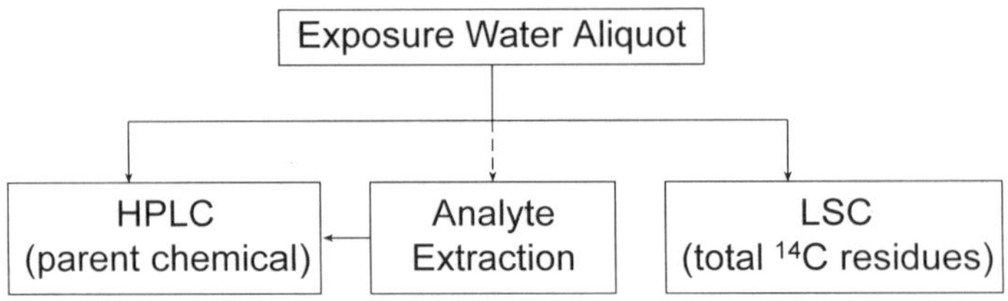

Figure 4. Analytical scheme for sampling water. Analyte extraction may include liquid:liquid or solid phase extraction. LSC:liquid scintillation counting.

the study, as well as provide tissue for analytical chemistry blanks. The dosing and exposure systems are equilibrated for at least 48 hours to assure equilibration with system components and to demonstrate a stable exposure concentration of the pesticide.

Typical fish species used in bioconcentration studies include juvenile rainbow trout and bluegill (e.g., 3–7 cm and 2–4 cm total lengths, respectively). At test initiation, fish are added to both control and exposure vessels (loading of 0.1 to 1.0 g-total wet weight fish $* L^{-1} * day^{-1}$ for each control or pesticide treatment). During the study, control and exposure water are monitored (e.g., measurement of dissolved oxygen, temperature, pH; 3 times/week) to assure adequate water quality (Table 2). Control and exposure water are also monitored for both parent (e.g., once/week) and total carbon-14 concentrations (e.g., 3 replicates per day) (Figure 4). Fish are sampled (e.g., 4–8 fish/sample time) and analyzed (parent, metabolites, and total carbon-14) during the uptake (≥5 sample times) and elimination (≥4 sample times) phases (Figure 5). The duration of the uptake phase is generally 28 days, although establishment of steady-state conditions may permit early termination. As noted in Table 2, the absence of steady-state conditions after 28 days may extend the exposure period to a maximum of 60 days (OECD, 1996). The duration of the elimination phase must generally be sufficient to demonstrate a 90 to 95% reduction in steady-state fish residues (see Table 2).

Metabolites present in fish at greater than 10% of the total residues at steady-state must be identified for those pesticides with a BCF ≥1,000. An example analytical strategy is presented in Figure 6. Typical analytical approaches include polar solvent extraction (e.g., 1% phosphoric acid in acetone) to remove both nonpolar parent chemical as well as polar metabolites. Metabolite identification is aided by analytical standards to determine the relative polarity of metabolites. The structural identity of metabolites can

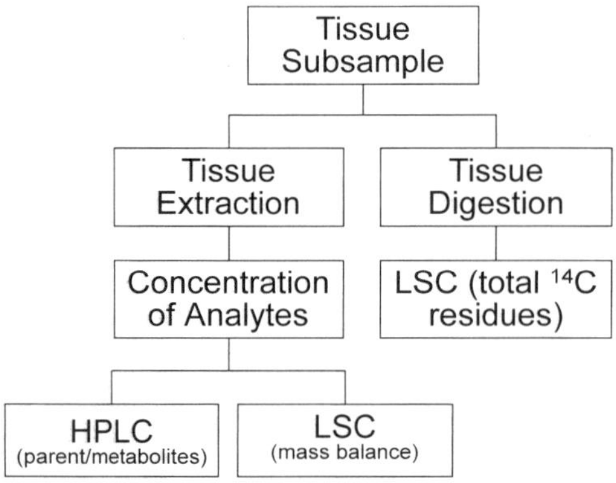

Figure 5. Analytical scheme for sampling fish. Tissue extraction of analytes may include solvent extraction (e.g., 3 × 10 mL of 1% H_3PO_4 in acetone), and determination of total carbon-14 residues by digestion and/or combustion and LSC (liquid scintillation counting).

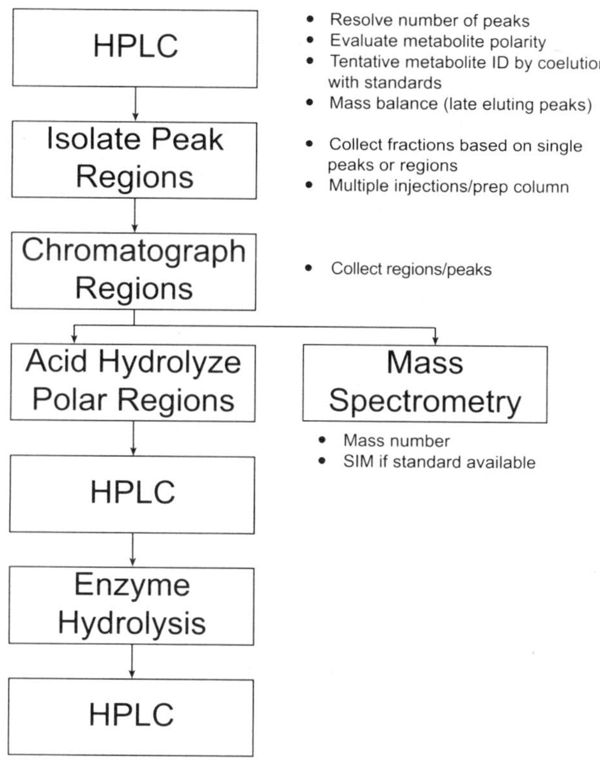

Figure 6. Strategy for metabolite identification.

be confirmed by co-elution with metabolite standards on two dissimilar high-performance liquid chromatography (HPLC) systems or by evaluating the mass spectra determined by selected ion monitoring (SIM) mode or full scan mass spectrometry (MS) (Figure 6).

4.2. Fish Metabolism Study

As discussed above, a fish metabolism study is required to support registration of any product directly applied to water to control aquatic pests such as weeds or insects (EPA, 1996b). Data from these studies are used in dietary risk assessments and to establish tolerances for enforcement purposes. The purpose of the fish metabolism study is to determine the qualitative metabolic fate of the pesticide in fish, so it differs in its objective from the fish bioconcentration study. The methodology of the study has general similarities to a bioconcentration study, including use of a radiolabeled pesticide, as well as fish and water sampling approaches and analytical chemistry procedures. Differences include study duration and exposure regime. The duration of a fish metabolism study is dictated by the environmental persistence of the pesticide (e.g., half-life in water), rather than the time required for fish to reach steady-state equilibrium with the chemical in water. Exposures may be static or static renewal if the test material is of adequate stability.

The principal experimental design aspects that are specific to the fish metabolism study are (1) the dose level (laboratory exposures are conducted at a concentration in excess of the maximum-proposed use rates), and (2) the rigorous identification of the residues accumulated in the fish, rather than assessing bioconcentration potential. The discussion below focuses on metabolite quantification requirements and experimental approaches. A listing of the recommended experimental conditions of a fish metabolism study is presented in Table 2.

The fish metabolism study must characterize and identify the total radioactive residue (TRR) in whole fish tissue. The TRR is defined as the sum of the parent pesticide, its degradation products, and any metabolites (free or bound) that may be of toxicologic interest. Properly characterizing the composition of the TRR often presents complex problems involving analytical methodology and proper quantification of residue data. Development of the study protocol and design may be done in conjunction with EPA chemists and toxicologists to determine the level of residue characterization required, which metabolites should be covered by tolerances, and the degree of residue identification/characterization that must be determined. The determination of whether the residue has been sufficiently characterized and/or identified will depend on the importance of the tissue containing the unidentified residue, the radioactivity remaining unidentified, the chemical structure of the active ingredient and identified metabolites, and the toxicology of compounds similar in structure to the metabolites. The terms characterization and identification have different meanings and should not be used interchangeably.

The term "identification" refers to the explicit structural determination of components of the TRR; this is typically accomplished by comparing chromatographic behavior to that of known standards and/or actual spectroscopic analyses [e.g., MS, nuclear magnetic resonance (NMR), etc.] (EPA, 1996b). Identification of metabolites must be established using two different analytical techniques, except when exact identification is made using a spectroscopic method such as gas or liquid chromatography/mass spectrometry (GC/MS or LC/MS), or the metabolite is determined to be of minimal importance because of its low absolute level (<0.05 ppm) or percentage of the TRR (<10% of the TRR). In the latter case, identification by one technique such as co-elution with standards is considered acceptable. In general, application of a chromatographic technique utilizing two different solvent systems and the same stationary phase is not considered to be an adequate two-method verification of metabolite identity (EPA, 1996b).

The term "characterization" refers to the structural description of the general nature/attributes of the radioactive residue without specific metabolite identification. The terms used to characterize residues include water or aqueous soluble, organosoluble, neutral, acidic or basic, polar, nonpolar, and nonextractable. Characterization may also involve conversion to a known structure through chemical treatment (e.g., acid hydrolysis). The degree of characterization refers to how close the assignment comes to structural identification. If identification of radioactive residues is not accomplished, the degree of characterization required will depend on several factors, including the amount of the TRR previously identified, the amount of residue present, the importance of the commodity, toxicological concern over the class of compounds, the suspected significance of the residue as determined by characterization already performed, and the capability of analytical methods to detect characterized but unidentified residues.

The application rate of the radiolabeled pesticide must be high enough to result in sufficiently high radioactivity levels to allow for characterization and/or identification of the residue. A rate of at least the maximum-registered application rate should be used, and for certain pesticides/uses an exaggerated rate (up to 10x) must be applied. The specific activity [i.e., number of disintegrations per minute (DPM) per amount of radiolabeled standard] of the pesticide should also be high enough to assure acceptable limits of detection for radioactive residues.

Figure 7 presents the preferred strategy for determining when identification or characterization of metabolites is required in a fish metabolism study (EPA, 1996b). If the total radioactivity in the fish is ≤10 ppb, then no further differentiation of radioactivity would be required unless there are toxicological concerns over residues at lower levels. For residues greater than 10 ppb, the sample should be extracted with organic and/or aqueous solvents and the levels of extractable and nonextractable (bound) residues should be measured to determine the degree of characterization required. If the extractable residues represent 10 ppb or less, additional characterization is not required. For extractable residues of 10–50 ppb, the partitioning behavior between aqueous and organic solvents should be determined, followed by chromatographic analysis (e.g., HPLC, GC, etc.) of the organosoluble residue. A comparison can then be made between the chromatographic behavior of the residue to the parent material and any possible metabolites. If the extractable activity exceeds 50 ppb, complete characterization and identification of activity in the organic and aqueous phases should be attempted; the "trigger" values for characterization and identification of residues in the aqueous portion

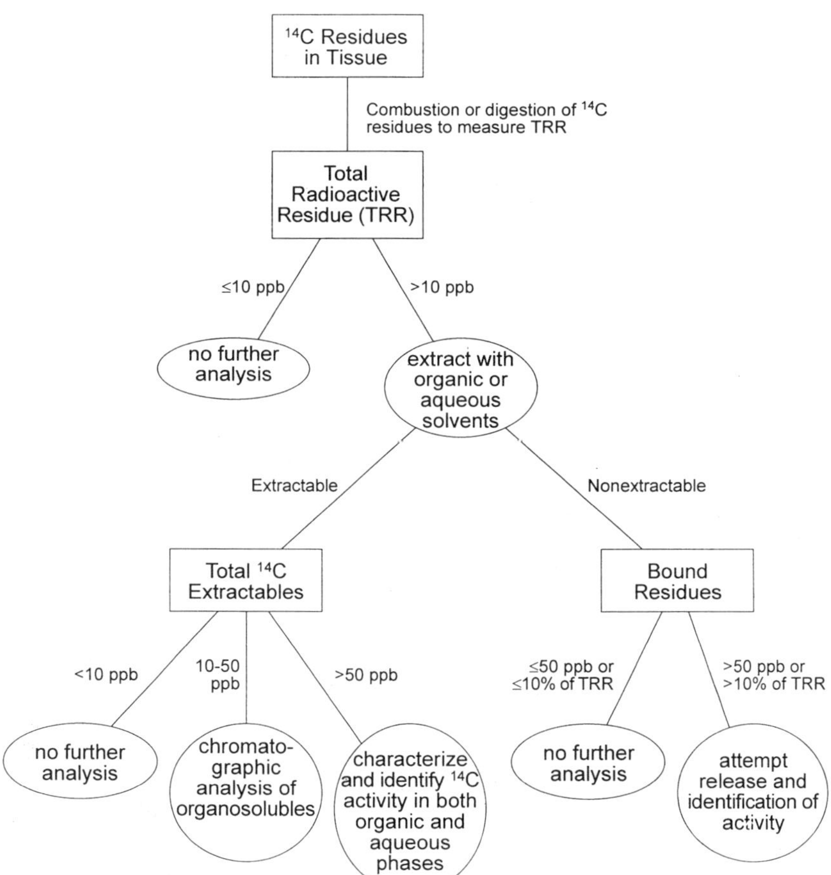

Figure 7. Strategy for identifying and characterizing residues in a fish metabolism study.

of the activity is levels down to 50 ppb or 10% of the TRR. While complete characterization and identification is required when extractable activity exceeds 50 ppb, this does not necessarily dictate that all individual components require identification; low level residues (both in terms of ppb and percent of TRR) do not require identification if a major portion of the residue has already been successfully identified. An example given by the EPA (1996b) suggests that if the TRR is 3 ppm and 75% of the residue has been identified, it is unlikely that further identification efforts would have to be expended on individual residues in the 50–100 ppb range. However, if the total activity was only 0.3 ppm, extensive efforts toward identification of 50–100 ppb would be required. The desired result of a fish metabolism study is identification of 90% of the TRR.

When bound residues are present at >50 ppb or >10% of TRR (whichever is greater), sample analysis must include an attempt at release and identification of residues. The sample treatments usually involve the following: acid and/or alkaline reflux, enzyme treatment, treatment with surfactants, and dilute acid and/or base treatment at ambient temperatures. If the bound residues are present at <50 ppb or <10% of TRR, no further analysis is required.

4.3. Aquatic Field Study

The aquatic field study should be conducted in at least two different sites that are representative of the areas where the pesticide is expected to be discharged or applied. The test substance should be a typical end-use product (same formulation), and it should be applied using the method of application specified on the product label. Samples of water, sediment, and plants should be taken, in addition to samples of representative fish species (samples of whole body and fillet) to measure the potential for residues of the pesticide to accumulate in fish tissue. Additionally, aquatic invertebrates may be sampled. The fate of the compound must be demonstrated with respect to the rate of dispersion downstream, degradation, volatilization (if appropriate), and sorption by plants or sediment. The aqueous data must show the highest level likely to occur in water at the maximum application rate. All samples should be analyzed for parent material and metabolites/degradation products.

5. DATA ANALYSIS AND STUDY EXAMPLES

5.1. Bioconcentration Study

Data analyses in a bioconcentration study include estimation of uptake (K_1; $mL{*}hr^{-1}{*}g^{-1}$) and elimination (K_2; hr^{-1}) rate coefficients, time to 90% steady state (t_{90}; hr), elimination half-life ($t_{1/2}$; hr), and BCF. The BCF is defined as the ratio of the chemical concentration in the fish $[C_f]$ to the chemical concentration in the exposure water $[C_w]$ at steady-state equilibrium (i.e., $BCF = [C_f]/[C_w]$). Estimation of BCF should be determined by both empirical measurement and pharmacokinetic methods. Empirical estimation of BCF requires that total residues in fish have reached apparent steady-state equilibrium with the exposure water. Steady-state is defined as total residues measured in three serial samples of fish (taken at least 2 days apart) that are within ±20%. Pharmacokinetic methods involve nonlinear modeling of fish residue values during uptake and elimination phases, and representing the fish as a one- or two-compartment system (Barron *et al.*, 1990). The simplest pharmacokinetic estimate of BCF is the ratio of uptake and

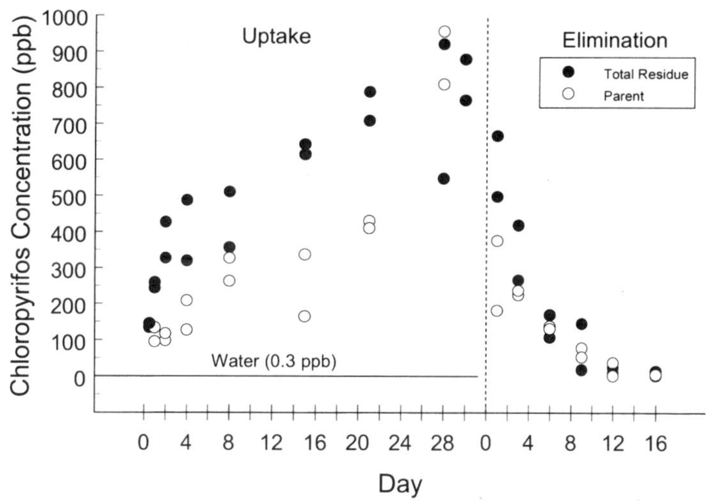

Figure 8. Example bioconcentration study data: chlorpyrifos.

elimination rate coefficients (i.e., BCF = k_1/k_2) in fish modeled as a one compartment system. Both t_{90} and $t_{1/2}$ are estimated from k_2, which is determined by the decrease in residues during the elimination phase, where C_t is concentration at time t and C_0 is concentration at time 0:

$$C_t = C_0 * e^{-k2t} \tag{1}$$

$$t_{90} = 1.1/k_2 \tag{2}$$

$$t_{1/2} = 0.693/k_2. \tag{3}$$

Data derived from a bioconcentration study with the insecticide chlorpyrifos [O,O-diethyl-O-(3,5,6-trichloro-2-pyridyl) phosphorothioate] is shown in Figure 8 (unpublished data; The Dow Chemical Company). Rainbow trout were continuously exposed for 30 days to an average chlorpyrifos concentration of 0.30 μg/L under flow-through conditions (uptake phase). Fish were then transferred to flowing clean water for 16 days (elimination phase). During the uptake and elimination phases, fish and exposure water were serially sampled and analyzed for parent chlorpyrifos and metabolites. The BCF of parent chlorpyrifos was estimated to be 1,400 mL/g in whole body and 725 mL/g in muscle fillet (data not shown). The estimated t_{90} for both parent and total carbon-14 residues was 7 to 9 days, and the $t_{1/2}$ was 2 to 3 days. Chlorpyrifos (40–80% of whole body residues) was metabolized to 3,5,6-trichloro-2-pyridinol (TCP) (5–20% of total residues) and two conjugates of TCP (each conjugate 10–30% of total residues).

5.2. Fish Metabolism Study

An example fish metabolism study of a pesticide intended for aquatic applications was conducted by Rick *et al.*, (unpublished data; The Dow Chemical Company) with the herbicide triclopyr (data not shown). In that study, bluegill sunfish were exposed to [^{14}C-2,6-pyridine ring labeled] triclopyr (3,5,6-trichloro-2-pyridinyloxyacetic acid) for 10 days under a static exposure regime of 7.4 mg/L (approximately three times the maximum-labeled application rate). Exposed bluegill were periodically sampled and analyzed for total radioactivity in whole fish, muscle, and viscera. The time required for triclopyr to achieve 90% of a whole fish steady-state concentration was approximately 2 days. Following the achievement of steady-state, the remaining fish were transferred after 10 days of exposure to clean flowing water for a 14-day elimination period.

Total carbon-14 residues increased rapidly to a steady-state concentration of 5.7 μg triclopyr equivalent/g-fish (wet), resulting in a calculated BCF value of 0.8 mL/g for TRR in whole fish. Extraction efficiencies averaged ~94%, with parent triclopyr constituting 25% of the TRR and 58% identified metabolites. Various minor metabolites constituted 7.5% of extracted carbon-14 and a characterized, unidentified metabolite made up 9%. The concentration of this latter metabolite may have been as low as 30 ng/g in whole fish tissue, making mass spectral identification extremely difficult. The study successfully identified or characterized 92% of the TRR in whole fish tissue.

5.3. Aquatic Field Study

An example aquatic field study of a pesticide intended for aquatic applications was described by Woodburn *et al.*, (1993a) for the herbicide triclopyr (Figure 9). This study was a joint U.S. Army Corps of Engineers/Dow AgroSciences aquatic dissipation study of an aquatic-use application of triclopyr in Lake Seminole, Georgia. The test and control plots consisted of three separate 4-ha (10 acre) plots. Each plot contained five sampling stations, one in the center of each equal-sized quadrant and one in the center of each plot. An upstream location was selected as the control plot, and the other two plots were used for either a helicopter or airboat application of herbicide. Sampling stations were also located on the perimeters of each plot, approximately 100 m from the center of the plot margins. A downstream sampling station was located approximately 1.5 km from the most downstream application plot.

Sampling of water, sediment, aquatic plants, clams, crayfish, and fish continued from the day of application until 42 days post-application. The sampling times were gen-

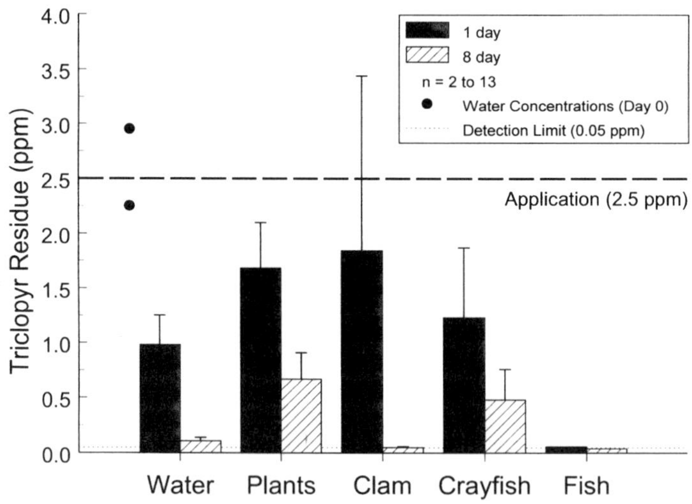

Figure 9. Example aquatic field study data: triclopyr.

erally pretreatment, 0, 1, 3, 8, 14, 21, and 42 days post-application. All samples were immediately frozen for later analysis for parent compound and metabolites. Water samples collected 1 and 4 hours post-application showed average triclopyr residues of 2.55 and 2.95 mg/L, respectively; the nominal application rate was 2.5 mg/L. The residue data indicated that degradation/dissipation of triclopyr was rapidly occurring in water, aquatic plants, clams, crayfish, and fish, as shown in Figure 9 for triclopyr residues on days 1 and 8 post-application. These field observations correlated well with previously-conducted environmental fate laboratory experiments with triclopyr (Barron *et al.*, 1991; Woodburn *et al.*, 1993b).

6. CONCLUSIONS

The intended use of a pesticide determines the bioaccumulation and metabolism study requirements for U.S. registration. Rotational and irrigated crop uses require a laboratory bioconcentration study for those chemicals with sufficient hydrophobicity ($K_{ow} > 1{,}000$) and persistence in water ($t_{1/2} > 4$ days). In contrast, fish metabolism and aquatic field studies are required for any pesticide intended for aquatic uses. The pesticide residue data generated in laboratory bioconcentration and fish metabolism experiments indicate the potential worst-case pesticide levels in food items such as fish. Bioconcentration data are used as part of an assessment of the potential for adverse effects of pesticide residues on nontarget organisms, whereas data from a fish metabolism study are primarily used in evaluating risks to humans and establishing allowable residues for enforcement purposes. The physicochemical properties of a pesticide are important in determining whether significant accumulation will occur in aquatic organisms. For example, the extent of bioconcentration is dependent on exposure time, the pesticide's persistence in water, and its hydrophobicity (K_{ow}) (Gillett, 1983; Veith and Kosian, 1983). In contrast to laboratory studies, aquatic field studies are conducted under actual-use conditions. Measurement of pesticide residues in water, sediments and biota are used in evaluating the environmental exposure and persistence of the pesticide.

7. REFERENCES

ASTM, 1996, Standard practice for conducting bioconcentration tests with fishes and saltwater bivalve mollusks, E 1022–94, *Annual Book of ASTM Standards,* Section 11, American Society of Testing and Materials, Philadelphia.

Barron, M.G., Hansen, S.C., and Ball, T., 1991, Pharmacokinetics and metabolism of triclopyr in the crayfish (*Procambarus clarki*), *Drug Metab. Dispos.* 19:163–167.

Barron, M.G., Stehly, G.R., and Hayton, W.L., 1990, Pharmacokinetic modeling in aquatic animals, I, Models and concepts, *Aquat. Toxicol.* 18:61–86.

Brazil, 1995, Brazilian Datalist for Pesticide Registration. SA-file: datatec2.doc, January 1995.

EPA, 1982a, Pesticide Assessment Guidelines Subdivision N-Chemistry: Environmental Fate. FIFRA Guideline 165-4. Laboratory studies of pesticide accumulation in fish, pp. 103–106. EPA 540/9-82-021.

EPA, 1982b, Pesticide Assessment Guidelines Subdivision E-Wildlife and Aquatic Organisms. FIFRA Guideline 72-6. Aquatic organism accumulation tests, pp. 81–83. EPA 540/9-82-024.

EPA, 1996a, Ecological Effects Test Draft Guidelines, OPPTS 850.1730: Fish BCF, Washington, D.C.

EPA, 1996b, Residue Chemistry Test Guidelines, OPPTS 860.1300, Nature of the Residue—Plants, Livestock, EPA 712-C-96-172, Washington, D.C.

EPA, 1996c, Residues Chemistry Test Guidelines, OPPTS 860.1400: Water, Fish, and Irrigated Crops, EPA 712-C-96-178, Washington, D.C.

Gillett, J.W., 1983, A comprehensive prebiological screen for ecotoxicologic effects, *Environ. Toxicol. Chem.* 2:463–476.

OECD, 1996, Test Guideline 305, Bioconcentration: flow-through fish test, Corrigendum to ENV/MC/CHEM (96) 2:26–48, Paris, France.

Official Journal of the European Communities, 1996, Commission Directive 96/12/EC of 8 March 1996, amending Council Directive 91/414/EC.

Touart, L.W., 1995, The federal insecticide, fungicide, and rodenticide act, in: *Fundamentals of Aquatic Toxicology, 2nd Ed.* G.M. Rand, ed. Taylor & Francis, Washington, D.C.

Veith, G.D. and Kosian, P., 1983, Estimating bioconcentration potential from octanol/water partition coefficients, in: *Physical Behaviour of PCBs in the Great Lakes,* Ann Arbor Science, Ann Arbor.

Woodburn, K.B., Green, W.R., and Westerdahl, H.E., 1993a, Aquatic dissipation of triclopyr in Lake Seminole, Georgia, *J. Agric. Food Chem.* 41:2172–2177.

Woodburn, K.B., Batzer, F.R., White, F.H., and Schultz, M.R., 1993b, The aqueous photolysis of triclopyr., *Environ. Toxicol. Chem.* 12:43–55.

5

CONSIDERATIONS IN COMPARTMENTAL PHARMACOKINETIC MODELING IN FISH

William L. Hayton

Division of Pharmaceutics
College of Pharmacy
The Ohio State University
Columbus, Ohio

1. ABSTRACT

Pharmacokinetic studies of xenobiotics are useful for quantitative characterization of their accumulation and elimination, bioavailability, bioconcentration and/or bioaccumulation, residue depletion, tissue distribution, and toxicity. Aspects of the study design that should be addressed in advance include identification of the desired information, test species (and size), route of administration, tissues and fluids to sample, and analytical methods for the test compound and metabolites. Small fish limit studies to a determination of the whole body concentration-time profile during and after bath or dietary administration. Whole body concentration-time and plasma concentration-time profiles may be determined in intermediate-size fish, and in large fish plasma and individual tissue concentration-time profiles are feasible using a variety of routes of administration. The principal pharmacokinetic parameters are clearance constants that characterize transfer of parent compound and metabolites among compartments, and the apparent volume of distribution of each compartment. Parameters derived from clearances and volumes of distribution include half-lives, bioavailability, and bioconcentration or bioaccumulation factor.

2. INTRODUCTION

Pharmacokinetic studies of chemicals in fish may be conducted to support the registration of chemicals for use in aquaculture and chemicals such as agrochemicals that have potential to enter surface water. As part of the hazard assessment process, the results of pharmacokinetic studies are used to quantitatively estimate the exposure of fish, and humans who consume fish, to chemicals. In addition, pharmacokinetic research in fish involves the development of techniques for extrapolation of the results of pharmacoki-

Xenobiotics in Fish, edited by *Smith et al.*
Kluwer Academic / Plenum Publishers, New York, 1999.

netic studies in one species under controlled environmental conditions to other species, body sizes, and environmental conditions.

Pharmacokinetic studies of chemicals in fish are useful for quantitative characterization of their rate and extent of accumulation, rate of elimination, bioavailability, and bioconcentration and/or bioaccumulation. In conjunction with pharmacokinetic studies, there may be closely related studies, such as residue depletion, tissue distribution, metabolite profile, and toxicity studies, that have similar data requirements. Achievement of each of these various purposes potentially requires different types of data, different sampling schedules, and use of radiolabeled test compound. If more than one purpose is intended for the study, careful planning will help to ensure that sufficient data are collected to achieve all the desired outcomes. The mode of administration of the chemical, the sample collection times, and the fluids and tissues to be sampled must be identified. The number of replicates must be decided. These issues may involve the use of pilot studies to gain preliminary information about the pharmacokinetics of the test chemical. In addition, the analytical methodology for quantification of the chemical and possibly its metabolites must be developed or implemented and validated.

Methods for analysis of the data also should be identified, preferably as part of the study planning. Computer software is available for compartmental pharmacokinetic analysis, as well as other types of pharmacokinetic analysis such as non-compartmental and physiologically based modeling (PBPK). Examples of programs include: WinNonlin, SAAM, ACSL, and Micromath SCIENTIST. Some of the packages have a number of preprogrammed models, and the user can simply select the desired model to analyze the data. As the packages are designed for analysis of data collected in humans and mammalian laboratory animals, the user may need to write programs for analysis of data collected in fish, and these packages generally allow for custom programs. This is particularly the case for modeling the accumulation of chemicals during a water-exposure study as the transfer of chemical from exposure water to the fish is an equilibrium rather than a one-way process as the preprogrammed models usually treat absorption.

For all but the simplest of models, the fitting of model-based equations to experimental data may require considerable effort. The fitting of multi-parameter models is frequently beset with the problems of estimation of initial values for the parameters, and multiple minima in the fitting criterion; e.g., weighted sum of squares. It may be desirable to perform *simulations* with the model, then fit the simulated data after introduction of random error, and compare the parameter values so obtained with those used in the simulation. This exercise may provide clues about the sensitivity of the model to random fluctuations in the data, to the degree of accuracy needed in the initial parameter estimates, and to the number of samples required as well as their schedule of collection. It is necessary to consider *identifiability analysis*, to ensure that a sufficient number of variables have been measured to support the complexity of the model; i.e., that the data will support the determination of a unique set of model parameters (Pohjanpalo, 1978). *Sensitivity analysis*, a series of simulations in which each model parameter is varied and the change in the model variables is observed, will assist in the identification of the parameters that require the most accurate initial estimates (Hetrick *et al.*, 1991). Some software packages will perform this analysis automatically. *Uncertainty analysis* may be used after the fitting of the model to show the expected range of model-predicted concentration-time profiles based on the variance that is found in the model parameters (Burmaster and Anderson, 1994). The adequacy of a model may be established by consideration of the following points (Gabrielsson and Weiner, 1994): does the model have biological relevance? does the fitted curve mimic the time course of the data and do the data scatter randomly about the fitted curve? do the model parameters have small uncertainty?

Compared with terrestrial animals, pharmacokinetic studies in fish present unique challenges. Procedures commonly performed on terrestrial animals, such as cannulation of blood vessels, are often considerably more difficult to perform in fish. Specialized laboratory facilities are needed to house fish. Attention must be devoted to water quality issues such as constant temperature, dissolved oxygen concentration, pH, hardness, and absence of chemical pollutants. The guidelines for acute and chronic toxicity tests in aquatic species of the American Society of Testing and Materials (ASTM, 1996) should generally be followed in pharmacokinetic studies. Among several test design parameters, loading densities of aquaria during studies are particularly relevant, as are minimum water flows in flow-through experiments. When the study results will support the registration of a drug or agrochemical, Good Laboratory Practice regulations usually must be followed (CFRa).

The test chemical may be administered to the fish via the water (bath exposure), orally via the diet or by oral gavage, or by injection into the blood, muscle, dorsal sinus, or peritoneal cavity. When water exposure is used, preliminary studies should characterize the stability of the chemical in the exposure water over the planned exposure period with no fish present. Potential loss of chemical by volatilization, binding to the exposure vessel wall, or chemical degradation should be studied and if present characterized. Such losses should be minimized, for example by using minimal aeration of exposure water and in the case of photodegradation by shielding the exposure vessel from light. These losses do not preclude pharmacokinetic study of the test chemical by water exposure, but it is important to establish their presence so that they can be incorporated into the pharmacokinetic model (Schultz and Hayton, 1993). The test chemical should be fully dissolved in the exposure water, and the use of solubilizers such as water miscible solvents and surfactants should be avoided. For lipophilic or water insoluble chemicals, achievement of the true solution state may require the use of specialized equipment, and the water should be free of substances that complex with the test chemical, such as dissolved organic materials (Suffet *et al.*, 1994).

In dietary exposure, the test compound is incorporated into the diet, which is usually pelleted. One approach is to dissolve the chemical in a nonaqueous solvent, place the solution and the pelleted diet into a round bottom flask, and then to remove the solvent using a rotary vacuum evaporator. Another approach is to mix the chemical and powdered food homogeneously and then to pellet the food. The concentration and stability of the chemical in the diet should be measured, as should the rate at which the chemical is leached from the diet after it is placed in the water for consumption by fish.

For all of the routes of administration, it is essential to know the amount of chemical administered to the fish; i.e., the dose. This should be measured rather than relying on the calculated concentration that should be present as a result of preparation of the dose. That is, the concentration obtained by dissolution of a weighed amount of chemical in a known volume of water should be checked by measurement of the concentration of the chemical in the exposure water.

3. SIZE OF FISH

The size of the fish selected for study constrains the number and types of samples that can be taken, and strongly determines the nature of the experiment and the approach to compartmental pharmacokinetic analysis of the data. Discussed in this section are the experimental designs that can be used for particular sizes of fish.

3.1. Small Size

Fish with a body weight below 50 g are more easily and cheaply housed than larger fish, and their exposure to waterborne chemicals is relatively easier and more economical. The usual approaches to bath exposure of small fish to waterborne chemicals are static exposure and flow-through exposure. Fish may be exposed individually or as a group, and generally the smaller the fish the greater the tendency to expose several individuals simultaneously in the same exposure vessel. The static approach involves exposure of fish to a known initial concentration and volume of test chemical. A variation of the static exposure is the static renewal exposure, where the fish are transferred to a fresh exposure solution periodically, which mitigates the problem of water quality degradation. Alternatively, part of the exposure solution may be periodically replaced.

The flow-through exposure involves the use of an exposure vessel that has a continuous inflow of exposure solution into the exposure vessel and an overflow tube to carry the excess solution from the exposure vessel. A specialized apparatus for continuous dilution of a concentrated stock solution of the test chemical to maintain a constant concentration in the flowing exposure solution is available (Ace Glass, Inc., Vineland, NJ). For very small fish, large volumes of the exposure solution may be prepared and delivered by hydrostatic pressure into the exposure vessel. Temperature should be controlled, for example by carrying out the study in a constant temperature room or by the use of a thermostatted water bath.

During the exposure the amount of chemical in the fish will exponentially increase. In the case of the flow-through exposure, the amount of chemical in the fish approaches a constant, or steady-state, value. At steady state, the rate of chemical uptake by the fish equals its rate of elimination, and the rate of change of the amount of chemical in the fish becomes zero. During the exposure, the concentration of test substance in the water remains at its initial value, due to the continuous renewal of the exposure solution. With the static exposure, the concentration of the test chemical in the water may also remain constant, at its initial value, if the amount of chemical accumulated by the fish is insignificant relative to the amount of chemical in the exposure water. This is generally not the case, however, and the water concentration of the chemical declines during the static exposure as the amount of chemical in the fish increases.

With small fish, the whole-body concentration of chemical is usually determined after slaughter or harvest and homogenization of the carcass. In the flow-through study, the exposure should be continued until the steady state is achieved. To provide both accumulation and wash-out profiles of chemical, additional fish may be exposed to steady state, placed in clean water, and the body concentration of chemical determined at various times. The length of the wash-out phase should be sufficient that the body concentration has declined to 10% of its initial value. When static or static renewal exposure maintains a constant exposure concentration, the wash-out profile may also be determined. If the water concentration declines during the static exposure as a result of accumulation of chemical by the fish, it meets a rising concentration in the fish. In the absence of loss of the chemical from this closed system, for example by metabolism by the fish, the concentration of the chemical in the water and fish will come to an equilibrium. When equilibrium is achieved, it may also be useful to characterize the wash-out profile. However, when there is loss of the chemical, the concentration in the fish peaks and then declines in parallel with the water concentration. If the loss is due to metabolism, the study may be designed to follow the declining concentrations of chemical in the fish and water and the total metabolite profile as well, giving a combined accumulation and wash-out profile. When the exposure water concentration declines and several fish are exposed in a single solution and removed periodically, it is desirable to maintain the fish weight to exposure water volume constant for the purpose of pharmacokinetic analysis. This can be accomplished at each sampling by removal of a fraction of the exposure water in proportion to the fraction of the initial weight of fish removed.

The approach to pharmacokinetic modeling of data collected from small fish is to represent the fish as one or two compartments, Figure 1. If the exposure water concentration is constant, and the fish concentration-time profile of the chemical appears to be monoexponential, the following model based differential equation applies, Figure 1A:

$$dX/dt = P(C_w - C_i) \tag{1}$$

where X is the amount of chemical, usually per gram of fish, C_w and C_i are the concentrations (amount/vol.) of chemical in the exposure water and in the body water of the fish, respectively, and P is a proportionality constant that has units of flow (volume $time^{-1}$ body $weight^{-1}$). X/V can substitute for C_i, where V is the apparent volume of distribution of the chemical referenced to the exposure water. The units of V are volume/unit body weight; e.g., mL/g if X had units of ng/g and C_w and C_i had units of ng/mL. Integration of Eq. 1 gives the following model-based equation for X as a function of time:

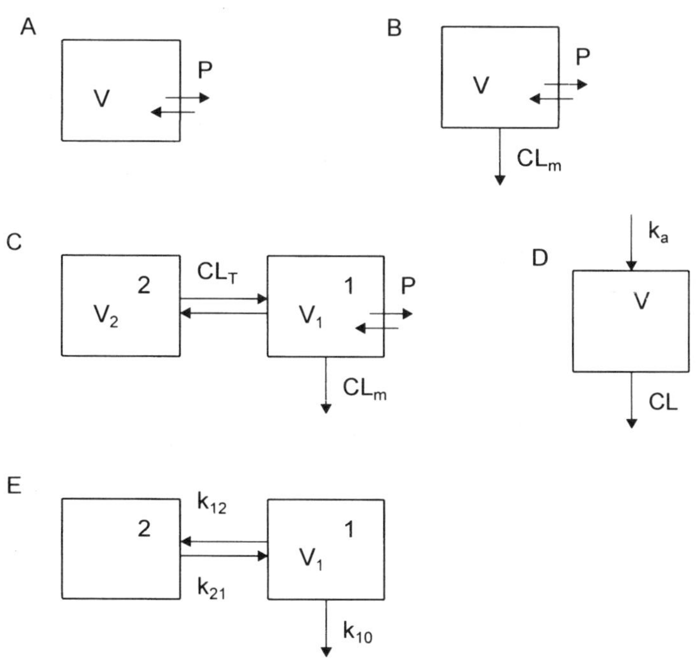

Figure 1. Examples of compartmental pharmacokinetic models used in fish. A, One compartment of volume V with uptake by the gill, P. B, Same as A except that elimination by metabolism, CL_m, is included. C, Two compartments of volume V_1 and V_2, with uptake by the gill, intercompartmental transfer, CL_T, and elimination by metabolism. D, One compartment with uptake by absorption, e.g., GI tract, controlled by a first-order absorption rate constant, k_a. E, Two compartments with intravascular administration of a bolus dose into Compartment 1, intercompartmental exchange controlled by first-order rate constants k_{12} and k_{21}, and elimination controlled by first-order rate constant k_{10}.

$$X = VC_w(1 - e^{-(P/V)t}) \tag{2}$$

This equation can be fitted to experimentally determined values of X at various times to determine values for P and V, and the half-life, for approach to equilibrium and also that would be observed if the fish were placed in chemical-free water, can be calculated using:

$$t_{1/2} = 0.693V/P \tag{3}$$

In the case of a declining exposure water concentration, the rate of decline is:

$$dC_w/dt = P(C_i - C_w)/V_w \tag{4}$$

where, V_w is the volume of the exposure solution. Equations 1 and 4 give the following equations, which may be fitted simultaneously to experimentally determined profiles of C_w and X for determination of values for P and V:

$$C_w = C_{w,o} - C_{w,o}V(1 - e^{-((P/V_w)+(P/V))t})/(V_w + V) \tag{5}$$

$$X = V \cdot V_w C_{w,o}(1 - e^{-((P/V_w)+(P/V))t})/(V_w + V) \tag{6}$$

$C_{w,o}$ is the initial water concentration. In the case of declining C_w, notice that the approach to steady state or equilibrium is faster than in the constant C_w case; i.e., the coefficient of time in the exponent of *e* is larger in Eq. 6 than in Eq. 2. This may be advantageous for long half-life chemicals, and the volume of the exposure solution may be decreased to shorten the time requirements for the experiment. When a metabolite profile (M) is available, Figure 1B, the following differential equations apply:

$$dX/dt = P(C_w - C_i) - CL_m C_i \tag{7}$$

$$dM/dt = CL_m C_i \tag{8}$$

where CL_m is the metabolism clearance for metabolite formation and M is the sum of metabolite in fish and water. The integrated forms of these equations, assuming that C_w is constant, are:

$$X = P \cdot VC_w(1 - e^{-((P+CL_m)/V)t})/(P + CL_m) \tag{9}$$

$$M = P \cdot CL_m C_w t/(P + CL_m) - P \cdot V \cdot CL_m C_w(1 - e^{-((P+CL_m)/V)t})/(P + CL_m)^2 \tag{10}$$

Equations 9 and 10 may be fitted simultaneously to X,t and M,t profiles to obtain values for P, V, and CL_m. In the case of a static exposure and declining C_w, the various time profiles (C_w, X, M) will be influenced by the size of the fish. Hence, the fish should either be about the same size so that their body weights may be assumed to be identical, or the equations must explicitly contain fish body weight, which becomes a second independent variable (Karara and Hayton, 1984).

Routes of administration other than via the exposure water are relatively uncommon in small fish. Administration by injection is rarely used, although administration by oral gavage is possible. Dietary administration is feasible, but quantification of the administered dose is a problem. If fish are fed individually, the amount consumed is small and difficult to quantify; if a group of fish is treated in the same vessel, there is a problem of fish-to-fish variability in the dosage, which would lead to interindividual variability in the measured values for X and M.

3.2. Medium Size

Compared with small fish, more routes of administration may be used in fish of intermediate body size (50–200 g); i.e., water exposure, oral gavage, diet, or injection. As with small fish, the whole body *versus* time profile is generally the variable followed as this size of fish is too small for removal of serial blood samples. Thus, characterization of the X,t and M,t profiles is carried out using a relatively large number of fish, with several replicates at each sampling time. This introduces substantially more variability into the profiles compared with serial concentrations determined in a single fish. With this size of fish it becomes feasible to remove a sample of blood at the time of sacrifice, which gives a profile of the plasma concentration along with the total amount of chemical in the body. This can be quite useful when a multicompartment pharmacokinetic model is used to analyze the data, as the plasma concentration (C_p) represents the central compartment concentration while X represents the average body concentration.

For example for water exposure and a two-compartment model for the fish, the rate equations would appear as, Figure 1C:

$$dC_p/dt = P(C_w - C_p)/V_1 + CL_T(C_2 - C_p)/V_1 - CL_m C_p/V_1 \quad (11)$$

$$dC_2/dt = CL_T(C_p - C_2)/V_2 \quad (12)$$

where C_p is the plasma concentration, and subscripts 1 and 2 refer to compartments 1 and 2; a metabolism clearance was included in Eq. 11. Upon setting $V_1 = X_1/C_p$, the model predicted profiles from integration of these equations are:

$$\begin{aligned} C_p = {} & P \cdot C_w/(P + CL_M) + P \cdot C_w(a - \alpha)e^{-\alpha t}/V_1\alpha(\alpha - \beta) \\ & + P \cdot C_w(\alpha - \beta)e^{-\beta t}/V_1\beta(\beta - \alpha) \end{aligned} \quad (13)$$

$$\begin{aligned} X = {} & P \cdot C_w(V_1 + V_2)/(P + CL_M) + P \cdot C_w(a + b - \alpha)e^{-\alpha t}/\alpha(\alpha - \beta) \\ & + P \cdot C_w(a + b - \beta)e^{-\beta t}/\beta(\beta - \alpha) \end{aligned} \quad (14)$$

$$\begin{aligned} M = {} & P \cdot CL_m C_w t/(P + CL_M) - P \cdot CL_m C_w(V_1 + V_2)/(P + CL_m)^2 \\ & + A(a - \alpha)e^{-\alpha t}/\alpha^2(\beta - \alpha) + A(a - \beta)e^{-\beta t}/\beta^2(\alpha - \beta) \end{aligned} \quad (15)$$

where

$$A = P \cdot CL_m C_w/V_1 \quad (16)$$

$$a = CL_T/V_2 \quad (17)$$

$$b = CL_T/V_1 \quad (18)$$

$$\alpha + \beta = (P + CL_m + CL_T)/V_1 + CL_T/V_2 \quad (19)$$

$$\alpha\beta = CL_T(P + CL_m)/V_1V_2 \quad (20)$$

Values for the rate constants α and β are the roots of the following quadratic equation:

$$X^2 - K_1X + K_2 = 0 \tag{21}$$

where K_1 and K_2 are the right-hand sides of Eqs. 19 and 20. It is then possible to simultaneously fit Eqs. 13–15 to the three profiles (Schultz and Hayton, 1993); without the measured C_p there would be little kinetic information to characterize the intercompartmental distribution of the chemical.

If after administration of a single oral dose the absorption of the chemical follows first-order, monoexponential kinetics; i.e., absorption is characterized by an absorption rate constant (k_a), then the rate of change in the total body level is, Figure 1D:

$$dX/dt = k_aX_g - CL \cdot C_p \tag{22}$$

where X_g is the amount of chemical remaining in the GI tract and CL is the total body clearance of the chemical, which includes branchial as well as renal, hepatic, and possibly other elimination pathways. This equation may be integrated to:

$$C_p = k_aF \cdot X_o(e^{-(CL/V)t} - e^{-k_at})/V(k_a - (CL/V)) \tag{23}$$

where F is the fraction of the dose that reaches the systemic circulation of the fish and X_o is the dose. The value of F may be estimated by comparison of the area under the C_p,t profile ($AUC = \int C_p dt$) with the area determined after an intravascularly administered dose where F is unity:

$$F = (AUC_{oral}/AUC_{iv})(X_{o,iv}/X_{o,oral}) \tag{24}$$

and Eq. 23 may be fitted to the experimentally determined C_p,t profile to estimate values for k_a, CL, and V. AUC values may be calculated using the trapezoidal or log trapezoidal method (Gibaldi and Perrier, 1982).

When the dietary exposure is continuous, the rate of change in the body level would be

$$dX/dt = F \cdot R_{in} - CL \cdot C_p \tag{25}$$

where R_{in} represents the rate of ingestion of chemical in the diet. In this case, the body level approaches a constant, steady state and the amount of chemical in the body at steady state is:

$$X_{ss} = F \cdot R_{in}/V \cdot CL \tag{26}$$

3.3. Large Size

In fish larger than 200 g, determination of the total body level of chemical by homogenization of the carcass becomes cumbersome. Pharmacokinetic studies in large fish are usually carried out using the plasma concentration as the measured variable. Plasma is generally obtained from blood that is removed from an indwelling cannula in the dorsal aorta, distal to the gills (Hoelton and Randall, 1967; Houston, 1990; Kleinow, 1991). Administration of chemical may be by any of the common routes. Administration directly into the blood is advantageous as the input kinetics can be precisely controlled and there is no uncertainty in the amount of chemical that is administered. This route of administration is ideal for characterization of the kinetics of distribution and elimination of the chemical. Generally, two or three compartment models are used to analyze the C_p,t profile after intravascular administration of a bolus dose; in the case of the two-compartment model, the rates of transfer of chemical from compartments one and two are, Figure 1E

$$dX_1/dt = k_{21}X_2 - k_{12}X_1 - k_{10}X_1 \tag{27}$$

$$dX_2/dt = k_{12}X_1 - k_{21}X_2 \tag{28}$$

where X_1 and X_2 are the amounts of chemical in compartments 1 and 2, and k_{12} and k_{21} ($time^{-1}$) are intercompartmental transfer rate constants, and k_{10} is the rate constant for elimination, which was assumed to occur in Compartment 1. These equations may be integrated to give X_1 and X_2 as a function of time:

$$X_1 = [X_o/(\alpha - \beta)][(k_{21} - \beta)e^{-\beta t} - (k_{21} - \alpha)e^{-\alpha t}] \tag{29}$$

$$X_2 = [X_o k_{12}/(\alpha - \beta)][e^{-\beta t} - e^{-\alpha t}] \tag{30}$$

where α and β are rate constants that are comprised of the model rate constants:

$$\alpha = 0.5\left\{(k_{12} + k_{21} + k_{10}) + \left[(k_{12} + k_{21} + k_{10})^2 - 4k_{21}k_{10}\right]^{1/2}\right\} \tag{31}$$

$$\beta = 0.5\left\{(k_{12} + k_{21} + k_{10}) - \left[(k_{12} + k_{21} + k_{10})^2 - 4k_{21}k_{10}\right]^{1/2}\right\} \tag{32}$$

The total amount of chemical in the body (X) is ($X_1 + X_2$):

$$X = [X_o/(\alpha - \beta)][(k_{12} + k_{21} - \beta)e^{-\beta t} - (k_{12} + k_{21} - \alpha)e^{-\alpha t}] \tag{33}$$

The plasma concentration is X_1/V_1 or:

$$C_p = [X_o/V_1(\alpha - \beta)][(k_{21} - \beta)e^{-\beta t} - (k_{21} - \alpha)e^{-\alpha t}] \tag{34}$$

which may be represented as:

$$C_p = Ae^{-\alpha t} + Be^{-\beta t} \tag{35}$$

To obtain estimates of the model parameters, this equation can be fit to the C_p,t profile to obtain values for A, α, B, and β. These values may then be used to calculate the model and associated parameters (Gibaldi and Perrier, 1982):

$$V_1 = X_o/(A + B) \tag{36}$$

$$k_{21} = (A\beta + B\alpha)/(A + B) \tag{37}$$

$$k_{10} = \alpha\beta/k_{21} \tag{38}$$

$$k_{12} = \alpha + \beta - k_{21} - k_{10} \tag{39}$$

$$V_2 = V_1(k_{12}/k_{21}) \tag{40}$$

$$V_{ss} = V_1 + V_2 \tag{41}$$

$$V_\beta = V_1(k_{12} + k_{21} - \beta)/(k_{21} - \beta) = V_1 k_{10}/\beta \tag{42}$$

$$CL = k_{10}V_1 = \beta V_\beta \tag{43}$$

$$t_{1/2,\alpha} = \ln 2/\alpha \tag{44}$$

$$t_{1/2,\beta} = \ln 2/\beta \tag{45}$$

V_{ss} and V_β are the steady state and β-phase volumes of distribution. The former when multiplied by C_p at steady state, e.g., after i.v. infusion, dietary or water exposure, gives the amount of chemical in the organism. The latter when multiplied by C_p during elimination; e.g., during the log-linear or β-phase, gives the amount of chemical in the organism. V_{ss} reflects only distribution whereas V_β is, in addition, a function of the kinetics of distribution and elimination. $V_\beta \geq V_{ss}$ because during the β-phase there is a positive gradient from Compartment 2 to Compartment 1, which results for a given plasma concentration in more chemical in the organism during the β-phase than during steady state. The $t_{1/2,\alpha}$ is characteristic of the time required for distribution, and the $t_{1/2,\beta}$ is the elimination half-life of the chemical, similar to the half-life for the one-compartment model.

4. SCALING PHARMACOKINETIC PARAMETERS

Given the diversity of aquatic species and the environmental conditions under which they live, it is frequently of interest to project the results of pharmacokinetic studies to species, body sizes, and environmental conditions different from those used in the studies. One approach involves the use the physiologically based pharmacokinetic (PBPK) type of model, which is presented in Chapter 7. When a compartmental model has been used, it is possible to project the results of the study to other conditions. Detailed examination of this topic is available (McKim and Erickson, 1991; Hayton and Schultz, 1991) and only a brief summary is presented here.

4.1. Body Size

Pharmacokinetic parameter values (pv) such as V and CL generally follow the empirical allometric equation (Schmidt-Nielsen, 1984; Calabrese, 1991; Hayton, 1989):

$$pv = aW^b \tag{46}$$

where W is body weight, and *a* and *b* are constants. If the *b* value is one, then the *pv* is directly proportional to W and its value for other body weights is calculated as a simple, weight-based proportionality. This is commonly the case for V values; exceptions are encountered; e.g., when the adiposity changes with body size and the chemical of interest is lipophilic.

The *b* value for clearance parameters such as CL and P is generally less than one, and commonly it is in the range 0.67–0.8. This means that as body weight increases, there is a less-than-proportional increase in the clearance value. The underlying reason for this relationship is that a similar relationship holds for W, organ weight (including gills, liver, and kidney), and cardiac output. With V values that increase in direct proportion to body size and clearance values that increase in a less-than-proportional fashion, the general trend is for the half-lives of processes such as time to steady state and time for elimination to increase as body size increases. In other words, the time span over which a chemical accumulates in and disappears from large fish is longer than it is for small fish.

If pharmacokinetic parameter values have been measured in fish of several different body weights, the validity of the allometric equation can be tested by a fit of Eq. 46 to the pv,W values. Whether a plot of pv versus W on log-log coordinates is linear can assist in this evaluation. If the plot is linear the *b* value may be obtained from the slope and used to estimate a *pv* value for any particular body weight. As with any extrapolation, projection beyond the limits of the body size range studied experimentally may yield an erroneous estimate. If the *pv* is available in only two body sizes, the *b* value may be estimated from:

$$b = \log(pv_1/pv_2)/\log(W_1/W_2) \tag{47}$$

where subscripts 1 and 2 indicate the two different sizes of fish studied. The *b* value could then be used to estimate the *pv* for any body weight, W_a, between W_1 and W_2:

$$pv_a = pv_1(W_a/W_1)^b \tag{48}$$

4.2. Environmental Conditions

Temperature affects the physiological and biochemical processes that control the uptake and elimination of chemicals by fish. Cardiac output and ventilation volume increase linearly with increasing temperature; e.g., cardiac output in rainbow trout increases linearly with an increase in temperature from 10 to 20°C (Barron *et al.*, 1987a). The temperature dependence of clearances that are blood or water flow limited will parallel the changes in flow. These include transport parameters such as gill uptake clearance (P), renal clearance, intercompartmental transport (CL_T) and hepatic metabolism when that is blood flow dependent.

The activity of metabolizing enzymes is also temperature dependent. While there may be a temperature of maximum activity and a decline in activity as temperature deviates from this optimum (James *et al.*, 1979), the maximum may lie outside the temperature range tolerated by the particular species. Also in temperature-acclimated fish the phenomenon of compensation may occur. A temperature-induced change in intrinsic activity (activity per unit amount of enzyme; i.e., altered K_M or V_{max}) may be diminished by an opposing change from factors such as the total amount of enzyme or the availability of co-substrate, with the result being that the effective activity of the enzyme changes relatively less than does the intrinsic activity (Ankley *et al.*, 1985).

The volume of distribution parameters may or may not change with a change in temperature. When the distribution of a chemical involves temperature-sensitive equilibria such as reversible binding to proteins and distribution between lipid and aqueous phases, the value of V may change with temperature. When distribution simply involves the total body water space, the V value may not change appreciably with a temperature change. An interesting example of temperature influence on the value of V is the lipophilic plasticizer di-2-ethylhexylphthalate (Barron *et al.*, 1987b). In rainbow trout between 6 and 18°C, its V value increased exponentially with temperature by a factor of 7.5 while its metabolic clearance increased linearly by a factor of 3.6. As a consequence, the half-life for elimination of the plasticizer increased from 79.5 to 130 hr between 6 and 18°C. In this case, the increased elimination clearance capacity was more than offset by the increased volume to be cleared, which resulted in increased persistence.

4.3. Species

Given the diversity among fish species, it is not surprising to find substantial interspecies differences in pharmacokinetic parameter values among fish, even after allowance is made for differences in body size and temperature. An indication of the interspecies variability in the bioconcentration factor (BCF, see below), which largely reflects variability in the volume of distribution, was shown in a study of 15 organophosphate insecticides (Tsuda *et al.*, 1997). The chemicals varied widely in their lipid solubilities and BCF values, which were measured in guppy (*L. reticulatus*), killifish (*O. latipes*), goldfish (*C. auratus*), and white cloud mountain fish (*T. albonubes*). The range of BCF values for each pesticide was less than a factor of 5 (ratio of highest to lowest) for 12 of the 15 chemicals and between 5 and 10 for two of them; one chemical showed a range of BCF values that was 15. For any pair of species, the BCF values were highly correlated ($r > 0.94$). Thus interspecies differences in body composition, such as lipid content, acted similarly on the BCF of these compounds.

Similar interspecies differences have been reported for activity of metabolizing enzymes. James' review concluded that there were "considerable interspecies differences in the ability of fish to conjugate xenobiotics" (James, 1986). The CYP450 content (nmol/mg protein) varied over a 5-fold range in 5 species of marine teleosts, and the hepatic microsomal oxidative activity with specific substrates (benzpyrene hydroxylase, benzphetamine N-demethylase, 7-ethoxycoumarin O-deethylase, analine hydroxylase) varied over the ranges (no. of species tested) of 25(8), 15(5), 38(6), and 6(5) (James *et al.*, 1979).

5. PURPOSE OF STUDY

5.1. Pharmacokinetics of Disposition

Large fish are advantageous for unambiguous characterization of the distribution and elimination kinetics of a chemical, after its intravascular administration and removal of serial blood samples over three to four elimination half-lives. The advantage of this approach is the knowledge of the amount of chemical that reaches the systemic circulation and the ability of the investigator to control the kinetics of administration; i.e., by use of a rapid bolus injection or an infusion of known rate and duration. While pharmacokinetic studies can be performed in small fish via water exposure, the amount of chemical that reaches the systemic circulation is generally not precisely known. Regardless of fish size, administration of the chemical by water or dietary exposure introduces uncertainty in the amount of chemical that reaches the systemic circulation and this uncertainty is transferred to the values determined for the pharmacokinetic parameters. In addition, these modes of administration may mask the kinetics of distribution if the absorption rate is slow compared with the rate of distribution. If the objective of the study is limited to an assessment of the persistence of the chemical, use of small fish and water or dietary exposure of the chemical may be preferred for simplicity and lack of invasiveness.

5.2. Bioavailability

For chemicals administered orally, including in the diet, and by injection into a site that involves absorption, e.g., intramuscular and intraperitoneal, bioavailability refers to the fraction of the administered dose of chemical that reaches the systemic circulation of the fish (Abdou, 1989). The U.S. Food and Drug Administration definition of bioavailability includes the rate of absorption, when this is relevant to the therapeutic effect of the drug (CFRb). When the fraction of the dose absorbed is the desired parameter, the study design generally involves intravascular administration as well as administration by the route of interest. Equation 24 may then be used to calculate the value of F. Since this calculation involves the plasma concentration-time profile and intravascular administration of the chemical, the fish have to be large enough to carry out these procedures.

The concept of bioavailability is different when the focus is on environmental toxicology and the question involves assessment of the availability of chemical present in the environment. For example, how bioavailable to fish is a chemical dissolved in water? The chemical is available, but fish do not generally accumulate all the chemical present. Bioavailability is this context is akin to bioaccessibility. It is sometimes referred to as environmental bioavailability (Landrum *et al.*, 1994), and its measurement may use a relative standard. For example, the effect of a water soluble polymer such as humic acid on the bioavailability of a particular waterborne chemical may be of interest. An approach in this case may be the use of small fish and static exposure. The area under the X,t profile could be used in place of AUC in Eq. 24, with the areas in the absence and presence of the polymer used in the denominator and numerator, respectively.

5.3. Bioconcentration/Bioaccumulation

An important component in environmental hazard assessment is the tendency of a chemical to accumulate in fish. When the chemical is present in the water, the accumulation tendency is characterized by the BCF, which is the ratio of the average body concentration of the chemical (X_{ss}, amount/g body weight) to its concentration in the exposure water at steady state:

$$BCF = X_{ss}/C_w \tag{49}$$

This ratio can be determined directly, or it can be calculated from the results of compartmental pharmacokinetic analysis, which may be advantageous in the case of a chemical that has a long half-life (Branson, Blau, *et al.*, 1975), or that degrades in the exposure water. From Eq. 2, it is apparent that BCF = V in the absence of nonbranchial elimination. When other elimination pathways are operative, the BCF value may be less than the value of V or V_{ss} (see below).

It is generally possible to estimate BCF from pharmacokinetic parameters when the V values are referenced to the exposure water, but not when they are referenced to plasma or some other region. In theory, it is possible to estimate BCF for plasma referenced pharmacokinetic parameter values if the extent of plasma protein binding is known:

$$\mathrm{BCF} = \mathrm{P} \cdot \mathrm{V}_{ss}/\mathrm{f}_{up}(\mathrm{P} + \mathrm{CL}_{nb}) \tag{50}$$

where P and CL_{nb} represent the branchial and nonbranchial clearances of the chemical, and f_{up} is the fraction unbound in plasma. Experimental verification of this approach is lacking, and it is recommended that the experimental design to obtain pharmacokinetic parameters for BCF calculation should include small fish and the approach to compartmental modeling for them described above.

Bioaccumulation, a more general term than bioconcentration, refers to the tendency of a chemical to accumulate from all environmental sources, including the water and the diet. Continuous dietary exposure to chemicals can lead to a much higher steady-state body concentration than observed for water exposure, even when the source of chemical in the diet is the water. This reflects the tendency of certain chemicals, particularly lipophilic chemicals, to concentrate in food and to further concentrate in fish that eat the food. Exposure to member species of a food chain thereby increases as the chain is ascended. For dietary exposure, the amount of chemical in the body at steady state is given by Eq. 26. If bioaccumulation results from both water and dietary exposure, the rate of accumulation would be

$$\mathrm{dX/dt} = \mathrm{F} \cdot \mathrm{R}_{in} + \mathrm{P} \cdot \mathrm{C}_{w} - (\mathrm{P} + \mathrm{CL})\mathrm{C}_{p} \tag{51}$$

At steady state, $dX/dt = 0$, and using the relationship $C_p = X/V$

$$\mathrm{X}_{ss} = (\mathrm{F} \cdot \mathrm{R}_{in} + \mathrm{P} \cdot \mathrm{C}_{w})\mathrm{V}/(\mathrm{P} + \mathrm{CL}) \tag{52}$$

5.4. Tissue Distribution Profile

While parameters such as the BCF and V provide information about the tendency for bioaccumulation of a chemical, it may be of interest to characterize the tendency of particular tissues to accumulate the chemical. This focus is relevant when assessment of risk from consumption of fish exposed to the chemical is being considered. The fish are exposed to the chemical using the dosage and route of administration that are representative of the exposure that fish will encounter in practice, generally via the water or the diet. At various times during and after the exposure, fish are sacrificed and samples of the tissues and fluids of interest are removed; e.g., blood, bile, muscle, kidney, liver, skin. The concentrations of the parent chemical and each metabolite are then determined in each sample. This type of study is similar in many respects to a pharmacokinetic study and when both studies will be done, it may be expedient to carry them out at the same time.

5.5. Residue Depletion

Assessment of risk involved in the consumption of fish exposed to a chemical must involve the consideration of toxicologically relevant metabolites of the chemical as well as the parent substance. *Residue* is the total amount of parent compound plus all of the metabolites that are present in the body of the fish, or in a particular tissue. The presence of residue is generally assessed using a nonspecific assay for the chemical, such as determination of the concentration of radioactivity in the total body or in a particular tissue after exposure of fish to radiolabeled chemical. Determination of pharmacokinetic parameters from a total residue concentration-time profile is generally not possible since the residue composition is heterogeneous and the profile reflects the kinetic behavior of the various components of the residue. It is usually the case that the residue concentration-time profile exhibits a terminal linear phase on semilogarithmic coordinates and this permits estimation of a half-life for the residue, and estimation of the time required to achieve a particular residue concentration. However, the uncertainty in extrapolation beyond the last measured concentration increases with the extent of the extrapolation, and projection beyond one half-life is generally unsound.

6. INTERPRETATION OF PHARMACOKINETIC PARAMETER VALUES

6.1. Volume of Distribution

The apparent volume of distribution (V) has units of volume or volume per unit body size; e.g., mL, liters, mL/g body weight, and liters/m^2 body surface area. The V value varies widely among a series of chemicals and it is a characteristic of the chemical and the species in which it is measured, as for example is the solubility of the chemical in a particular solvent. The numerical value of V_{ss} provides some insight into the distribution characteristics of the chemical. The V_{ss} is a proportionality constant that relates the amount of chemical in a compartment to its concentration in the reference region; *viz.*, blood plasma or exposure water. A V_{ss} value that is small (<1 mL/g) indicates that the chemical generally has less affinity for the fish tissues than it does for the reference region. In the case of a plasma referenced V_{ss}, this may occur as a result of plasma protein binding or an inability of the chemical to penetrate cell membranes, resulting in extracellular distribution. For plasma referenced V_{ss} values, the lower limit is the plasma volume (0.04 mL/g) for a chemical that did not leave the circulatory system. A V_{ss} value $\cong$ 1 mL/g indicates similar affinity of the chemical for compartment-associated tissues as for the reference region. A chemical that distributed throughout the body water and did not bind to tissues or accumulate in fat would have a V_{ss} value of the total body water volume, about 0.7 mL/g. Large V_{ss} values (>>1 mL/g) indicate that the chemical has a much greater affinity for the fish tissues that for the reference region. This affinity may be for fat in the case of a lipophilic chemical, or for tissue binding sites; e.g., a chemical such as acriflavine that binds avidly to DNA.

6.2. BCF

While the apparent volume of distribution is a determinant of BCF, there may be other determinants as well. For a water referenced V, and in the absence of nonbranchial elimination of the chemical, the values of V and BCF are equal. When there is nonbranchial elimination of the chemical, this may cause the BCF to be less than the value of V. In the case of a water referenced V:

$$BCF = P \cdot V_{ss}/(P + CL_{nb}) \tag{53}$$

Eqs. 50 and 53 show that V_{ss} (or V_{ss}/f_{up} for the plasma referenced case) is equivalent to BCF only when $CL_{nb} \ll P$. When $CL_{nb} \gg P$, then nonbranchial elimination causes BCF $\ll V_{ss}$. This can cause poor correlation between BCF and measures of lipophilicity such as $K_{o/w}$. It can also cause large interspecies variability in BCF values when there is interspecies variability in any of the three determinants of BCF; viz., V_{ss}, P, or CL_{nb}.

6.3. Elimination Clearance

For water exposure studies, the exchange of the chemical between the exposure water and the gill is modeled as an event that is separate from the elimination processes such as metabolism; e.g., Eq. 7. For dietary and parenteral routes of administration, the branchial elimination is considered to be a part of the total body clearance; e.g., Eqs. 22 and 43. In the latter case, the total body clearance represents the sum of the clearances for the elimination pathways; e.g.,

$$CL = CL_H + CL_R + P \tag{54}$$

where the subscripts H and R refer to the hepatic and renal pathways. Each pathway specific CL value is determined by organ-specific physiological and biochemical determinants. For example, the hepatic clearance is controlled by the plasma protein binding of the chemical, the rate of blood flow to the liver, and by the activity of metabolizing enzymes in the liver toward the chemical. Thus, hepatic clearance may be blood flow limited, or limited by the activity of the metabolizing enzymes. In the former case, interspecies differences in hepatic blood flow would introduce similar interspecies differences in CL_H values whereas in the latter case, differences among species in the activity of metabolizing enzymes would introduce similar interspecies differences in CL_H values. Renal clearance is controlled by the renal blood flow, by the glomerular filtration rate, and by the activity of the active tubular secretion pathway when that is involved. Branchial clearance is considered under the gill uptake clearance section.

The upper limit for the clearance value for any particular pathway is generally the blood flow to the organ. Exception to this generality occurs when the chemical spontaneously degrades in the blood at a relatively rapid rate. A check on the validity of CL values may include a comparison with the known blood flow rates to the clearing organs. The total body clearance should generally not exceed the cardiac output.

6.4. Elimination Half-life

The elimination half-life of a chemical is determined by the total body clearance and by the apparent volume of distribution:

$$t_{1/2} = 0.693V/CL \tag{55}$$

In other words, the half-life reflects both the capacity of the fish to clear the chemical from its blood and the size of the space to be cleared. This mathematical representation does not fully convey this important concept. The implication is that a change in $t_{1/2}$ may result from a change in *either* V or in CL. Furthermore, V and CL are independent parameters; a change in one of them does not produce a change in the other, but a change in either one of them produces a change in the $t_{1/2}$. To understand why $t_{1/2}$ changes as a result of a change in some environmental condition or why it varies among species, it is essential to examine the behavior of the V and CL values that make up the $t_{1/2}$.

6.5. Gill Clearance

The gill is an important site for movement of chemicals into and out of the fish body. The key physiological determinants of the rate of chemical transport at the gill are the permeability of the gill epithelium, the rate of blood flow through the secondary lamellae (about 75% of cardiac output), and the rate of water flow across the secondary lamellae (5–15 times the cardiac output). Any of these determinants may be the rate controlling factor, depending on the properties of the chemical. As a general rule chemicals with low log $K_{o/w}$ values are limited by epithelial permeability, those with intermediate log $K_{o/w}$ values are limited by blood flow, and those with high log $K_{o/w}$ values are limited by water flow across the gills (Hayton and Barron, 1990).

7. REFERENCES

Abdou, H.M., 1989, *Dissolution, Bioavailability & Bioequivalence*, Mack Publishing Co., Easton, PA.

Ankley, G.T., Reinert, R.E., Wade, A.E., and White, R.A., 1985, Temperature compensation in the hepatic mixed-function oxidase system of bluegill, *Comp. Biochem. Physiol.* 81C:125–129.

ASTM, 1996, Standard practice for conducting bioconcentration tests with fishes and saltwater bivalve mollusks. E 1022–94, in: *Annual Book of ASTM Standards, Section 11*, American Society of Testing and Materials, Philadelphia.

ASTM, 1996, Standard practice for conducting acute toxicity tests with fishes, macroinvertebrates, and amphibians, E 729–88a, in: *Annual Book of ASTM Standards, Section 11*, American Society of Testing and Materials, Philadelphia.

Barron, M.G., Tarr, B.D., and Hayton W.L., 1987a, Temperature-dependence of cardiac output and regional blood flow in rainbow trout, *Salmo gairdneri* Richardson, *J. Fish Biol.* 31:735–744.

Barron, M.G., Tarr, B.D., and Hayton W.L., 1987b, Temperature-dependence of di-2-ethylhexyl phthalate (DEHP) pharmacokinetics in rainbow trout, *Toxicol. Appl. Pharmacol.* 88:305–312.

Branson, D.R., Blau, G.E., Alexander, H.C., and Neely, W.B., 1975, Bioconcetration of 2,2′,4,4′-tetrachlorobiphenyl in rainbow trout as measured by an accelerated test. *Trans. Am. Fish. Soc.* 104:785–792.

Burmaster, D.E., and Anderson, P.D., 1994, Principles of good practice for the use of Monte Carlo techniques in human health and ecological risk assessment, *Risk Analysis* 14:477–481.

Calabrese, E.J., 1991, *Principles of Animal Extrapolation*, Lewis Publishers, Inc., Chelsea, MI.

CFRa, *Code of Federal Regulations*, Title 21, Volume 1, Part 58, Good laboratory practice for nonclinical laboratory studies.

CFRb, *Code of Federal Regulations*, Title 21, Volume 5, Part 320, Bioavailability and bioequivalence requirements, Section 320.1, Definitions.

Gabrielsson, J., and Weiner, D., 1994, *Pharmacokinetic and Pharmacodynamic Data Analysis*, Swedish Pharmaceutical Press, Stockholm.

Gibaldi, M., and Perrier, D., 1982, *Pharmacokinetics*, 2nd ed., Marcel Dekker, Inc., New York.

Hayton, W.L., 1989, Pharmacokinetic parameters for interspecies scaling using allometric techniques, *Health Physics* 57(Suppl. 1):159–164.

Hayton, W.L., and Barron, M.G., 1990, Rate-limiting barriers to xenobiotic uptake by the gill, *Environm. Toxicol. Chem.* 9:151–157.

Hayton, W.L., and Schultz, I.R., 1991, Scaling bioconcentration and pharmacokinetic parameters for body size and environmental variables in fish, in: *Aquatic Toxicology and Risk Assessment: Fourteenth Volume, ASTM STP 1124*, M.A. Mays and M.G. Barron, eds., American Society for Testing and Materials, Philadelphia, pp. 149–65.

Hetrick, D.M., Jarabek, A.M., and Travis, C.C., 1991, Sensitivity analysis for physiologically based pharmacokinetic models, *J. Pharmacok. Biopharm.* 19:1–20.

Hoelton, G.F., and Randall, D.J., 1967, The effect of hypoxia upon the partial pressure of gasses in blood and water afferent and efferent to the gills of rainbow trout, *J. Exp. Biol.* 46:317–322.

Houston, A.R., 1990, Blood and circulation, in: *Methods for Fish Biology*, C.B. Schreck and P.B. Moyle, eds., American Fisheries Society, Bethesda, MD, pp. 273–322.

James, M.O., 1986, Conjugation of organic pollutants in aquatic species, in: *Xenobiotic Conjugation Chemistry*, J. Caldwell, D.H. Hutson, J.J. Menn, and G.D. Paulson, eds., American Chemical Society, Washington, D.C., pp. 29–47.

James, M.O., Khan, M.A.Q., and Bend, J.R., 1979, Hepatic microsomal mixed-function oxidase activities in several marine species common to costal florida. *Comp. Biochem. Physiol.* 62C:155–164.

Karara, A.H., and Hayton, W.L., 1984, Pharmacokinetic model for the uptake and disposition of di-2-ethylhexyl phthalate in sheepshead minnow *Cyprinodon variegatus, Aquatic Toxicol.* 5:181–195.

Kleinow, K.M., 1991, Experimental techniques for pharmacokinetic data collection in free-swimming fish, in: *Aquatic Toxicology and Risk Assessment: Fourteenth Volume, ASTM STP 1124*, M.A. Mays and M.G. Barron, eds., American Society for Testing and Materials, Philadelphia, pp. 131–138.

Landrum, P.F., Hayton, W.L., Lee, H., II, McCarty, L.S., Mackay, D., and McKim, J.M., 1994, Synopsis of discussion session on the kinetics behind environmental bioavailability, in: *Bioavailability: Physical, Chemical, and Biological Interactions*, P.F. Landrum, H.L. Bergman, W.H. Benson, and J.L. Hamelink, eds., Lewis Publishers, Boca Raton, FL, pp. 203–219.

McKim, J.M., and Erickson, R.J., 1991, Environmental Impacts on the Physiological Mechanisms Controlling Xenobiotic Transfer Across Fish Gills, *Physiol. Zool.* 64:39–67.

Pohjanpalo, H., 1978, System identifiability based on the power series expansion of the solution, *Math. Biosci.* 41:21–33.

Schmidt-Nielsen, K., 1984, *Scaling Why Is Animal Size So Important?*, Cambridge University Press, Cambridge.

Schultz, I.R., and Hayton, W.L., 1993, Toxicokinetics of trifluralin in rainbow trout. *Aquatic Toxicol.* 26:287–306.

Suffet, I.H., Jafvert, C.T., Kukkonen, J., Servos, M.R., Spacie, A., Williams, L.L., and Noblet, J.A., 1994, Synopsis of discussion session: influences of particulate and dissolved material on the bioavailability of organic compounds. in *Bioavailability: Physical, Chemical, and Biological Interactions*, P.F. Landrum, H.L. Bergman, W.H. Benson, and J.L. Hamelink, eds., Lewis Publishers, Boca Raton, FL, pp. 93–108.

Tsuda, T., Kojima, M., Harada, H., Nakajima, A., and Aoki, S., 1997, Relationships of bioconcentration factors of organophosphate pesticides among species of fish, *Comp. Biochem. Physiol. Part C: Pharmacol. Toxicol. Endocrinol.* 116:213–218.

6

PHARMACOKINETIC MODELING IN AQUATIC ANIMALS

Bioconcentration and Bioavailability

Guy R. Stehly[1], Mace G. Barron[2] and William L. Hayton[3]

[1]U.S. Geological Survey
Biological Resources Division
Upper Midwest Environmental Sciences Center
2630 Fanta Reed Road
La Crosse, Wisconsin
[2]Hagler Bailly, Inc., Boulder, CO
[3]Department of Pharmaceutics and College of Pharmacy
The Ohio State University
Columbus OH

1. ABSTRACT

Pharmacokinetic models used for estimating and interpreting the bioconcentration and bioavailability of xenobiotics in fish and other aquatic organisms are critically reviewed. Equilibrium partitioning, the simplest model for determining and interpreting bioconcentration, is based on the concept that xenobiotic distribution is thermodynamically driven. Although a number of factors can result in deviation from the model, the model works well for a number of lipophilic xenobiotics that are not metabolized. The accelerated bioconcentration model was developed as a method to determine bioconcentration factor for xenobiotics that require a long time to reach equilibrium. Model limitations include differing units for uptake and elimination, and representation of the fish as a single compartment. Clearance-volume models of bioconcentration are interpretable in terms of animal physiology. These models measure uptake and elimination in the same units and allow prediction of their relative importance to bioconcentration. Physiologically based pharmacokinetic models are complex models of xenobiotic distribution based on the anatomy and physiology of the fish. These models are most useful in their predictive value for different conditions and in interpretation of processes controlling accumulation of xenobiotic. Several alternate methods are available to determine bioconcentration factor without exposure to a constant concentration of a xenobiotic.

Xenobiotics in Fish, edited by *Smith et al.*
Kluwer Academic / Plenum Publishers, New York, 1999.

In clinical pharmacokinetics, bioavailability most frequently is a measure of the extent that an absorbed xenobiotic reaches the systemic circulation, but it can also be a measure of the rate. The term "bioavailability" is sometimes ambiguous because environmental toxicologists use it as a measure of the uptake clearance of unbound contaminants in water. Food web models of xenobiotic transfer in the environment use the basic kinetic concepts of equilibrium partitioning, bioconcentration, and bioavailability. Application of these models to some environmental contaminants has successfully predicted their kinetic behavior in the environment. However, simplifying assumptions may restrict the application of food-web models to a limited number of xenobiotics.

2. INTRODUCTION

Bioconcentration and bioavailability of a xenobiotic in fish are important factors in the environmental risk assessment of water- or sediment-borne xenobiotics. Determination of bioconcentration and bioavailability requires an understanding of pharmacokinetic principles to define the controlling mechanisms. General pharmacokinetic principles important to bioconcentration and bioavailability were previously reviewed (Barron, 1990). The models reviewed here will mostly focus on use with fish, however these pharmacokinetic principles can be used to describe the disposition of xenobiotics in other aquatic organisms.

Bioconcentration is the accumulation of a waterborne xenobiotic by fish and does not include uptake via the diet (Veith *et al.*, 1979). Bioconcentration factor (BCF) is defined as the ratio of the concentration of a xenobiotic in a fish (mass$*$body weight^{-1}) at steady-state to the average concentration (mass$*$volume^{-1}) of the xenobiotic in water (Rand and Petrocelli, 1985). The BCF of a xenobiotic is a major consideration in risk assessment because it is an estimate of the xenobiotic's propensity to accumulate in the fish. The accelerated bioconcentration test or similar models frequently are used to estimate the BCF of xenobiotics in aquatic organisms (Branson *et al.*, 1975).

In clinical pharmacokinetics, bioavailability refers to the fraction of a dose which reaches the systemic circulation unchanged, and the rate at which this occurs (Gibaldi and Perrier, 1982). Thus, bioavailability refers to the rate and the extent of a chemical entering the fish. The use of the term bioavailability in environmental toxicology refers to the uptake of xenobiotics from sediments or associated pore water and should not be confused with the definition used in clinical pharmacokinetics.

Food-web models describe the movement of xenobiotics within the environment based on knowledge of input of contaminants and the biotic and abiotic factors affecting transfer. These models are useful in predicting the fate of a xenobiotic in the environment and for determining its bioaccumulation in organisms (accumulation from food and water sources).

3. BIOCONCENTRATION

3.1. Equilibrium Partitioning

Conceptually equilibrium partitioning (EqP) is the simplest method to determine BCF. A group of fish is exposed to a constant concentration of xenobiotic in water. Periodically the concentration of xenobiotic in the fish (mass$*$body weight^{-1}) is measured until

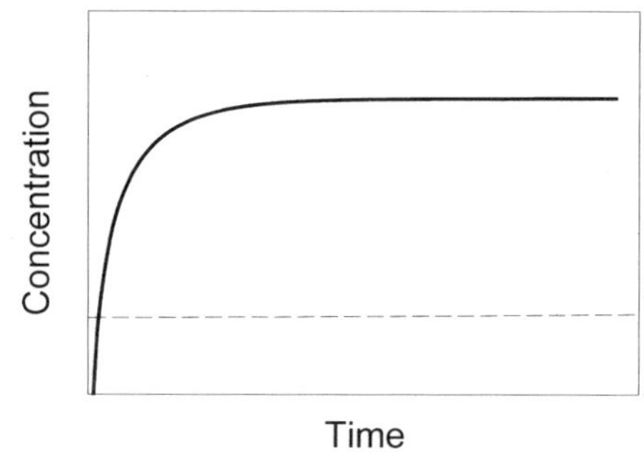

Figure 1. Graphical representation of the concept of equilibrium partitioning or accumulation to steady state. A waterborne xenobiotic (dotted line denotes concentration) will bioconcentrate in fish based on characteristics of the xenobiotic and on the lipid content of the fish. After some time, the concentration of xenobiotic in the fish will reach steady-state. The ratio of the concentration of xenobiotic in the fish to the concentration in the water is its bioconcentration factor.

a steady-state concentration of xenobiotic is accumulated; the ratio of the steady-state concentration in the fish to the concentration of xenobiotic in the water is defined as the BCF (Figure 1). Typically BCF experiments are conducted in a flow-through system to maintain a constant concentration of the xenobiotic in water.

Although EqP is most often applied to distribution of xenobiotic between sediment, associated pore water, and sediment dwelling organisms, the same basic principle applies to fish exposed to a waterborne xenobiotic. The EqP theory assumes that the driving force for bioconcentration is the thermodynamic partitioning of xenobiotic between the lipid component of an organism and water (Mackay and Paterson, 1981). Additionally, an implicit assumption of EqP theory is that only the freely dissolved concentration of a xenobiotic is available for uptake.

The EqP theory was originally applied to xenobiotics that were relatively lipophilic and had insignificant metabolism. As a result, the BCFs of many xenobiotics were shown to be correlated to its 1-octanol:water partition coefficient, (K_{OW}; Barron, 1990). The linear relationship between lipophilicity and bioconcentration was later shown to be curvilinear for log K_{OW} above 6 (Bruggeman *et al.*, 1984; McKim *et al.*, 1985). Factors that result in the curvilinear nature of the log K_{OW} vs bioconcentration curve have been critically reviewed (Barron, 1990) and include stearic hindrance, concentration dependent uptake, and significant metabolism or degradation of the contaminant.

3.2. Accelerated Bioconcentration Test

The accelerated bioconcentration test is an alternative method to determine BCF for xenobiotics requiring long times (usually >1 month) to reach steady-state (Branson *et al.*, 1975). Experimentally, determination of BCF is conducted with exposure of a group of fish to a constant concentration of waterborne xenobiotic for a short period (several days) followed by a depuration period. The concentration of xenobiotic is determined in individual (or small groups of) fish during the experiment. Uptake (k_1) and depuration (k_2) rate constants are estimated from the uptake and depuration curves (Figure 2).

An underlying assumption of this model and other models of accumulation is that the process can be modeled kinetically. The accelerated bioconcentration test also assumes that the aquatic organism and water each act as a single compartment for the xenobiotic (Figure 3):

The change in the concentration of xenobiotic in the organism (mass*weight of organism^{-1}) with respect to time is:

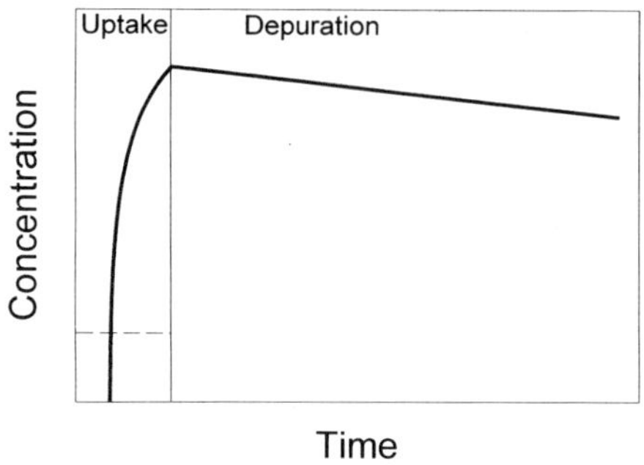

Figure 2. Bioconcentration is determined in the accelerated bioconcentration test by exposure of the fish to a constant concentration of waterborne xenobiotic (dotted line) for a short period (several days) followed by a depuration period. The concentration of the xenobiotic in the fish is determined periodically to obtain uptake and depuration curves. Estimates for the uptake (k_1) and depuration (k_2) are obtained with computer software, based on the fit to the observed data, and are used to calculate the bioconcentration factor.

$$dC_F/dt = k_1 * C_w - k_2 * C_F \tag{1}$$

where C_F is the concentration in the organism (mass*body weight^{-1}), C_W is the concentration of xenobiotic in water (mass*volume of water^{-1}), k_1 is the uptake rate constant (volume of water*body weight^{-1}*time^{-1}) and k_2 is the elimination rate constant (time^{-1}). Note that the units of k_1 are those of a clearance (volume of water*body weight^{-1} *time^{-1} rather than time^{-1}) and that it represents the volume of exposure solution cleared of toxicant per unit time normalized to body weight (e.g., mL*g^{-1}*h^{-1}) and it thus should be differentiated from a rate constant. Rate constants represent a fractional transfer of xenobiotic and have units of reciprocal time. If k_1 did represent a true rate constant, then it would not make sense in a flow through system; i.e., k_1 would vary with the volume passing through the exposure vessel and would be extremely small because an insignificant quantity of total xenobiotic is absorbed when constant concentration in water is assumed (Barron *et al.*, 1990; Wilkinson, 1987). Clearances have most frequently been used in pharmacokinetics to represent elimination but should not be restricted to this definition. A general feature of clearance is its independence with the associated transfer volume (rate constant = clearance*volume^{-1}). For example, clearance should be constant regardless of exposure volume; the rate constant however will decrease as the exposure volume increases. Uptake clearance is independent of the volume of the exposure solution, unless for example, changes in the physiology of the animal occur as volume is changed.

When the waterborne concentration of the xenobiotic is assumed constant,

$$C_F = k_1/k_2 * C_w(1 - e^{-k_2 t}). \tag{2}$$

At steady-state, time (t) is large and equation 2 reduces to:

$$BCF = C_F/C_W = k_1/k_2 \tag{3}$$

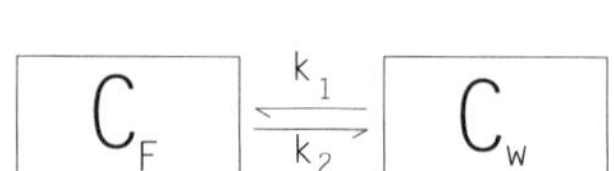

Figure 3. Model description for the accelerated bioconcentration test, where C_F is the concentration in the organism (mass *body weight^{-1}), C_W is the concentration of xenobiotic in water (mass*volume of water^{-1}), k_1 is the uptake rate constant (volume of water*body weight^{-1}*time^{-1}) and k_2 is the elimination rate constant (time^{-1}).

Bioconcentration factors were originally thought to be unitless; units for k_1 and k_2 canceled using the assumption that tissue density was $1\,g*mL^{-1}$ (Spacie and Hamelink, 1985). However examination of equation 1 indicates that dCF/dt has units of mass of xenobiotic*body weight^{-1}*time^{-1}. Therefore, units in the right side of equation 1 (k_1*C_W or k_2*C_F) must be similar. If C_W has units of mass of xenobiotic*volume^{-1}, then k_1, the uptake rate constant must have units of volume water*body weight^{-1}*time^{-1}. Therefore uptake clearance is a weight normalized measure of the ability to remove the xenobiotic from water. Therefore, BCF is not a comparison of the specific gravity or density of water to fish but rather the equilibrium ratio of the amount of xenobiotic in a volume of water per gram of fish. Intuitively this reasoning makes sense, because in this case, apparent volume of distribution is a partition coefficient for the xenobiotic between fish and water.

3.3. Additional Methods to Interpret BCF

Although the accelerated BCF test is frequently used to calculate BCF, a more complex model may be required to adequately describe the accumulation of a xenobiotic with multicompartmental characteristics (Barron *et al.*, 1990; Stehly and Hayton, 1989a). Multicompartmental pharmacokinetic models to describe xenobiotic disposition in fish are probably more realistic in many cases than a single compartment model. A model that describes the organism as two compartments, with the equation developed analogous to the accelerated BCF, predicts that BCF would be:

$$BCF = k_1(k_{21} + k_{12})/k_{21}k_2 = k_1/k_2k_{21} + k_{12}/k_{21}. \tag{4}$$

where k_{12} and k_{21} are intercompartmental transfer rate constants (Stehly and Hayton, 1989a). Therefore, xenobiotics with multicompartmental characteristics are dependent on intercompartment rate constants (equation 4) and an increase in the number of compartments results in an increase in equation complexity.

Alternatively, BCF based on a clearance-volume model can be derived for a simple two compartment system with the animal and water each acting as a single compartment (Figure 4).

Note in the clearance-volume model that the unit of X_F is the same as C_F used in the accelerated BCF (Stehly *et al.*, 1990). The change in the amount of xenobiotic in the animal normalized to body weight (X_F) with respect to time for this model is written as:

$$dX_F/dt = CL_U * C_W - CL_U * X_F/V \tag{5}$$

where where CL_U is the uptake clearance (volume of water*body weight^{-1}*time^{-1}) from the water, which is equal to the elimination clearance from the gills, and V is the

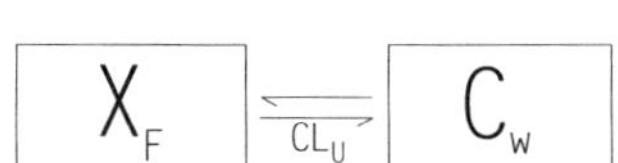

Figure 4. Model description for a clearance-volume model, where CL_U is the uptake clearance (volume of water*body weight^{-1}*time^{-1}) from the water, which is equal to the elimination clearance from the gills, X_F is amount of xenobiotic in the organism normalized on a weight basis (mass*body weight^{-1}) and C_W is the concentration of xenobiotic in water (mass*volume of water^{-1}).

apparent volume of distribution (volume of water*body weight^{-1}) of the organism (Spacie and Hamelink, 1985). Assuming a constant C_W, equation 5 can be integrated.

$$X_F = (C_W * V)(1 - e^{-(CL_U/V)*t}) \tag{6}$$

As time approaches infinity, the xenobiotic in the animal reaches steady-state, and equation 6 reduces to:

$$X_F = C_W * V \tag{7}$$

and

$$BCF = X_F/C_W = V \tag{8}$$

The assumptions used for the accelerated BCF test are also used for this model; i.e., the fish and water each acted as a single compartment and elimination is only at the site of absorption (gills). The example of equation 8 is, by definition, equivalent to equilibrium partitioning. An additional term can be included to account for routes of elimination such as metabolism, urinary excretion, or biliary excretion. When the term CL_B (total body clearance) is substituted in equation 8 to cover a more general case where elimination also occurs at sites other than the respiratory surface, then:

$$BCF = X_F/C_W = CL_U * V/CL_B \tag{9}$$

If the pharmacokinetics of a xenobiotic display multicompartmental characteristics with any number of peripheral compartments then:

$$BCF = CL_U * V_{ss}/CL_B \tag{10}$$

where V_{SS} is the steady-state apparent volume of distribution and is equal to the sum of the apparent volumes of distribution of all compartments that represent the organism (Stehly and Hayton, 1989b; Schultz and Hayton, 1991). Branchial and metabolic elimination are generally the two most important routes of elimination in fish. Consider the case where $CL_B = CL_U + CL_M$, where CL_M is the metabolic clearance; two limiting categories of BCF are possible, metabolism and uptake-dependent or metabolism and uptake-independent. When $CL_M << CL_U$, then BCF is equal to V_{SS} (i.e., $CL_B \cong CL_U$) and again is equivalent to EqP. In this case, changes in metabolism (induction or inhibition) or uptake are not likely to change BCF. When the magnitude of metabolic clearance approaches or exceeds uptake clearance, then the BCF of a xenobiotic, as described by equation 8 or 9, will be a fraction of the apparent volume of distribution (CL_U/CL_B, or $CL_U/(CL_U + CL_M)$). Changes in metabolism and uptake can result in significant changes in the BCF of these xenobiotics.

Thus, if uptake and elimination are described in clearance terms their importance to BCF can be assessed by the relative magnitudes of elimination and uptake clearances (Stehly and Hayton, 1989b). The clearance based derivation of BCF (Equation 10) indicates that regardless of the model, BCF is simply dependent upon three physiological variables: uptake, partitioning (i.e., distribution), and elimination. The importance of distribution in the accelerated BCF model is not as obvious and illustrates dependence of the elimination rate constant, k_2, on clearance and volumes. Interpretation of rate constant-clearance models is more difficult in multicompartment systems and uptake and elimination are compared by parameters with different units of measure (Eq. 4). Clearance-volume models also allow direct interpretation of uptake and elimination in terms of the physiology of the animal because clearances may be explicitly connected with processes such as ventilation volumes and blood flows, which have the same units. For example, because uptake clearances, ventilation volumes, and blood flows have units of $mL{*}min^{-1}$, assumptions concerning the rate limiting barriers to uptake can be hypothesized based on their relative magnitudes (Hayton and Barron, 1990). These models also define the limits when uptake and elimination are important to steady-state level of xenobiotic in the organism. Finally, clearances are preferred because of the independence of apparent volumes of distribution and rate of removal (Barron *et al.*, 1990).

Equations 2 and 6 are equivalent as previously demonstrated (Stehly *et al.*, 1990). In these models C_F and X_F are equal (note same units). In pharmacokinetic models concentrations in a compartment are described by the amount of xenobiotic in a compartment (mass or weight normalized mass) divided by an apparent volume of distribution in terms of the reference compartment (Gibaldi and Perrier, 1982; Barron *et al.*, 1990). When plasma is the reference the concentration in a compartment is the amount of xenobiotic divided by the apparent volume of distribution (in terms of plasma). For exposure of aquatic organisms to waterborne xenobiotic, the water is the "reference" compartment; therefore, the "concentration" in the organism should be the amount of xenobiotic in the compartment divided by the apparent volume of distribution of the organism in terms of the exposure solution (Barron *et al.*, 1990). Dividing mass of xenobiotic (amount) in the organism by weight is therefore a normalization for the weight of the organism in an analogous manner to the dosing of animals by amount per body weight. The animals are dosed an amount of xenobiotic not a concentration. While the use of the notation C_F as a concentration does not seem confusing to those using it for environmental toxicology, this is contrary to the use of the term as used in kinetic models in pharmacology, toxicology, and physiology.

In environmental toxicology, an evaluative model of the distribution of a xenobiotic between the phases of the environment (air, water, sediment and biota) can be interpreted by fugacity, the "escaping tendency" of the xenobiotic (Mackay and Patterson, 1981). Fugacity models use principals similar to the pharmacokinetic models presented and assume equilibrium distribution between phases. However, the xenobiotic in air (partial pressure in atmosphere) is the reference compartment in fugacity models and the fugacity capacity (Z) is analogous to the apparent volume of distribution. Therefore, as in clearance models, concentrations in compartments will have a net flux toward compartments with lower concentrations and concentrations will be equal at steady-state (Mackay and Paterson, 1981).

More complex models have been used to describe the disposition of xenobiotics in aquatic organisms. Physiologically based pharmacokinetic models (PBPK) have had a limited use in aquatic organisms, however descriptions with three compounds are presented in this Symposium. These models are generally composed of a number of compartments that are important to describe the disposition of a xenobiotic (Figure 5). These compartments can be classified into several categories; storage compartments (poorly perfused tissues, adipose tissue), elimination compartments (liver, kidney), and in fish an important uptake compartment (gills). Physiologically based pharmacokinetic models require a large amount of data obtained from the literature or experimentally, such as, physiological variables (clearances and metabolism) transport variables (absorption, permeability), thermodynamic (tissue: blood partition coefficients), and anatomical values

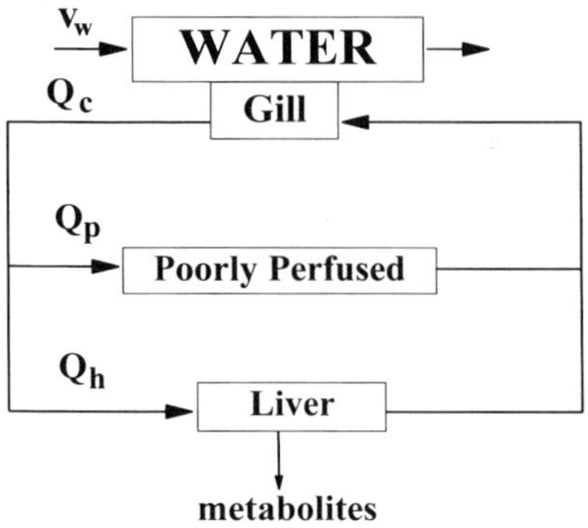

Figure 5. A simple physiologically based pharmacokinetic model for fish. This model contains a component for transfer of xenobiotic between the water and fish through the gill, a storage compartment (poorly perfused) and an eliminating organ (liver). The model shown requires information on physiological (e.g., metabolism), transport (e.g., permeability), thermodynamic (e.g., tissue: blood partition coefficients), and anatomical variables (e.g., Q = blood flows, V = ventilation volume). Subscripts refer to a component of the model, W = water, c = cardiac, p = poorly perfused, and h = hepatic.

(blood flows, tissue volumes) (Lutz and Dedrick, 1985). In PBPK models, the sum of the individual tissue/water partition coefficients multiplied by the volume of the respective compartments is, by definition, equal to the apparent volume of distribution, or the BCF at steady-state. Calculation of BCF with the PBPK model based on partition coefficients and volumes of the compartments is equivalent to EqP in a much more detailed model. Because PBPK models are based on anatomy and physiology they are useful for extrapolation to other conditions, doses, and species. In addition, evaluation of the effect of physiological, transport, thermodynamic, and anatomical variables on BCF can be determined using PBPK models. A two compartment clearance-volume model can be viewed as a simplified version of the PBPK model presented in Figure 5, where the apparent volume of distribution is an estimated parameter rather than an experimentally determined value (partition coefficients). Uptake and metabolism are essentially the same in these two models while the blood flows and permeability of the PBPK model are incorporated in the intercompartmental clearance of the clearance-volume model.

3.4. Experimental Methods to Determine BCF

In addition to determination of BCF by EqP and the accelerated bioconcentration test, several other methods are available. One difficulty frequently encountered during measurement of BCF by the accelerated BCF test is the maintenance of a constant C_W. Gobas and Zhang (1992) described a method using a first degree approximation of the concentration in the aquatic organism with respect to time, taking into account the change in the C_W. Values for k_1 and k_2 are determined by an iterative procedure with numerical integration of the first degree approximation. Gobas and Zhang found that their method of estimating BCF was more reliable compared with using a mean or time weighted average of the C_W.

The derivation of BCF equations (equations 1–10) and parameter estimation is simplified in a flow through system because C_W is constant. An exposure can be conducted in a closed static system, but C_W could decrease if the organism is large relative to the volume of water or if the xenobiotic has a large capacity to accumulate in the fish. If the fish absorbs a significant quantity of xenobiotic, an equation for the change in C_W with respect to time can be written. The differential for change in the amount of xenobiotic in the fish is integrated with C_W as a variable rather than a constant and the equations are fitted simultaneously (Stehly and Hayton, 1989b; Karara and Hayton, 1984). If significant metabolism occurs, a third equation, to describe total metabolites within the system (fish and water) is fitted simultaneously (Stehly and Hayton, 1989b; Karara and

Hayton, 1984). Simultaneous fitting of the change in C_W and the xenobiotic in fish within a static system provides a dynamic system for estimating model parameters. This methodology allows better estimates of model parameters because the xenobiotic in fish and water can rapidly reach equilibrium, without approaching steady-state, unlike constant concentration exposures. Further, the fish can be represented as any number of compartments the data will support.

4. BIOAVAILABILITY

In clinical pharmacokinetics, bioavailability refers to the rate and extent that an absorbed dose reaches the systemic circulation (Gibaldi and Perrier, 1982). Although the rate of absorption is infrequently measured, it has important implications concerning the effectiveness or toxicity of a xenobiotic. Compounds or formulations having relatively rapid rates of absorption have larger maximum concentrations, which occur sooner, than one with a slow absorption rate. The most common means of determining absorption kinetics is to fit the plasma concentration (C) vs time data to the equation appropriate for the selected model. For a one compartment model with first order absorption and elimination;

$$C = \frac{k_a F X_0}{V(k_a - k)}\left(e^{-kt} - e^{-k_a t}\right) \tag{11}$$

where k_a is the absorption rate constant (time^{-1}), k is the elimination rate constant (time^{-1}), F is the extent of bioavailability of the dose (X_0), and V is the apparent volume of distribution. Care must be taken in the interpretation of data if absorption is complex or if the absorption rate is slower than the initial elimination rate (Gibaldi and Perrier, 1982). More complex models and alternate methods used to estimate absorption rates in clinical pharmacokinetics are reviewed in Gibaldi and Perrier (1982).

The extent of bioavailability (F) is the most determined and reported bioavailability term. Bioavailability is a unitless term referencing one dosage form or route of exposure to another, and has a value from 0 to 1. Incomplete absorption due to low membrane permeability or slow dissolution, and metabolism (intestinal or site of absorption and hepatic first pass) are major factors reducing F (Klaassen and Rozman, 1991). Bioavailability is most often determined as a ratio of the total area under the drug concentration in the plasma versus time curve (AUC) after oral administration and intravascular dosing, i.e.,

$$F = \frac{\left(\int_0^{\infty} C\,dt\right)_{oral}}{\left(\int_0^{\infty} C\,dt\right)_{iv}} = \frac{AUC_{oral}}{AUC_{iv}} \tag{12}$$

for the same dose and in a range where no dose-dependent kinetics occur (Gibaldi and Perrier, 1982). The majority of determinations of oral bioavailability (F) in fish are for therapeutants and were recently reviewed (Stoffregren *et al.*, 1996). Because determination of bioavailability is common in clinical pharmacokinetics, additional details on experimental methods are not presented here. Specific examples of determination of bioavailability can be obtained from Gibaldi and Perrier (1982) or from Stoffregren *et al.*, (1996).

Although determination of F in fish is most frequently determined after oral administration, it has been calculated for intramuscular (Stoffregren *et al.*, 1996) and waterborne administration (Barron *et al.*, 1989). Determination of the extent of bioavailability after waterborne exposure requires the weight of the fish and volume of the waterborne exposure. For example, F of the plasticizer DEHP after waterborne exposure was determined by,

$$F = \frac{AUC_W * D_{iv} * FW}{AUC_{iv} * C_i * V_W} \tag{13}$$

where AUC_W and AUC_{iv} are the AUCs for water administration and intravascular dosing respectively, D_{iv} is the intravascular dose, FW is the fish weight, C_i the initial water concentration, and V_W the volume of the exposure solution. Although not used to date, the bioavailability of waterborne therapeutants that act systemically would be useful in evaluating a proper therapy.

In contrast, the term bioavailability is frequently used in environmental toxicology in reference to waterborne or sediment associated contaminants and should not be confused with the description given above. In environmental toxicology the term bioavailability refers to the availability of xenobiotic to aquatic organisms and is a measure of the uptake clearance of unbound xenobiotic in the water column or sediment (Niimi and Oliver, 1983; Schrap and Opperhuizen, 1990).

The measurement in environmental toxicology that corresponds to F is the dietary absorption efficiency (Niimi and Oliver, 1983; Niimi and Dookhran, 1989). Residue concentrations after a single feeding are plotted versus time. Absorption efficiency is estimated from the log linear extrapolation to day 0 of the residue concentration versus time plot (Niimi and Oliver, 1983). Alternately, absorption efficiency is estimated after several days of oral administration if elimination is taken into account during the time of administration (Niimi and Dookhran, 1989). Measurement of F (dietary absorption efficiency) by these methods may be a good estimate for relatively hydrophobic xenobiotics. However, for more polar xenobiotics, particularly when the kinetics are best described by multicompartmental models, F will be underestimated. Underestimates of F result because some elimination occurs before distribution is complete, and a portion of the elimination is quicker than the terminal phase. Long-term dosing tends to decrease the multicompartmental characteristics of a compound (Gibaldi and Perrier, 1982).

5. FOOD WEB MODELS

Food web bioaccumulation models incorporate the mass balance of contaminant bioaccumulation in biota exposed by multiple pathways (Thomann, 1989; Thomann *et al.*, 1992a). The models are composed of a series of equations that calculate contaminant concentrations in biota from EqP (e.g., benthos:sediments; plankton:water), uptake from water, or intake of contaminated food, and processes which reduce contaminant concentrations (i.e., growth, excretion). Contaminant concentrations in source compartments (water, sediment) are generally assumed to be in steady-state equilibrium with the biota composing the food web. Food web bioaccumulation models allow estimation of contaminant residues in component trophic groups, species, or life stages, but have typically not included biotransformation or reproductive processes of biota (Thomann *et al.*, 1992a; Gobas, 1992). Despite their utility, the models generally incorporate a large number of parameters, many with high levels of uncertainty.

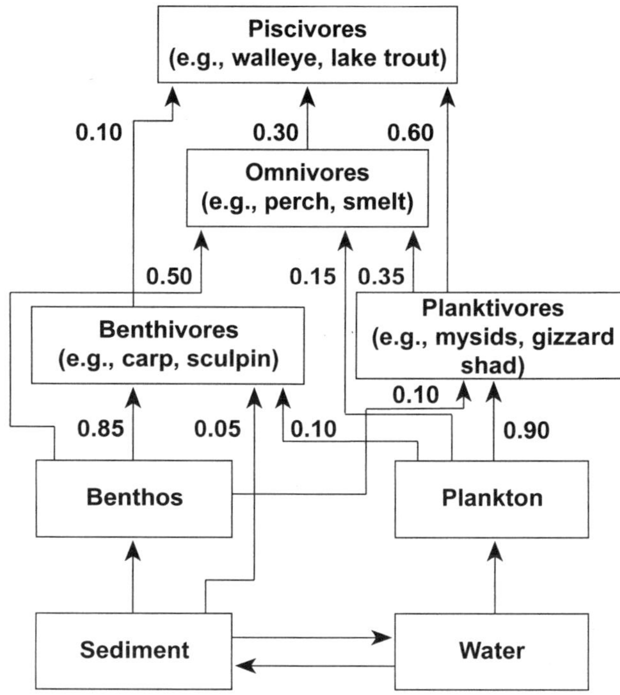

Figure 6. Schematic representation of a food web bioaccumulation model. Numbers are dietary fractions consumed by each trophic group. Arrows without numbers designate equilibrium partitioning.

This section summarizes the model based equations of a generalized food web model adapted from Thomann *et al.*, (1992b) and Gobas (1992). The model is based on a Great Lakes food web and can be viewed conceptually as a series of compartments linked by transport pathways (Figure 6). Compartments of the model are defined based on functional characteristics (e.g., feeding habits), in addition to source compartments (i.e., water, sediment) (Figure 6). In addition to the pathways shown in Figure 6, the model also calculates residue concentrations in fish tissues resulting from exposure to a contaminant dissolved in the water column. More complex formulations of food web models are available, such as those based on species and age-specific bioaccumulation (Thomann *et al.*, 1992b).

5.1. Source Compartments: Sediment and Water

The food web bioaccumulation model is coupled to sediment and water source compartments (Figure 6). Generally sediment and water concentrations of a contaminant are specified as input parameters.

5.2. Plankton and Benthos

Contaminant concentrations in plankton and benthos are calculated from EqP. Equilibrium partitioning theory predicts that the contaminants concentration in a benthic organism ($C_{benthos}$) is determined by the lipid content of the organism (L_{biota}), the xenobiotic's concentration in sediment (C_s) and the fraction of organic carbon (OC_s) in the sediment:

$$C_{\text{benthos}} = L_{\text{biota}} * (C_s / OC_s) \tag{14}$$

Assumptions relevant to the EqP theory were discussed above. The equilibrium residue concentration accumulated by aquatic organisms is assumed to be independent of the number and types of exposure routes (e.g., sediment ingestion and/or exposure via pore water). Food web bioaccumulation models use the EqP theory to predict the contaminant concentration in phytoplankton (C_{plankton}) from lipid content (L_{biota}), the xenobiotic's K_{ow}, and the contaminant concentration in water (C_w):

$$C_{\text{plankton}} = L_{\text{biota}} * K_{ow} * C_w \tag{15}$$

As a simplification, the model illustrated in Figure 6 assigns the same contaminant concentration in both phytoplankton and zooplankton. This simplification is reasonable where contaminant concentrations in phytoplankton and zooplankton are closely coupled because of density dependent population regulation, rapid partitioning with the xenobiotic dissolved in water, and when there is limited biomagnification between these two compartments (Connolly *et al.*, 1992).

5.3. Benthivores, Planktivores, Omnivores, Piscivores

Predator trophic groups are defined by their prey selection (Figure 6). Contaminant concentrations in predator trophic groups (C_v) are calculated from the ratio of xenobiotic intake (uptake from water and food) to elimination (excretion and egestion) and growth dilution according to:

$$C_V = \frac{[(K_{UW} * C_W) + (K_{UF} \sum P_i * C_i)]}{K_{EX} + K_{EG} + K_G} \tag{16}$$

where dietary intake is the sum of the proportion of each prey trophic group selected (P_i) (Figure 6) multiplied by the contaminant concentration in the dietary item (C_i). Biotransformation was assumed to be negligible. This equation is a generalized relationship for estimating contaminant concentrations in each trophic group, using trophic group-specific parameters (e.g., body weight, lipid content, prey selection, assimilation efficiency).

The uptake rate (K_{UW}) from water is dependent on initial transfer of xenobiotic from water to the aqueous phase of the organism, then to the lipid phase of the organism (Gobas, 1992). K_{UW} is calculated by:

$$K_{UW} = \frac{1}{[(W_V / Q_W) + (W_V / (K_{OW} * Q_W / 100))]} \tag{17}$$

which is dependent on the weight of the organism (W_V) and Q_W (xenobiotic transport parameter). Qw is calculated by:

$$Q_W = 88.3 * W_V^{\,0.6} \tag{18}$$

The uptake rate from food (K_{UF}) is dependent on the assimilation efficiency ($\propto$; fraction of dietary xenobiotic absorbed), the feeding rate (FR) and organism weight (W_v). K_{UF} is calculated by:

$$K_{UF} = \propto * FR/W_V \tag{19}$$

FR is dependent on biota weight and ambient temperature (T; annual average), and is calculated by:

$$FR = 0.022 * W_V^{0.85} * e^{0.06*T} \tag{20}$$

The excretion rate of xenobiotic across the gills (K_{EX}) is calculated by:

$$K_{EX} = \frac{1}{[(W_V * L_{biota*}K_{OW}/Q_W) + (W_V/L_{biota*}100/Q_W)]} \tag{21}$$

K_{EX} is dependent on initial transfer of xenobiotic from the lipid to the aqueous phase of the fish, then to the ambient water (Gobas, 1992). The term 100/Qw is equal to one divided by the fish lipid phase transport parameter (Gobas, 1992).

The fecal egestion rate (K_{EG}) is calculated as 0.25 of the uptake rate from food (Gobas, 1992):

$$K_{EG} = K_{UF}/4 \tag{22}$$

Contaminant concentrations in fish are reduced by 'growth dilution'; e.g., increasing tissue mass of a fish results in a decrease in xenobiotic concentration. Growth dilution is estimated by calculating K_G as a weight dependent coefficient:

$$K_G = 0.000502 * W_V^{-0.2} \tag{23}$$

Food web models are typically calibrated using existing data on contaminants in similar ecological systems to the one being modeled. Parameter inputs such as body sizes, temperature, prey selection, and lipid content can be adjusted to obtain better agreement between observed and model estimated concentrations of contaminants in biota.

6. REFERENCES

Barron, M.G., 1990, Bioconcentration. Environ. Sci. Technol. 24:1612–1618.

Barron, M.G., Stehly, G.S., and Hayton W.L., 1990, Pharmacokinetic modeling in aquatic animals. I. Models and concepts. Aquat. Toxicol. 17:187–212.

Barron, M.G., Schultz, I.R., and Hayton, W.L., 1989, Presystemic branchial metabolism limits di-2-ethylhexyl phthalate accumulation in fish. Toxicol. Appl. Pharmacol. 98:49–57.

Branson, D.R., Blau, G.E., Alexander, H.C., and Neely, W.B., 1975, Bioconcentration of 2,2′, 4,4′-tetrachlorobiphenyl in rainbow trout as measured by an accelerated test. Trans. Am. Fish. Soc. 4: 785–792.

Bruggeman, W.A., Opperhuizen, A., Wybenga, A., and Hutzinger, O., 1984, Bioaccumulation of superlipophilic chemicals in fish. Toxicol. Environ. Chem. 7:173–189.

Connolly, J.P., Parkerton, T.F., Quadrini, J.D., Taylor, S.T., and Thumann, A.J., 1992, Development and application of a model of PCBs in Green Bay, Lake Michigan walleye and brown trout and their food webs. Unpublished.

Gibaldi, M., and Perrier, D., 1982, Pharmacokinetics. Marcel Dekker, Inc., New York.

Gobas, F.A.P.C., 1992, Modeling the accumulation and toxicity of organic chemicals in aquatic food chains. pp. 129–151, In: Chemical Dynamics in Fresh Water Ecosystems (Eds: F.A.P.C. Gobas and J.A. McCorquodale). Lewis Publishers, Boca Raton.

Gobas, F.A.P.C., and Zhang, X., 1992, Measuring bioconcentration factors and rate constants of chemicals in aquatic organisms under conditions of variable water concentrations and short exposure time. Chemosphere. 25:1961–1971.

Hayton, W.L., and Barron, M.G., 1990, Rate-limiting barriers to xenobiotic uptake by the gill. Environ. Toxicol. Chem. 9:151–157.

Karara. A.H., and Hayton, W.L., 1984, Pharmacokinetic model for the uptake and disposition of di-2-ethylhexyl phthalate in sheepshead minnow *Cyprinidon variegatus*. Aquat. Toxicol. 5:181–195.

Klaassen, C.D., and Rozman, K., 1991, Absorption, distribution, and excretion of toxicants , in: Casarett and Doull's Toxicology, The Basic Science of Poisons, M.O. Amdur, J. Doull, and C.D. Klaassen, ed., Pergamon Press, New York.

Landrum, P.F., Eadie, B.J., and Faust, W.R., 1992, Variation in the bioavailability of polycyclic aromatic hydrocarbons to the amphipod *Diporeia* (spp.) with sediment aging. Environ. Toxicol. Chem. 11:1197–1208.

Lutz, R.J., and Dedrick, R.L., 1985, Physiological Pharmacokinetics: Relevance to Human Risk Assessment. In A.P. Li, eds., New Approaches in Toxicity Testing and Their Application in Human Risk Assessment. Raven Press, New York, pp. 129–149.

Mackay D., and Paterson, S., 1981, Calculating fugacity. Environ. Sci. Technol. 15:1006–1014.

McKim, J., Schmeider, P., and Veith, G., 1985, Absorption dynamics of organic chemical transport across trout gills as related to octanol-water partition coefficient. Toxicol. Appl. Pharmacol. 77:1–10.

Niimi, A.J., and Oliver, B.G., 1983, Biological half-lives of polychlorinated biphenyl (PCB) congeners in whole fish and muscle of rainbow trout (*Salmo gairdneri*). Can. J. Fish. Aquat. Sci. 40:1388–1394.

Niimi, A.J., and Dookhran, G.P., 1989, Dietary absorption efficiencies and elimination rates of polycyclic aromatic hydrocarbons (PAHs) in rainbow trout (*Salmo gairdneri*). Environ. Toxicol. Chem. 8:719–722.

Rand, G.M., and Petrocelli, S.R., 1985, Fundamentals of Aquatic Toxicology. Hemisphere Publishing Corporation, Washington.

Reuber, B.,McKay, D., Paterson, S., and Stokes, P., 1987, A discussion of chemical equilibria and transport at the sediment-water interface. Environ. Toxicol. Chem. 6:731–739.

Schrap, S.M., and Opperhuizen, A., 1990, Relationship between bioavailability and hydrophobicity: reduction of the uptake of organic chemicals by fish due to the sorption on particles. Environ. Toxicol. Chem. 9:715–724.

Schultz, I.R., and Hayton, W.L., 1991, Experimental design for pharmacokinetic data analysis using compartmental pharmacokinetic models. In M.A. Mayes and M.G. Barron, eds., Aquatic Toxicology and Risk Assessment. American Society for Testing and Materials, Philadelphia, pp. 139–147.

Spacie, A., and Hamelink, J.L., 1985, Fundamentals of Aquatic Toxicology. Hemisphere Publishing Corporation, Washington, pgs 495–525.

Stehly, G.R., and Hayton, W.L., 1989a, Errors in the use of the accelerated bioconcentration test. Aquatic toxicology and environmental fate: Eleventh Volume, ASTM STP 1007, edited by G.W. Suter, II and M.A. Lewis, American Society for Testing and Materials, Philadelphia, pp. 573–584.

Stehly, G.R., and Hayton, W.L., 1989b, Disposition of pentachlorophenol in rainbow trout (*Salmo qairdneri*): Effect of inhibition of metabolism. Aquat. Toxicol. 14:131–148.

Stehly, G.S., Landrum, P.F., Henry, M.G., and Klemm, C., 1990, Toxicokinetics of PAH in Hexigenia. Environ. Toxicol. Chem. 9:167–174.

Stoffregren, D.A., Bowser, P.R., and Babish J.G., 1996, Antibacterial Chemotherapeutants for finfish aquaculture: A synopsis of laboratory and field efficacy studies. J. Aquat. Anim. Health 8:181–207

Thomann, R.V., 1989, Bioaccumulation model of organic chemical distribution in aquatic food chains. Environ. Sci. Technol. 23:699–707.

Thomann, R.V., Connolly, J.P., and Parkerton, T.F., 1992a, An equilibrium model of organic chemical accumulation in aquatic food webs with sediment interaction. Environ. Toxicol. Chem. 11:615–629.

Thomann, R.V., Connolly, J.P., and Parkerton, T., 1992b, Modeling accumulation of organic chemicals in aquatic food webs. pp. 153–186, In: Chemical Dynamics in Fresh Water Ecosystems (Eds: F.A.P.C. Gobas and J.A. McCorquodale). Lewis Publishers, Boca Raton.

Vieth, G.D., Defoe, D.L., and Bergstedt, B.V., 1979, Measuring and estimating the bioconcentration factor of chemicals in fish. J. Fish. Res. Board Can. 36:1040–1048.

Wilkinson G.R., 1987, Clearance approaches in pharmacology. Pharmacol. Rev. 39:1–47.

7

RECENT ADVANCES IN THE DEVELOPMENT AND USE OF PHYSIOLOGICALLY BASED TOXICOKINETIC MODELS FOR FISH

John W. Nichols

U.S. Environmental Protection Agency
National Heath and Environmental Effects Laboratory
Mid-Continent Ecology Division
6201 Congdon Boulevard
Duluth, MN, 55804

1. ABSTRACT

Existing physiologically based toxicokinetic (PBTK) models for fish accurately predict the uptake and disposition of waterborne organic chemicals of low to intermediate hydrophobicity. These models have been used to evaluate factors that control chemical kinetics in fish and to extrapolate data among fish species. The utility of these models is limited, however, by a lack of understanding of the factors that control chemical uptake from dietary sources and an inability to measure, describe, and predict metabolic biotransformation. In this chapter, critical questions related to dietary uptake and metabolic biotransformation are identified and discussed. Research with fish that addresses some of these questions is presented. Guidance is also provided from modeling efforts with mammalian species. It is recommended that future research be directed towards developing a relatively simple gut sub-model description. In vitro methods for characterizing chemical biotransformation must be complemented by new techniques for measuring metabolism in the intact animal.

2. MODELING APPROACH

2.1. Introduction

Increasingly, there is a need to extrapolate chemical kinetic information obtained from one or a few species of fish to a much larger number of untested species. One example is the need to relate toxicity information from laboratory test species to animals

Xenobiotics in Fish, edited by *Smith et al.*
Kluwer Academic / Plenum Publishers, New York, 1999.

present in the environment. A second is the need to provide guidance on the registration and use of chemical products in aquacultural practice. A promising approach to these and similar problems is the development and use of physiologically based kinetic models. Physiologically based models are developed from information about the anatomy, physiology and biochemistry of the exposed animal, and physical-chemical attributes of the compound of interest. As such, these models possess biological and chemical integrity and are therefore well suited for extrapolation to untested species and compounds by appropriate reparameterization of model inputs. An important advantage of this modeling approach is that it yields chemical time-course predictions for specific tissues of interest. In toxicity testing, this information provides a measure of dose that can be compared directly with tissue-level effects. The ability to predict chemical kinetics in selected tissues may also yield improved exposure estimates for animals (including humans) that consume contaminated fish.

The goal of this chapter is to describe recent efforts to develop and use PBTK models for fish. The models described herein have been developed primarily for the purpose of extrapolating toxicity information among fish species. It should be emphasized, however, that the basic principles of this modeling approach are "chemically generic" and may be applied to any compound of interest. An effort is made to review basic elements of model structure with an emphasis on underlying assumptions. Particular attention is given to the development of physiological models for chemical uptake from the diet and the incorporation of data pertaining to metabolic biotransformation.

2.2. Historical Development

The mathematical equations that are used to describe the chemical mass-balance on a tissue exchanging with blood were first developed by Teorell (1937). The computing power needed, however, to simultaneously solve mass-balance differential equations for several compartments was not available until the 1960s. Early use of the physiological modeling approach consisted primarily of chemotherapeutic and other pharmacological applications (reviewed by Jerlowski and Jain, 1983). In this regard, the pioneering efforts of Bischoff, Dedrick and co-workers are particularly noteworthy (Bischoff and Brown, 1966; Dedrick and Bischoff, 1968). Subsequently, this approach was adopted by toxicologists interested in extrapolating animal testing data to humans. Andersen, Gargas, Clewell and co-workers employed PBTK models to estimate apparent rate and capacity parameters for compounds that undergo metabolic biotransformation (Andersen *et al.*, 1984; Gargas *et al.*, 1986). These efforts were facilitated by the use of *in vitro* methods to estimate chemical partitioning relationships for volatile compounds (Gargas *et al.*, 1989). Increasingly, PBTK dosimetry models are being linked with biologically-based models of chemical effect, including multistage models for carcinogenesis (Conolly *et al.*, 1988). These linked models are now being employed to augment more traditional methods in human health risk assessment (Reitz *et al.*, 1996).

The first physiological models for fish were adapted from existing rodent models and incorporated intravascular routes of exposure. Based on previous experience with mice, Zaharko *et al.*, (1972) developed a model for methotrexate in stingrays. Bungay *et al.*, (1976) adapted a rat model to describe the disposition of phenol red in the dogfish shark. The first fish model incorporating the branchial route of exposure was published by Nichols *et al.*, (1990). This model was made possible by the prior development of a physiologically based gill submodel which incorporated both flow and diffusion limitations (Erickson and Mckim, 1990). PBTK models for fish have been used to extrapolate kinetic information among species (Nichols *et al.*, 1993) and evaluate *in vitro* metabolic rate and capacity parameters (Law *et al.*, 1991). An effort has also been made to link dosimetry information with measures of toxic effect in target tissues (Abbas and Hayton, 1997).

2.3. Model Structure

A PBTK model is composed of compartments that correspond to individual groups of tissues and organs with similar kinetic characteristics. The transfer of material between compartments is mediated strictly by blood flow. Figure 1 shows a schematic of an inhalation model for fish (Nichols *et al.*, 1990). The right-hand side of this schematic represents the arterial circulation. Q_F, Q_P, Q_R, Q_K and Q_L refer to adipose tissue, poorly perfused tissue (mostly white muscle), richly perfused tissue (consisting of the stomach, intestinal tract, spleen and gonads), kidney, and liver blood flows as proportions of Q_C, the cardiac output. Q_W is the flow rate of inspired water that is available to exchange with blood and C_{INSP} is the chemical concentration in this water. Chemical concentrations in expired water (C_{EXP}) and arterial blood (C_{ART}) are calculated from the gill submodel (Erickson and McKim, 1990). C_{VF}, C_{VP}, C_{VR}, C_{VK}, and C_{VL} refer to chemical concentrations in venous blood leaving these tissues, and C_{VEN} is the chemical concentration in mixed venous blood. Consistent with the circulatory anatomy of fish, venous blood draining the well perfused tissue compartment becomes a portal blood supply to the liver, while a portion of venous blood draining the poorly perfused compartment flows directly to the kidney. The model also incorporates kinetic constants for saturable metabolism in the liver compartment. In theory, this or some other kinetic description of chemical elimination could be ascribed to any tissue compartment.

2.4. Mass Balance on Each Compartment

The model itself consists of a set of mass-balance differential equations that describe the rate of change of the amount of chemical within each of the tissue compartments. A relatively simple model may be formulated if one assumes that diffusion does not limit chemical flux between blood and tissues. In this case, it is assumed that chemical partitioning proceeds to equilibrium with the result that chemical concentration in venous blood exiting the tissue (C_{Vi}) is equal to that in the tissue (C_i) divided by a tissue:blood equilibrium partition coefficient (P_i):

$$C_{Vi} = C_i / P_i \tag{1}$$

A mass-balance on the compartment may then be written as:

$$dA_i/dt = Q_i(C_{ART} - C_{Vi}) \tag{2}$$

where A_i is the amount of chemical in the tissue compartment and Q_i is blood flow rate to the compartment. This type of chemical distribution is said to be blood flow-limited, because the rate of chemical accumulation in tissue is controlled by blood flow rate. Importantly, if both sides of the equation are divided by tissue volume, one obtains an equation in which the rate of change of tissue concentration is determined as the product of a concentration term and the ratio of tissue blood flow rate to tissue volume. Dividing blood flow by tissue volume yields a term with units of reciprocal time,

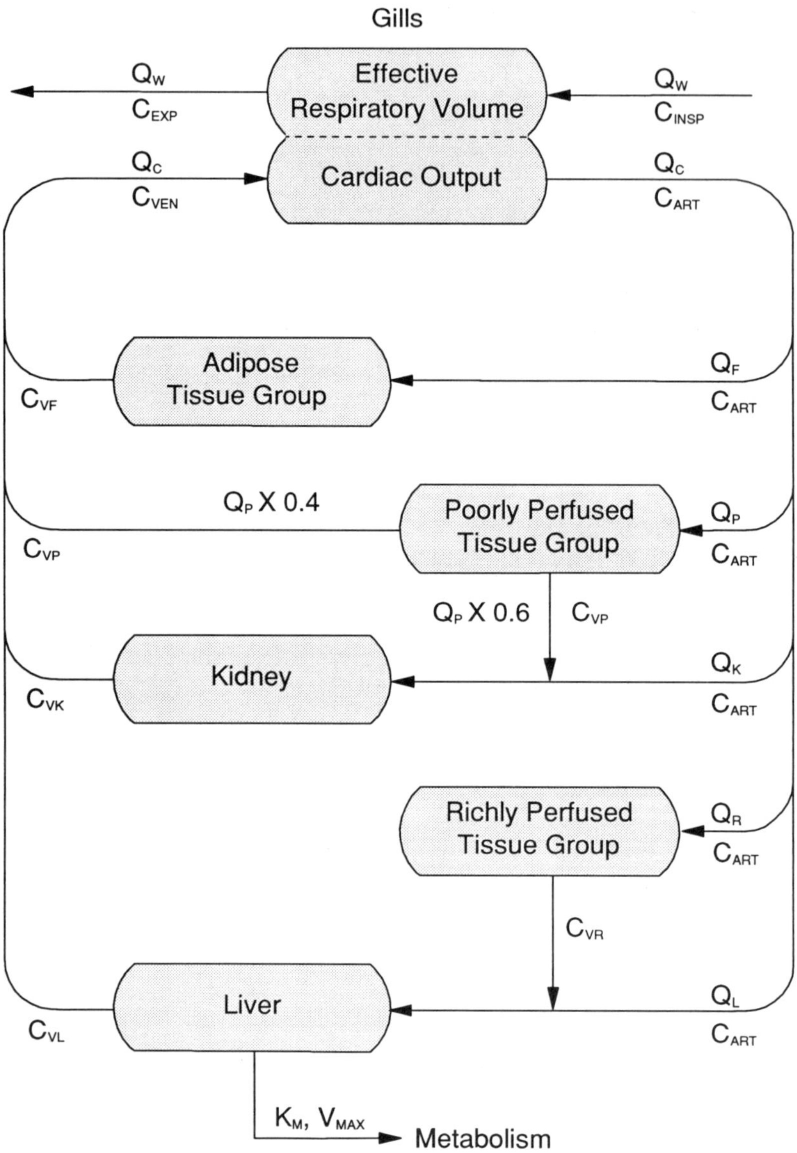

Figure 1. Schematic of a PBTK model for the uptake and disposition of waterborne organic chemicals by fish. Symbols and abbreviations are described in the text.

which are the same units employed for a first-order rate constant in a traditional "data-based" compartmental model. Thus, the physiological description is mathematically similar to that of a data-based model with fitted rate constants. In a physiological model, however, the rate constants derive from the animal's anatomy and physiology.

Diffusion limitations may prevent a chemical equilibrium from being established between blood and tissues. Typically, this is indicated when the rate of chemical accumulation in a tissue is slower than that predicted by a flow-limited description. Contributing factors include the volume-weighted tissue blood flow rate (in general, low perfusion rates are associated with large intercapillary distances), ultrastructural features of the capillary endothelium (e.g., tight junctions at the blood: brain barrier), and attributes of the compound (e.g., molecular volume and charge character). In such cases, mass-balance equations must be written for both the blood and extravascular tissue spaces, and a permeability coefficient is needed to relate chemical flux to the magnitude of the chemical gradient (McKim and Nichols, 1994). Remarkably, the assumption of blood flow-limitation has been shown to approximate the kinetic behavior of many chemicals and is frequently used as a default assumption.

2.5. Data Needs for Model Parameterization

Briefly, the information needed to develop a PBTK model for fish includes: 1) estimates of compartment volumes and blood perfusion rates, 2) equilibrium tissue: blood chemical partition coefficients, 3) metabolic and excretory routes and rates, and 4) a mathematical description of the route(s) of exposure. A discussion of these data needs, as well as related issues such as allometric scaling and growth, is beyond the scope of this presentation. A detailed review of the physiologically based modeling approach and its application to fish is provided by McKim and Nichols (1994). This review also describes the gill submodel developed by Erickson and McKim (1990). A physiological description of chemical uptake across the skin is given in a recent paper by Nichols *et al.*, (1996).

3. DEVELOPMENT OF A PHYSIOLOGICAL GUT MODEL

3.1. Importance of Dietary Uptake

From bioenergetics-based considerations and knowledge of the extent to which compounds bioconcentrate in prey species, it can be shown that dietary uptake is the predominant route by which large piscivorous fish accumulate environmental contaminants with log K_{OW} values greater than about 5 (Bruggeman *et al.*, 1984). Special cases also exist, resulting in dietary uptake of low log K_{OW} compounds by fish. One example is the use of various food additives in aquaculture such as oxytetracycline.

3.2. Approaches to Modeling Dietary Uptake

Dietary uptake may be described very simply as a first-order absorptive process. If the dose is modeled as the total amount of compound instead of a concentration there is no need to define a gut compartment volume. There is still a requirement, however, to specify at what point in the circulatory system the dose enters the bloodstream. A disadvantage of this approach is that it does not provide for fecal elimination of unassimilated compound (i.e., oral bioavailability <100%). A second approach, therefore, is to define a gut lumen compartment and to model ingestion and fecal egestion as bulk flow rates or periodic events. A time lag may also be required, due to dissolution, digestion (if the chemical is incorporated into a prey item), or passage of the dose to the absorptive gut segment.

A highly detailed PBTK model was developed by Bungay *et al.*, (1981) to describe the dietary uptake of chlordecone by rats. Observed kinetics were reproduced using a gut sub-model that consisted of six segments corresponding to the stomach, small intestine (three subcompartments), cecum and lower intestine. Diffusion limitations were considered to exist at both the tissue:blood and tissue:gut lumen interfaces, thereby requiring that twelve different rate constants (permeability × area products) and equilibrium partition coefficients be fitted based on measured residues in tissues and gut contents. Transit times and exit flow rates for each compartment were developed from previous studies with non-absorbable gut markers. Unfortunately, very little of this type of information is available for fish. There is also some question whether a model incorporating this level of detail would be useful for extrapolation, given the large differences in gastrointestinal anatomy and physiology that exist among fish species.

A need exists to identify principles that can be used to develop PB-TK models for dietary uptake by fish without a requirement for excessive detail. In this regard, a relatively simple description of the gut was developed by Barber *et al.*, (1991) to model PCB accumulation by lake trout. In this gut sub-model, the assumption was made that at the conclusion of digestion chemical in feces is in equilibrium with that in the rest of the fish. From this "fecal partitioning" assumption it is possible to calculate assimilation efficiency and the chemical egestion rate in feces. Moreover, because uptake is based on the existence of a concentration gradient between the gut and the fish, this approach provides for the possibility of a steady-state condition with no net uptake of chemical.

3.3. Kinetics of TCB Uptake From the GI Tract of Fish

Until recently, data required to test the fecal partitioning assumption did not exist. In particular, chemical time course data was needed for both tissues and lumenal contents of the GI tract following the administration of a contaminated meal. A study of this type was performed using [^{14}C] 2,2′,5,5′-tetrachlorobiphenyl (TCB) as a model for hydrophobic compounds generally (Fitzsimmons *et al.*, 1996). The study was conducted by loading 60-d old fathead minnows with TCB in water and then feeding the minnows to 100 g rainbow trout. TCB levels in the stomach contents of trout declined rapidly after feeding and peaked in the upper intestinal contents at 12 hr (Figure 2). The ratio of TCB in upper intestinal tissues to that in blood declined from 6 to 48 hr (Figure 3). These obsevations are suggestive of an early liquid phase emptying of stomach contents to the upper intestine followed by rapid coassimilation of lipid and TCB. Tissue:blood concentration ratios for the stomach, lower intestine, muscle, liver, and kidney were constant and probably represented near equilibrium conditions.

Peak chemical assimilation efficiencies were greater than 95%. Based on total counts, however, the chemical kinetics in feces did not follow those of the lower intestinal tissues. A peak of radioactivity was instead observed passing through the lower GI tract at a time when blood and lower intestinal tissue counts were steady or declining (Figure 4). Presently, it is not known whether this radioactivity consisted of TCB or some combination of TCB and metabolites. On a chemical mass basis, the total amount of counts in feces during the peak of radioactivity was about 1% of the dose. This corresponds to levels of metabolism previously reported for TCB in rainbow trout by Melancon and Lech (1976). A definitive test of the fecal partitioning assumption will require the separation and identification of parent TCB and metabolites in gut contents and tissues. This work is presently ongoing.

3.4. Development of a Physiological Gut Model: Dietary Uptake of TCDD by Brook Trout

A variation on the fecal partitioning assumption was subsequently used to describe the dietary uptake of 2,3,7,8-tetrachlorodibenzo-p-dioxin (TCDD) by brook trout

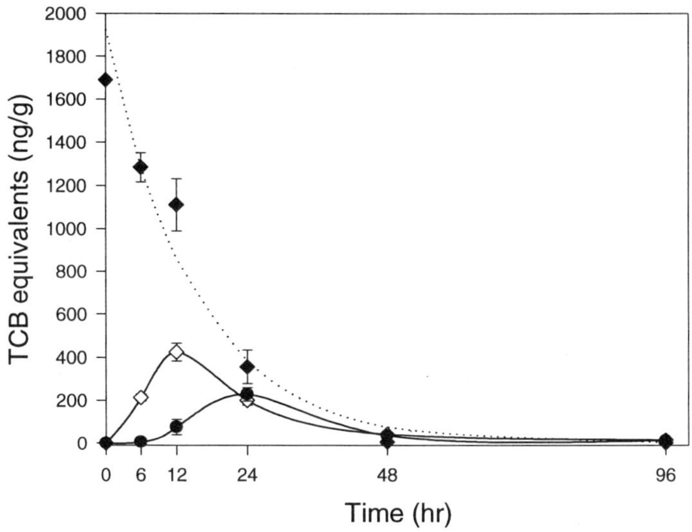

Figure 2. ^{14}C-TCB equivalents in the lumenal contents of the stomach (◆) and upper (◇) and lower (●) intestine of rainbow trout fed a single meal of contaminated prey. Each point represents the mean ± SE. TCB concentrations in the stomach contents were fitted to an exponential model: $Y = 1922.2 \cdot 0.94^X$ (dotted line, $r^2 = 0.934$). Solid lines were fitted using a cubic spline interpolation procedure.

(Nichols *et al.*, 1998). Brook trout were exposed to TCDD spiked into a prepared diet. Briefly, the study consisted of a 28-d loading phase, a 77-d maintenance phase, and a 77-d depuration phase. Dose levels were reduced during the maintenance phase to achieve constant whole-body concentrations. This study design provided the opportunity to study uptake and depuration, as well as a condition close to steady-state.

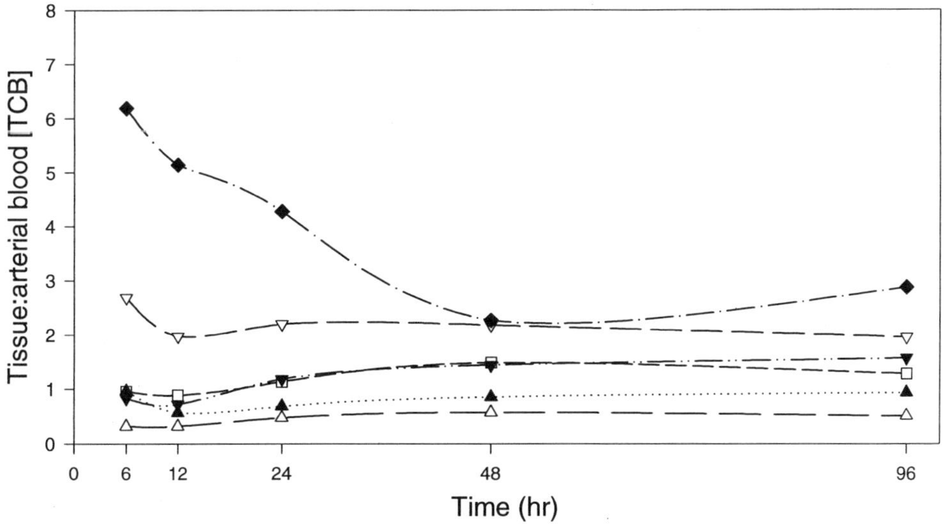

Figure 3. Tissue:blood TCB concentration ratios in lean tissues. Concentration ratios for muscle (△), liver (▽), kidney (□), stomach (▲), upper intestine (◆), and lower intestine (▼) were calculated from total TCB residue equivalents. All lines were fitted using a cubic spline interpolation procedure.

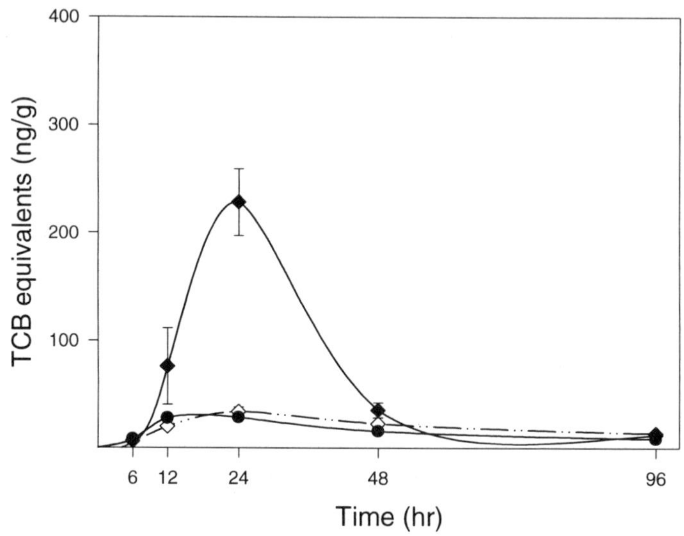

Figure 4. ^{14}C-TCB equivalents in the lower intestinal contents (◆), lower intestinal tissue (◇), and arterial blood (●) of rainbow trout fed a single meal of contaminated prey. Each point represents the mean ± SE. All lines were fitted using a cubic spline interpolation procedure.

Uptake was modeled initially by assuming that chemical in gut contents following the uptake of dietary lipid was in equilibrium with that in tissues that comprise the intestinal tract. The intestinal tract was modeled as a single compartment, and TCDD exchange between blood and tissues of the intestinal tract was assumed to be blood flow-limited. Three repeating features of the data provided the strongest test of the model: 1) TCDD assimilation efficiencies ranged from 72% to 82% during the first three weeks of the study; 2) there was little change in assimilation throughout the two exposure phases; and 3) the elimination rate during the depuration phase was too slow to estimate accurately. Observed assimilation efficiencies were consistently lower than those predicted by the fecal partitioning assumption. A diffusion-limited gut sub-model was therefore employed to retard chemical transport between the gut contents and the bloodstream. Exposure data were then used to fit the value of a diffusion rate constant (Figure 5).

The origin of this "diffusion" limitation is unknown. TCDD accumulation in all tissues except adipose tissue was adequately described by assuming that internal distribution was blood flow-limited (Figure 6). Thus, any diffusion limitation that may have existed in the gut did not appear to limit mass transfer from blood to these tissues. Alternatively, some portion of the TCDD spiked into food may have remained strongly bound throughout digestion, limiting bioavailability. Whatever its origin, this kinetic limitation appears to be a general feature of feeding studies with intermediate to high log K_{OW} compounds (Gobas *et al.*, 1988). The challenge that remains is to develop a PBTK model for dietary uptake by fish that is flexible enough to handle a broad range of compounds as well as variations in gastrointestinal anatomy and physiology.

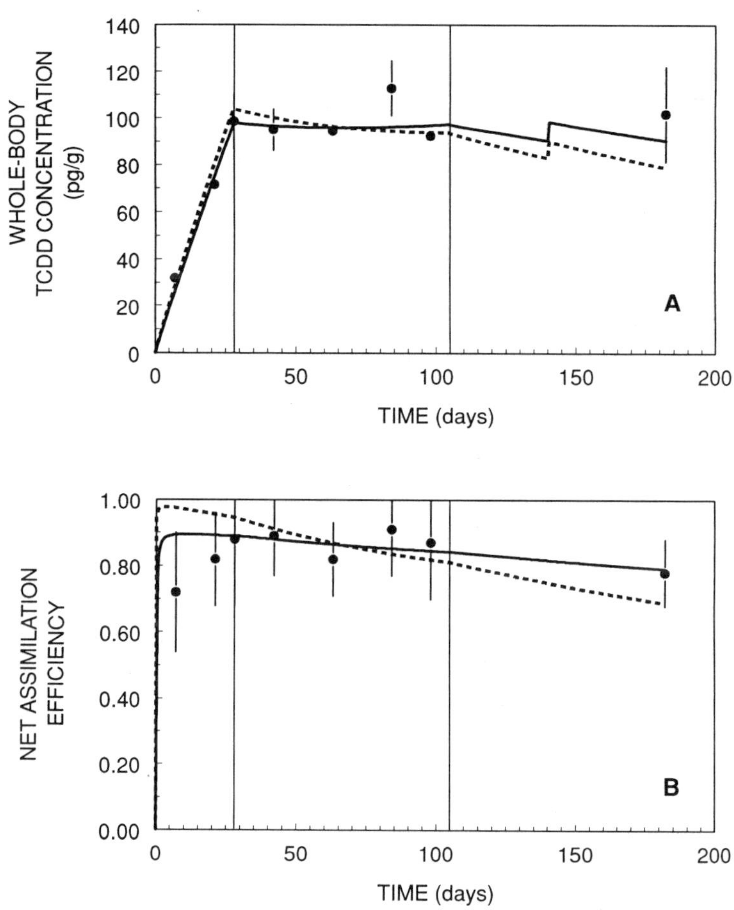

Figure 5. A. Whole-body TCDD concentrations simulated using a PBTK model for brook trout. The dashed line was obtained using a fecal partitioning gut sub-model. The solid line was generated using a gut sub-model with diffusion limitations. Measured residues are shown as individual points (mean ± SD). Vertical lines delineate the three exposure phases. B. Net assimilation efficiencies simulated using the fecal partitioning (dashed line) and diffusion-limited (solid line) gut sub-models. Calculated assimilation efficiencies are shown as individual points (mean ± SD). Adapted from Nichols *et al.*, (1998).

4. PHYSIOLOGICAL MODELING OF CHEMICAL ELIMINATION

4.1. Branchial, Dermal, and Dietary Elimination

Current physiological descriptions of gill, skin and gut exchange surfaces are symmetric with respect to chemical uptake and elimination (Erickson and McKim, 1990; Nichols *et al.*, 1996; Nichols *et al.*, 1998). Models employing these descriptions can therefore be used to simulate changing exposure concentrations and depuration. Some caution is advised, however, since the assumption of symmetry has not been rigorously evaluated. Moreover, these exchange surface descriptions are based entirely on factors related to bulk flow and passive diffusion. The existence of facilitated or active transport systems for a compound of interest may require the development of more detailed descriptions.

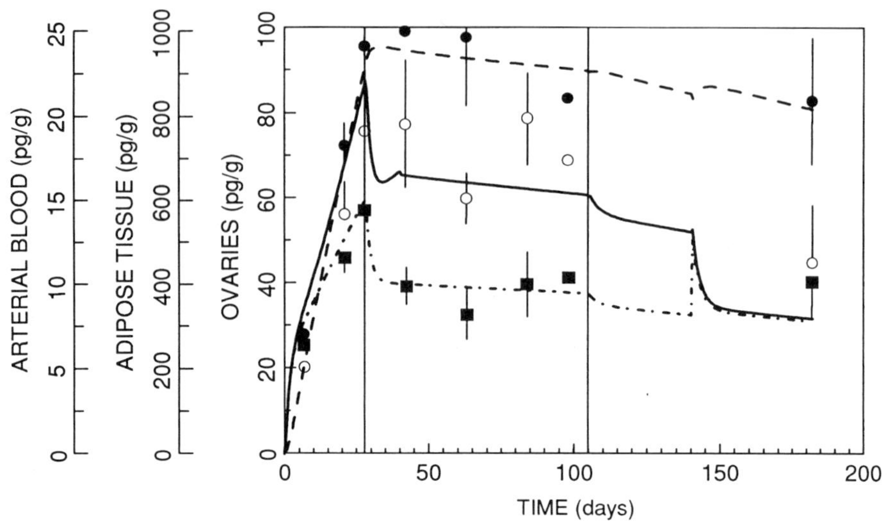

Figure 6. TCDD kinetics in adipose tissue (solid line), ovaries (dashed line), and blood (dotted line) simulated using a PBTK model for brook trout. Measured residues are shown as dots (adipose tissue), open circles (ovaries), and solid squares (blood)(mean ± SD). TCDD accumulation in the ovaries was assumed to be blood flow-limited. TCDD accumulation in adipose tissue was modeled using a diffusion-limited tissue description. Functionally, the diffusion limitation in adipose tissue was equivalent to multiplying blood flow rate by 0.25. From Nichols *et al.*, (1998), with permission.

4.2. Hepatic Clearance

Hepatic clearance may contribute substantially to the elimination of a compound, particularly if metabolic pathways (e.g., glucuronidation) promote its secretion directly to bile. As in mammals, a compound's molecular weight tends to influence its disposition; high molecular weight compounds are more likely to be eliminated by the liver than by the kidney due to limitations on glomerular filtration. The molecular weight cutoff in fish is not as well known as that in mammals, but is probably around 500 daltons. The kinetics of active chemical secretion to bile are poorly known. In mammals, however, hepatic clearance often exhibits first order kinetics. A PBTK model with hepatic secretion would therefore include a first order loss term from the liver compartment. Hydrophobic compounds may partition passively to bile and would then be passed to the GI tract. This route of elimination could be modeled from existing knowledge of bile flow rates in fish (0.1 ml/kg/h in rainbow trout, Schmidt and Weber, 1973; 0.04 to 0.17 ml/kg/h in the skate and dogfish shark, Ballatori and Boyer, 1986) and an estimate of the relative affinity of a compound for bile and liver tissue. To the extent, however, that fecal partitioning determines chemical uptake from the gut, most or all of this compound may be reabsorbed.

4.3. Renal Clearance

Data on renal clearance by fish are available for several compounds and species (Pritchard and Bend, 1984). In general, a distinction can be made between fresh and saltwater-adapted fish due to differences in kidney structure and function. The kidney of a saltwater fish tends to possess fewer functional glomerulii. This and other adaptations result in the production of small quantities of concentrated urine, and limit the potential for chemical elimination by glomerular filtration. In contrast, freshwater fish produce substantial quantities of dilute urine (3.0 to 6.0 ml/kg/hr, McKim *et al.*, 1999). Renal clearance by glomerular filtration may therefore be important for some hydrophilic compounds, provided that they are not bound substantially in plasma. The kinetics of renal clearance by glomerular filtration are likely to be first order with respect to chemical concentrations in blood. It is important, however, to recognize that glomerular filtration rate is limited by arterial blood flow to the kidney, which is perhaps 20% of total renal blood flow in fish (Pritchard and Bend, 1984). A PBTK model for a compound that is filtered at the kidney would therefore require segregation of arterial and portal blood supplies.

Organic cations and anions may be actively transported to urine by renal tubular secretion. In fish, as in mammals, this clearance can greatly exceed that due to glomerular filtration. It is likely, therefore, that a chemical in portal blood flowing to the kidney is available to this pathway. The kinetics of active secretion are more likely to exhibit complex (e.g., saturable) kinetics than those of glomerular filtration. It should be possible, however, to model this elimination in the same manner as metabolic biotransformation in the liver; that is, by using a kinetic description referenced to the chemical concentration in venous blood exiting the kidney.

Current work in our laboratory is directed toward better understanding the renal handling of Phase II metabolic products (McKim *et al.*, 1999). In a recent study, rainbow trout were exposed to phenol in water. A microdialysis probe was used to monitor levels of phenol, hydroquinone, phenyl glucuronide and phenyl sulfate in plasma. The advantage of this sampling technique is that it directly measures the "free" (unbound) chemical concentration, or that which is available for glomerular filtration. The same compounds were also analyzed in urine, and for each compound the ratio of renal clearance to glomerular filtration was characterized (Figure 7). Urine clearance ratios greater than 1.0 were observed for phenyl glucuronide and phenyl sulfate. These values suggest the existence of renal tubular secretion. Ratios less than 1.0 were determined for phenol and hydroquinone. Such values are usually interpreted as evidence for glomerular filtration followed by passive reabsorption.

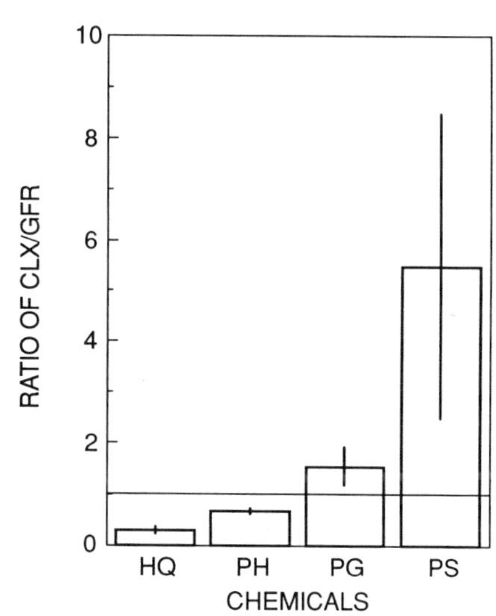

Figure 7. Ratios of renal clearance (CLX) to glomerular filtration rate (GFR) for hydroquinone (HQ), phenol (PH), phenyl glucuronide (PG), and phenyl sulfate (PS) in rainbow trout that were continuously exposed to waterborne PH for 144 hr. Ratios are expressed as the mean ± SE. Adapted from McKim *et al.* (1999).

4.4. Metabolic Biotransformation

4.4.1. General Considerations. A need exists to develop PBTK models for compounds that are metabolically transformed by fish. As these models are developed it will necessary to determine whether the primary toxicological concern is for the parent compound or its metabolites. If the concern is for the parent compound and metabolism is simply a detoxification pathway, it may be sufficient to group all metabolism without regard for the identity of metabolites and to localize this activity to a tissue of convenience (e.g., the liver). If, on the other hand, one or more metabolites represent bioactivated products (i.e., more toxic than the parent compound) it may be necessary to model the formation of these metabolites and to localize this metabolism to the tissues where it occurs. This is especially important if a metabolite has attributes, such as high reactivity, which make it likely that damage will occur at or near the site of formation. For metabolites that diffuse readily and can distribute within the organism it may be necessary to develop linked models for the parent chemical and its metabolic products.

4.4.2. Literature. A large and growing literature details the metabolic biotransformation of xenobiotic compounds by fish. This information is particularly useful for identification of primary metabolic products and may aid in identifying important species differences. In the absence of fish studies for a particular compound, mammalian information may exist which would provide guidance on principal metabolic products and metabolite disposition. It may then be possible to compare this information with data from fish to determine whether the operative enzymatic pathways are likely to be present. It should be noted that in mammals, the enzyme affinity constant (Km) for a given reaction tends to be relatively constant across species, while the capacity constant (Vmax) tends to scale to a fractional exponent of body weight (Travis *et al.*, 1990). Existing data are insufficient to determine whether these generalizations can be applied to fish, Moreover, in fish it may be necessary to consider temperature effects on enzyme activity.

4.4.3. Whole Organism Approaches to Metabolism.

4.4.3.1. Parent/Other Mass-Balance. Several whole organism experimental approaches are available to obtain metabolism information from fish. One is the parent/other mass-balance approach (McKim *et al.*, 1986). The goal of this approach is to characterize in a general way the disposition of a parent compound and it's metabolites. It is not necessary to know the identity of the metabolites, provided that they differ in polarity from the parent. Solvents of different polarity may then be used to extract samples of tissue, bile, urine and gill water. This approach is facilitated by using a radiolabeled parent compound. The dose is generally administered as a bolus i.v. injection; however, it is important to keep in mind that the metabolic fate of a compound may vary greatly with the route of exposure.

4.4.3.2. Modeling the Disappearance of Parent. A second whole organism approach is to assume that a PBTK model is "correct" with respect to the uptake and distribution of parent compound. The model can then be used to solve for the value of rate and capacity parameters required to simulate observed kinetics. Depuration data are frequently preferred for this purpose because of the absence of complications associated with uptake and the rapid phase of internal distribution. By modeling to data across a range of exposure concentrations it is possible to assess the concentration dependence of the system. Importantly, metabolic parameters determined in the way are "apparent"; that is, they may or may not reflect real parameters in the organ designated in the model as the site of metabolism. Using this approach, metabolism data has been obtained for a large number of compounds in several rodent species (Gargas *et al.*, 1986). The utility of this approach for studies with fish has not been evaluated. However, because rates of metabolism in fish are frequently quite low, parent chemical kinetics may be insufficient in many cases to determine whether metabolism has occurred.

4.4.3.3. Measuring Metabolites *In Vivo.* A third whole organism approach is to analyze blood, tissues, and the principal elimination products for both the parent compound and it's metabolites. Generally, this approach is used only when one or more metabolites are of specific interest, and preliminary work has identified the site(s) of metabolism and the general fate of metabolic products. With sufficient information, it is possible to develop linked models for the parent compound and metabolites. For the metabolite model, fitted rate and capacity parameters comprise the "input" description. Metabolite levels in the fish depend, however, on both rates of formation and elimination. Thus, it is possible for Phase I metabolic products to reach high levels if rate limitations exist for subsequent events such as Phase II metabolism. In mammals, Phase II conjugation pathways tend to exhibit first-order kinetics due to high levels of enzyme and the availability of substrate (e.g., glucose, glutathione). Prolonged production of metabolite may, however, result in substrate depletion, particularly at the site of formation. In fish, the kinetics of Phase II metabolism and the fate of Phase II metabolic products remain poorly understood.

4.4.4. Incorporation of In Vitro Data. *In vitro* experimental systems can be used to identify major metabolites and to estimate the value of enzymatic rate and capacity parameters. With respect to the identification of metabolic products, it is important to recognize the limitations of different types of *in vitro* systems. In general, microsomal protein systems can only yield Phase I metabolic products and glucuronide conjugates. Isolated cell preparations and precision-cut tissue slices have the potential to yield the full range of Phase I and Phase II products.

Metabolic rate and capacity parameters determined using these systems can in theory be incorporated into PBTK models. Doing so, however, requires that the information be scaled up to the organ or tissue level. For example, microsomal enzyme activity is typically expressed as moles of product produced per unit time per milligram of microsomal protein. This value must be scaled up using a factor which converts milligrams of microsomal protein to milliliters of tissue. The scale-up of metabolism data from isolated cell preparations or tissue slices is more straightforward. Problems may arise, however, due to changes in cell viability during the period over which the information was collected. Additional complications exist with tissue slices due to the time required for substances to diffuse in (parent) and out (metabolite). For this reason, it is often recommended that tissues be homogenized along with the bathing medium. A separate analysis of the medium may aid, however, in characterizing the diffusing properties of metabolic products. Finally, any *in vitro* system will yield results of questionable value if, as is routinely done, the system is optimized for maximum product yield and/or the animal from which the tissues are derived are pretreated with an enzyme inducing agent.

4.4.5. In vivo Estimates of Metabolism. An alternative to the use of *in vitro* systems is to obtain direct *in vivo* estimates of metabolism. The whole organism approach to measuring metabolites has already been discussed. A second method involves the use of a technique called microdialysis (Figure 8). Dialysis probes can be implanted into a blood

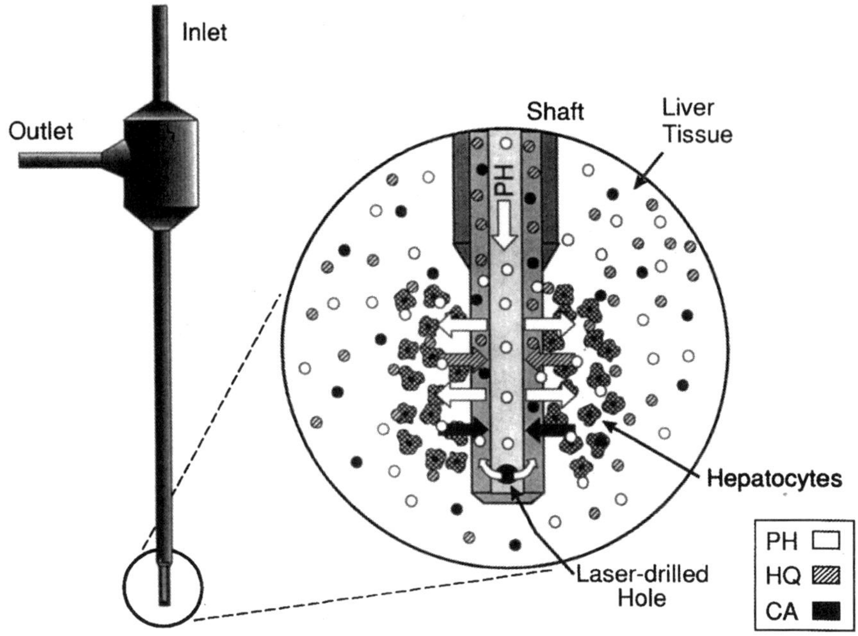

Figure 8. Operation of a microdialysis probe implanted into the liver of a rainbow trout and used for delivery dialysis of phenol (PH). Metabolites of PH, including hydroquinone (HQ) and catechol (CA), are formed in hepatocytes and diffuse back into the probe for analysis by HPLC.

vessel or placed directly within a tissue. The probes are continuously perfused with a saline solution, and free metabolites (Phase I and Phase II) as well as the parent compound diffuse across the dialysis membrane. The dialysate is then sent to an HPLC for separation and analysis. In this manner it is possible to characterize metabolite kinetics *in vivo* in close to real time. This technique has been employed extensively in mammalian neuroscience investigations and is particulary useful for the analysis of labile products that may be difficult to extract and purify using conventional methods. The scale-up of microdialysis results to the whole tissue level requires the development and validation of mathematical models to describe the diffusion, binding and capillary washout of test compounds (Morrison *et al.*, 1991; Stenken *et al.*, 1997). Mammalian researchers are currently working on microdialysis models for hepatic metabolism (Stenken *et al.*, 1998).

Limited studies have demonstrated the utility of microdialysis for toxicokinetic studies with fish. Initially, the technique was used to measure phenol and a major metabolite, phenyl glucuronide, in blood (Figure 9) (McKim, Jr., *et al.*, 1993). The goal of ongoing studies is to measure phenol and its metabolites in fish liver (McKim *et al.*, 1998). These studies are being conducted using a method called delivery dialysis wherein the parent compound is infused via the probe itself. This technique minimizes the potential for systemic toxicity and maximizes control of chemical concentration in the tissue of interest.

4.4.6. In Vitro/in *Vivo Comparisons.* Comparisons of *in vitro* and *in vivo* metabolism data have received a great deal of attention among mammalian toxicology researchers. Clearly, if *in vitro* kinetic parameters consistently yielded good estimates of *in vivo* chemical behavior, the utility of PBTK models would be greatly enhanced. Wilkinson (1987)

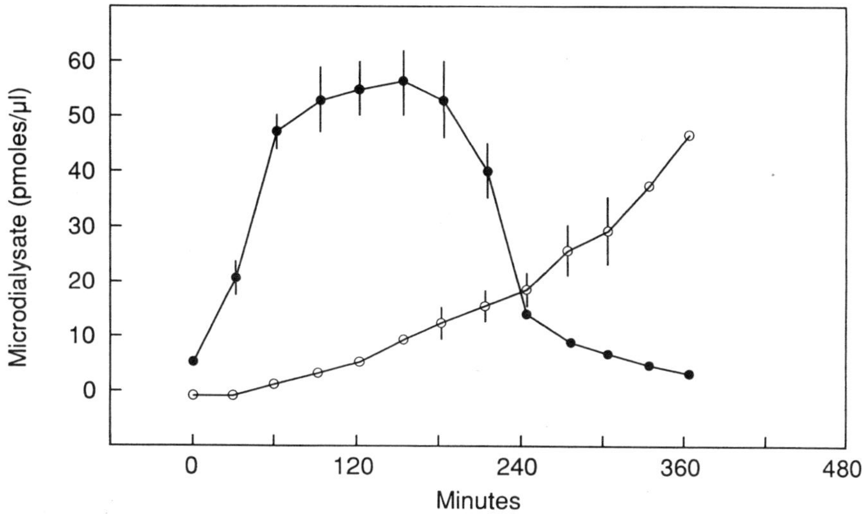

Figure 9. Free plasma concentrations of phenol (●) and phenyl glucuronide (○) in rainbow trout measured by microdialysis (mean ± SE). The phenol exposure was terminated at 180 min. Samples collected during the depuration phase provided evidence, however, for a continued increase in glucuronide levels. Adapted from McKim, Jr., *et al.*, (1993).

found that microsomal and hepatocyte preparations gave good correspondence with *in vivo* liver metabolism predicted using a venous equilibration model. Similarly, Houston (1994) determined that rat hepatocytes accurately predicted chemical clearance by isolated rat livers. Liver slice preparations also have potential as a predictive model, although relatively few comparisons have been made with *in vivo* or isolated organ systems (Olinga *et al.*, 1997). On the strength of these results, numerous authors have used *in vitro* information as inputs to PBTK models. The success of such efforts has been mixed (Andersen *et al.*, 1995; Fiserova-Bergerova, 1995). This is not surprising. In addition to problems associated with the *in vitro* systems themselves, substantial metabolism may occur in tissues other than those modeled. A variety of additional factors, including differences among individuals, sexes, and animals of different age, contribute to the difficulty in making *in vitro*/*in vivo* comparisons.

In a pioneering study, Law *et al.*, (1991) used Km and Vmax data from a trout hepatocyte system to parameterize a PBTK model for pyrene. Total pyrene clearance greatly exceeded that predicted by hepatic metabolism, however, requiring the addition of two fitted clearance terms. The need to make *in vitro*/*in vivo* metabolism comparisons in fish remains a critical area for future research. In particular, a systematic effort should be made to identify chemical classes and/or categories of metabolic activity for which *in vitro* data are predictive of *in vivo* disposition. In time it may be possible to extend these observations even further using Quantitative Structure Activity Relationships (QSARs) or other structure-based approaches.

5. REFERENCES

Abbas, R., and Hayton, W.L., 1997, A physiologically based pharmacokinetic and pharmacodynamic model for paraoxon in rainbow trout, *Toxicol. Appl. Pharmacol.* 145:192.

Andersen, M.E., Gargas, M.L., and Ramsey, J.C., 1984, Inhalation pharmacokinetics: Evaluating systemic extraction, total in vivo metabolism, and the time course of enzyme induction for inhaled styrene in rats based on steady-state blood:air concentration ratios, *Toxicol. Appl. Pharmacol.* 73:176.

Andersen, M.E., Clewell III, H.J., and Frederick, C.B., 1995, Applying simulation modeling to problems in toxicology and risk assessment—A short perspective, *Toxicol. Appl. Pharmacol.* 133:181.

Ballatori, N., and Boyer, J.L., 1986, Slow biliary elimination of methyl mercury in the marine elasmobranchs, *Raja erinacea* and *Squalus acanthias*, *Toxicol. Appl. Pharmacol.* 85:407.

Barber, M.C., Suarez, L.A., and Lassiter, R.R., 1991, Modelling bioaccumulation of organic pollutants in fish with an application to PCBs in Lake Ontario salmonids, *Can. J. Fish. Aquat. Sci.* 48:318.

Bischoff, K.B., and Brown, R.G., 1966, Drug distribution in mammals, *Chem. Eng. Progr. Symp. Ser.* 62:32.

Bruggeman, W.A., Opperhuizen, A., Wijbenga, A., and Hutzinger, O., 1984, Bioaccumulation of superlipophilic chemicals in fish, *Toxicol. Environ. Chem.* 7:173.

Bungay, P.M., Dedrick, R.L., and Guarino, A.M., 1976, Pharmacokinetic modeling of the dogfish shark (*Squalus acanthias*): distribution and urinary and biliary excretion of phenol red and its glucuronide, *J. Pharmacokinet. Biopharm.* 4:377.

Bungay, P.M., Dedrick, R.L., and Matthews, H.B., 1981, Enteric transport of chlordecone (Kepone®) in the rat, *J. Pharmacokinet. Biopharm.* 9:309.

Conolly, R.B., Reitz, R.H., Clewell III, H.J., and Andersen, M.E., 1988, Pharmacokinetics, biochemical mechanism and mutation accumulation: a comprehensive model of chemical carcinogenesis, *Toxicol. Lett.*, 43:189.

Dedrick, R.L., and Bischoff, K.B., 1968, Pharmacokinetics in applications of the artificial kidney, *Chem. Eng. Progr. Symp. Ser.* 64:32.

Erickson, R.J., and McKim, J.M., 1990, A model for exchange of organic chemicals at fish gills: flow and diffusion limitations, *Aquat. Toxicol.* 18:175.

Fiserova-Bergerova, V., 1995, Extrapolation of physiological parameters for physiologically based simulation models, *Toxicol. Lett.* 79:77.

Fitzsimmons, P., Whiteman, F., and Nichols, J, 1996, Kinetics of dietary uptake of 2,2′,5,5′-tetrachlorobiphenyl in rainbow trout, Poster presented at the 17th Annual Meeting of the Society of Environmental Toxicology and Chemistry, November 17–21, Washington, DC.

Gargas, M.L., Andersen, M.E., and Clewell III, H.J., 1986, A physiologically based simulation approach for determining metabolic constants from gas uptake data, *Toxicol. Appl. Pharmacol.* 86:341.

Gargas, M.L., Burgess, R.J., Voisard, D.E., Cason, G.H., and Andersen, M.E., 1989, Partition coefficients of low molecular weight volatile chemicals in various liquids and tissues, *Toxicol. Appl. Pharmacol.* 98:87.

Gobas, F.A.P.C., Muir, D.C.G., and Mackay, D, 1988, Dynamics of dietary bioaccumulation and faecal elimination of hydrophobic organic chemicals in fish, *Chemosphere* 17:943.

Houston, J.B., 1994, Utility of in vitro drug metabolism data in predicting in vivo metabolic clearance, *Biochem. Pharmacol.* 47:1469.

Jerlowski, L.E., and Jain, R.K., 1983, Physiologically based pharmacokinetic modeling: principles and applications, *J. Pharm. Sci.* 72:1103.

Law, F.C.P., Abenini, S., and Kennedy, C.J., 1991, A biologically based toxicokinetic model for pyrene in rainbow trout, *Toxicol. Appl. Pharmacol.* 110:390.

McKim, Jr., J.M., McKim, Sr., J.M., Naumann, S., Hammermeister, D.E., Hoffman, A.D., and Klaassen, C.D., 1993, *In vivo* microdialysis sampling of phenol and phenyl glucuronide in the blood of unanesthetized rainbow trout: Implications for toxicokinetic studies, *Fund. Appl. Toxicol.* 20:190.

McKim, J.M., Schmieder, P.K., and Erickson, R.J., 1986, Toxicokinetic modeling of [^{14}C]pentachlorophenol in the rainbow trout (*Salmo gairdneri*), *Aquat. Toxicol.* 9:59.

McKim, J.M., and Nichols, J.W., 1994, Use of physiologically based toxicokinetic models in a mechanistic approach to aquatic toxicology, in: *Aquatic Toxicology. Molecular, Biochemical, and Cellular Perspectives*, D.C. Malins and G.K. Ostrander, eds., Lewis, Boca Raton.

McKim, J.M., Kolanczyk, R.C., and Solem, L.E., 1998, In vivo liver microdialysis of phenol in rainbow trout, Poster presented at the 37th annual meeting of the Society of Toxicology, March 1–5, Seattle, Washington.

McKim, J.M., Kolanczyk, R.C., Lien, G.J., and Hoffman, A.D., 1999, Dynamics of renal excretion of phenol and major metabolites in the rainbow trout (*Oncorhynchus mykiss*), *Aquat. Toxicol.* 45:265–277.

Melancon, M.J., and Lech, J.J., 1976, Isolation and identification of a polar metabolite of tetrachlorobiphenyl from bile of rainbow trout exposed to ^{14}C-tetrachlorobiphenyl, *Bull. Environ. Contam. Toxicol.* 15:181.

Morrison, P.F., Bungay, P.M., Hsiao, J.K., Mefford, I.N., Dykstra, K.H., and Dedrick, R.L., 1991, Quantitative microdialysis, in: *Microdialysis in the Nuerosciences*, J.B. Justice and T.E. Robinson, eds., Elsevier, London.

Nichols, J.W., McKim, J.M., Andersen, M.E.,Gargas, M.L., Clewell III, H.J., and Erickson, R.J., 1990, A physiologically based toxicokinetic model for the uptake and disposition of waterborne organic chemicals in fish, *Toxicol. Appl. Pharmacol.* 106:433.

Nichols, J.W., McKim, J.M., Lien, G.J., Hoffman, A.D., Bertelsen, S.L., and Gallinat, C.A., 1993, Physiologically-based toxicokinetic modeling of three waterborne chloroethanes in channel catfish, *Ictalurus punctatus, Aquat. Toxicol.* 27:83.

Nichols, J.W., McKim, J.M., Lien, G.J., Hoffman, A.D., Bertelsen, S.L., and Elonen, C.M., 1996, A physiologically based toxicokinetic model for dermal absorption of organic chemicals by fish, *Fund. Appl. Toxicol.* 31:229.

Nichols, J.W., Jensen, K.M., Tietge, J.E., and Johnson, R.D., 1998, A physiologically based toxicokinetic model for maternal transfer of 2,3,7,8-tetrachlorodibenzo-p-dioxin in brook trout (*Salvelinus fontinalis*), *Environ. Toxicol. Chem.* 17:2422–2434.

Olinga, P., Meijer, D.K.F., Slooff, M.J.H., and Groothuis, G.M.M., 1997, Liver slices in *in vitro* pharmacotoxicology with special reference to the use of human liver tissue, *Toxicol. In Vitro* 12:77.

Pritchard, J.B., and Bend, J.R., 1984, Mechanisms controlling the renal excretion of xenobiotics in fish: effects of chemical structure, *Drug Metab. Rev.* 15:655.

Reitz, R.H., Gargas, M.L., Andersen, M.E., Provan, W.M., and Green, T.L., 1996, Predicting cancer risk from vinyl chloride exposure with a physiologically based pharmacokinetic model, *Toxicol. Appl. Pharmacol.* 137:252.

Schmidt, D.C., and Weber, L.J., 1973, Metabolism and biliary excretion of sulfobromophthalein by rainbow trout (*Salmo gairdneri*), *J. Fish. Res. Board Can.* 30:1301.

Stenken, J.A., Lunte, C.E., Southard, M.Z., and Ståhle, L., 1997, Factors that influence microdialysis recovery. Comparison of experimental and theoretical microdialysis recoveries in the rat liver, *J. Pharm. Sci.* 86:958.

Stenken, J.A., Ståhle, L., Lunte, C.E., and Southard, M.Z., 1998, Monitoring in situ liver metabolism in rats using microdialysis. Comparison of microdialysis mass-transport model predictions to experimental metabolite generation data, *J. Pharm. Sci.* 87:311.

Teorell, T., 1937, Kinetics of distribution of substances administered to the body, *Arch. Int. Pharmacodyn. Ther.* 57:205.

Travis, C.C., White, R.K., and Ward, R.C., 1990, Interspecies extrapolation of pharmacokinetics, *J. Theoret. Biol.* 142:285.

Wilkinson, G.R., 1987, Prediction of *in vivo* parameters in drug metabolism and distribution from *in vitro* studies, in: *Pharmacokinetics in Risk Assessment*, Vol. 8, Drinking Water and Health, National Academic Press, Washington, D.C.

Zaharko, D.S., Dedrick, R.L., and Oliverio, V.T., 1972, Prediction of the distribution of methotrexate in the sting rays *Dasyatidae sabina* and *sayi* by use of a model developed in mice, *Comp. Biochem. Physiol.* 42A:183.

8

A PHYSIOLOGICALLY BASED PHARMACOKINETIC MODEL FOR PREDICTING THE WITHDRAWAL PERIOD OF OXYTETRACYCLINE IN CULTURED CHINOOK SALMON (*ONCORHYNCHUS TSHAWYTSCHA*)

F. C. P. Law

Environmental Toxicology Program
Department of Biological Sciences
Simon Fraser University
Burnaby, British Columbia V5A 1S6. CANADA

1. ABSTRACT

Oxytetracycline (OTC) is an antibiotic used widely to prevent or treat bacterial diseases in cultured salmon. In the present study, a physiologically based pharmacokinetic (PBPK) model of OTC was coupled with the Monte Carlo technique in predicting the withdrawal period (WP) of OTC for a population of cultured chinook salmon (*Oncorhynchus tshawytscha*). Since the PBPK model was implemented with ranges rather than fixed parameter values, it also could be used to account for the inter-individual variability across the population. The model-predicted WP, defined as the post-dosing time when the OTC residue was at or below 0.1 ppm with 95% certainty for the 99th percentile salmon population, were 60 days and 49 days at 9°C and 15°C seawater, respectively. These WP were very close to those determined by a currently used empirical method for the same data sets. However, the population based physiological model is a more useful tool than the empirical method of WP determination since treatment specific information such as fish weight, bioavailability, dose regime, and water temperature can be incorporated into the simulation.

2. INTRODUCTION

Oxytetracycline (OTC), a broad-spectrum antibiotic, is used widely as a prophylactic or treatment for bacterial diseases in aquaculture (Ahmad and Matty, 1989). OTC

Xenobiotics in Fish, edited by *Smith et al.*
Kluwer Academic / Plenum Publishers, New York, 1999.

is often administered to cultured salmon as medicated feed at a rate of 50–100 mg/kg body weight per day for 3–14 days depending on the severity of infection. After treatment with OTC, the salmon must be kept for a minimum time before harvesting of the stock can take place. Observation of the withdrawal period (WP) ensures that edible tissues of salmon are free of residues and are safe for human consumption.

Several empirically based methods are currently used to determine the WP of drugs in cultured finfish. These include the statistical tolerance limit method of the U.S. Food and Drug Administration (FDA, 1994), the 90% upper prediction limit method (Salte and Liestφl, 1983), and the degree day method of the European Union (Schnick, 1992). Because the empirical method derives its WP directly from the drug residue elimination data, it has at least three major limitations: First, a drug depletion study must be performed at several water temperatures (Brown, 1992) since fish are poikilothermic and the elimination of the drug from fish varies with the water temperature. Second, the estimated WP is useful only for interpolation *i.e.*, the WP is valid only for the fish species and exposure conditions used in the depletion study. Third, the effects of inter-individual variability on the WP are not considered although it is assumed that the variability associated with the fish used in the depletion profiles of the tolerance method (FDA, 1994) may provide an adequate approximation of the inter-individual variability of a population. The limitations of the empirically based approach can be minimized or overcome by a physiologically based pharmacokinetic (PBPK) model which describes the absorption, distribution and elimination of a drug in terms of identifiable anatomical spaces and physiological and biochemical processes of the fish.

A single PBPK model has been used successfully to describe the pharmacokinetics of OTC in both trout and salmon (Law, 1992; Broklebank *et al.*, 1997) because the salmonids have similar anatomy, physiology (Thorarensen, 1994) and OTC pharmacokinetic profiles (Abedini *et al.*, 1998; Namdari *et al.*, 1998; Namdari *et al.*, 1999). The same PBPK model also has been used to describe the pharmacokinetics of OTC in the tissues of cattle after *i.v.* or *i.m.* administration (Achenbach *et al.*, 1998). Currently, the PBPK model of OTC is deterministic *i.e.*, a set of fixed values are used as model input parameters and the outputs of the model also are fixed values. Therefore, the model does not have the benefits of a formal statistical procedure and is not designed to describe or understand individual variability in a fish population.

The purposes of this paper were (a) to describe a novel, theoretical approach of WP determination by implementing the PBPK model with the Monte Carlo technique (Rubenstein, 1981) for a population of cultured chinook salmon, (b) to validate the population based physiological model with experimental data, (c) to identify key PBPK model parameters to which the predicted WP is particularly sensitive, and (d) to study the extent to which inter-individual variability in the population confers variability on the predicted WP.

3. MATERIALS AND METHODS

3.1. Model Concepts

3.1.1. Model Structure. We reported a 9-compartment PBPK model which described closely the pharmacokinetics of OTC in the tissues of salmon after multiple OTC-medicated feed treatments (Law, 1992; Brocklebank *et al.*, 1997). The original PBPK model was developed with little or no basic information on the skin and the bone com-

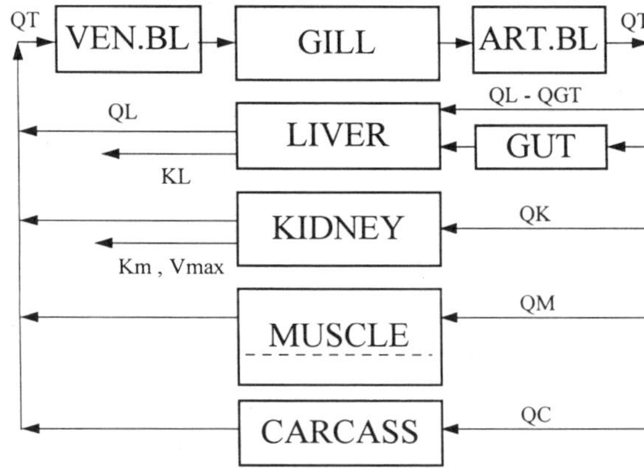

Figure 1. A population based physiological model for predicting the WP of OTC in the salmonids.

partments. To minimize uncertainty in the WP determination, the PBPK model was modified by combining the skin, bone and carcass compartments into a single compartment. The structural modification did not affect significantly the model outputs (data not shown).

Figure 1 shows a schematic diagram of the modified PBPK model which is comprised of the liver, gut, kidney, muscle, blood, gills and carcass compartments. Each of these compartments represented a main site of OTC accumulation and/or elimination. The gills, liver, gut, kidney and carcass were blood flow-limited compartments. The muscle was a membrane-limited compartment. The gut compartment included the highly perfused organs such as the esophagus, stomach, intestine, spleen, and heart. The carcass was a residual compartment including all tissues and organs not separately incorporated into the model.

The modified model has a similar structure as the PBPK model of pyrene (Law *et al.*, 1991). Therefore, the assumptions underlying the mathematical description followed closely those of the pyrene model with the following exceptions: (a) The OTC model (Law, 1992) was developed for salmon exposed to multiple, oral OTC doses whereas the pyrene model was designed for trout injected intraarterially with the chemical (Law *et al.*, 1991); (b) although both chemicals were assumed to be eliminated by the kidney and the liver compartments of the models, OTC was eliminated unchanged by salmon (Björklund and Bylund, 1990; Namdari *et al.*, 1999) whereas pyrene was eliminated mainly as metabolites by trout (Law *et al.*, 1991). Because the equations used to described the OTC model had been presented in preceding papers (Law *et al.*, 1991; Law, 1992), they were not repeated here.

3.1.2. Gastrointestinal Absorption of OTC. Gastrointestinal absorption of OTC was assumed to be a first-order rate process. It was modeled by adding a gain term (GAIN) to the gut compartment description:

$$\mathrm{VGT}(d\mathrm{CGT}/dt) = \mathrm{QGT}(\mathrm{CBA} - \mathrm{CGT}/\mathrm{RGT}) + \mathrm{GAIN} \quad (1)$$

$$\mathrm{GAIN} = \mathrm{Ka} \cdot \mathrm{DA} \quad (2)$$

where Ka (day^{-1}) is the first-order absorption rate constant for OTC; DA (μg) is the amount of OTC absorbed and is calculated by multiplying the administered dose (μg) with the fraction of dose absorbed (%); VGT is the gut volume (ml); QGT ($\mathrm{ml \cdot day^{-1}}$) is the blood flow to the gut; CBA ($\mu g \cdot ml^{-1}$) is the concentration of OTC in the arterial blood; CGT ($\mu g \cdot g^{-1}$) is the concentration of OTC in the gut; and RGT is the gut : blood partition coefficient.

3.1.3. Elimination of OTC. No information was found on the branchial elimination of OTC by salmon. However, both the liver and the kidney were involved in the elimination of OTC by salmon and trout (Namdari *et al.*, 1999; Björklund and Bylund, 1991). Moreover, Björklund and Bylund (1991) suggested that hepato-biliary elimination of OTC was more important than renal elimination because a large amount of OTC was found in the bile of rainbow trout after a single oral dose. These observations were confirmed in the salmon (Namdari *et al.*, 1999).

OTC was assumed to be eliminated unchanged in the model because OTC was largely excreted unchanged by mammals (Baggot, 1977; Ritchel, 1980) and no OTC metabolites were found in the tissues of finfish (Björklund and Bylund, 1991). Therefore, hepatic and renal clearances of OTC were modeled by first-order and zero-order rate processes, respectively:

$$\mathrm{VL}(d\mathrm{CL}/dt) = (\mathrm{QL} - \mathrm{QGT})(\mathrm{CBA} - \mathrm{CL}/\mathrm{RL}) - \mathrm{Kl}(\mathrm{CL}) \tag{3}$$

$$\mathrm{VK}(d\mathrm{CK}/dt) = \mathrm{QK}(\mathrm{CBA} - \mathrm{CK}/\mathrm{RK}) - (\mathrm{V\,max} \cdot \mathrm{CK}/\mathrm{RK})\mathrm{VK}/(\mathrm{Km} + \mathrm{CK}/\mathrm{RK}) \tag{4}$$

where Kl ($ml \cdot day^{-1}$) is the hepatic clearance of OTC; Vmax ($\mu g \cdot day^{-1} \cdot g^{-1}$) and Km ($\mu g \cdot ml^{-1}$) are the Michaelis-Menten constants of the renal clearance; VL (ml) is the liver volume; QL ($ml \cdot day^{-1}$) is the total blood flow to the liver; CL is the concentration of OTC in the liver ($\mu g \cdot g^{-1}$); RL is the liver : blood partition coefficient; VK (ml) is the kidney volume; QK ($ml \cdot day^{-1}$) is the blood flow to the kidney; CK is the concentration of OTC in the kidney ($\mu g \cdot g^{-1}$); and RL is the kidney : blood partition coefficient.

3.2. Estimation of Parameter Variability

3.2.1. Parameter Variability. Nineteen model parameters were assigned either a normal or a uniform probability distribution based on the on the variability of the physiological and anatomical data of trout and salmon (Thorarensen, 1994; Barron *et al.*, 1987; Gingerich *et al.*, 1987; Daxboeck, 1981; and Stevens, 1968). The mean values of the probability distributions were taken directly from the deterministic PBPK model (Law, 1992) because they had been chosen from either the literature or through experimentation. The remaining 6 model parameters were assigned fixed values because there was little or no basic information on these parameters (section 3.2.2.).

In order to represent the inter-individual variability of the population as realistically as possible, subjective or expert opinion sometimes were also used in combination with relevant data to assign the frequency distributions and their range values. The normal distributions were slightly truncated at 0.001 and 0.999 quantiles in order not to restrict the inter-individual variability of the population too much.

The parameter distributions also were subjected to biological constraints. Therefore, the total weight of the tissues was equal to the body weight of salmon; the sum of compartment blood flows was equal to the cardiac output; and the tissue : blood partition coefficients of OTC for different salmonid tissues decreased in the order of: liver > kidney > gill > carcass > gut > muscle, which reflected the level of fat in these tissues. The variability of parameters and the relevant data were summarized as follows:

(a) The variability of tissue: blood partition coefficients was assigned normal distributions. The mean and standard deviation (STD) of the normal distributions were determined experimentally with trout tissues using *in vitro* and *in vivo* methods (section 3.4).

(b) The variability of tissue volume and blood flows was based on inspection and analysis of data reported by Thorarensen (1994), Barron *et al.*, (1987), Gingerich *et al.*, (1987), Daxboeck (1981), and Stevens (1968). The gills were assumed to receive the entire cardiac output. For compartmental blood flows, values for 4 of the 5 parameters were determined by probabilistic methods and the fifth by subtraction (Table 1). For compartmental volumes, values for 7 of the 8 parameters were determined by probabilistic methods and the eighth by subtraction (Table 1).

(c) The variability of gastrointestinal absorption was assigned a uniform distribution with a minimal value of 1% and a maximal value of 5% (Table 1) because Elema *et al.*, (1996) reported a 2% oral bioavailability for OTC-medicated feed in the Atlantic salmon and Rogstad *et al.*, (1991) reported a 2.6% oral bioavailability for trout given an OTC suspension in gelatin capsules. An oral bioavailability of 1.2% also was reported in trout dosed with OTC in mixed feed pellets and water (Nouws, 1992). In contrast, about 5.6% of an orally administered OTC suspension was bioavailable to the trout (Björklund and Bylund, 1991). The oral bioavailability of OTC (7–9%) reported by Cravedi *et al.*, (1987) for

Table 1. Parameter probability distributions used to simulate the withdrawal period of OTC in chinook salmon acclimated at 15 ± 1°C seawater.

Abbreviation	Parameter	Distribution	Mean	STD
BW	Body weight (g)	Normal	71.4	16.3
T	Water temperature (°C)	Uniform	15[a]	
F	Fraction of dose absorbed (%)	Uniform	2.5[b]	
	Tissue volumes (% Body weight)[c]			
VGL	Gill volume (%)	Normal	4.0	0.65
VL	Liver volume (%)	Normal	1.3	0.32
VK	Kidney volume (%)	Normal	0.85	0.18
VGT	Gut volume (%)	Normal	11.5	1.78
VEM	Extracellular muscle volume (%)	Normal	0.15	0.016
VIM	Intracellular muscle volume (%)	Normal	46.0	0.97
VB	Blood volume (%)	Normal	4.0	0.65
	Blood flows (% cardiac output)[d]			
QL	Liver blood flow (%)	Normal	24.0	1.9
QK	Kidney blood flow (%)	Normal	10.0	1.8
QGT	Gut blood flow (%)	Normal	14.0	1.3
QM	Muscle blood flow (%)	Normal	42.0	1.9
	Partition coefficients (unitless)			
RGL	Gill: blood	Normal	3.0	0.16
RL	Liver: blood	Normal	9.16	1.34
RK	Kidney: blood	Normal	3.75	0.41
RGT	Gut: blood	Normal	1.40	0.07
RM	Muscle: blood	Normal	0.7	0.16

[a]range of uniform distribution, 14–16°C.
[b]range of uniform distribution, 1–5%.
[c]Body weight in g; values for VC were computed at each run so that the sum of the volumes was equal to the body weight.
[d]Cardiac output in ml·day^{-1}; values for QC were computed at each run so that the sum of the blood flows was equal to the total flow.

trout was not used in this study because it overpredicted OTC residues in the tissues (data not shown).

(d) The variability of cardiac output (QT, ml·day^{-1}) was assumed to be a function of body weight (BW, g) and water temperature (T, °C). This assumption was based on the study of Barron *et al.*, (1987). The following equation was used to calculate QT:

$$QT = (3.95T - 12.9)(BW/1{,}000)(60)(24) \tag{5}$$

3.2.2. Parameter Optimization. Little or no information was available for 6 of the model parameters including Vmax, Km, Kl, Ka, H, and RC. Ka was estimated initially by fitting the serum OTC concentration data of trout at 10 °C (Björklund and Bylund, 1990) to a one-compartment pharmacokinetic model (Gibaldi and Perrier, 1975); it was found to be 0.11 day^{-1}. The reported 0.24 h^{-1} Ka for salmon (Abedini *et al.*, 1998) was not used in this study since it was obtained by administering a methanolic solution of OTC to the fish. No information was available on individual Vmax, Km and Kl values. However, an elimination rate constant of 0.17 day^{-1} was reported in the blood of chinook salmon after a single intraarterial injection of 50 mg·kg^{-1} OTC (Abedini *et al.*, 1998). Elimination rate constants ranging from 0.06 day^{-1} to 0.11 day^{-1} also were reported for other tissues of chinook salmon after a 21-day OTC medicated feed treatment (Namdari *et al.*, 1996). RC was assumed to be identical to RGL. H was assumed to be similar to that used in the pyrene PBPK model (Law *et al.*, 1991).

These 6 parameter estimates were optimized simultaneously by fitting the 15 °C data set (Namdari *et al.*, 1996) to the PBPK model (Figure 1) as follows: (a) a coefficient of variation (CV) of 25% was assigned to each of the 6 parameters; (b) all the parameters of the PBPK model were selected randomly from their distributions for each of the 500 Monte Carlo runs; (c) the system equations of the model were solved numerically; and (d) the quality of fit was assessed by the log-likelihood (LL) technique (Kalbfleisc, 1985) according to Eq. 6:

$$LL = \underset{i=1}{\overset{N}{S}} \left[-\frac{n_i}{2} \ln\left(1 + \frac{(y_{ei} - y_{pi})}{S_i^2}\right)\right] \tag{6}$$

Where N is the total number of mean experimental data points; y_{ei} is the value of the experimental point and y_{pi} the corresponding model simulated value; n_i is the number of salmon sampled at each time point and S_i^2 is the variance (with n_i degree of freedom) for each data point. In this procedure, the squared differences between log-transformed model predictions and corresponding log-transformed reference data were minimized and weighed by the variance of each experimental data point. The set of input parameters with the highest absolute LL value represents the "best-fit" values for the model.

The "best-fit" parameter set of the 500 Monte Carlo simulations was found to have a log-likelihood of –42.74. Figure 2 shows a plot of the "best-fit" model predictions *versus* the observed, *in vivo* data at 15 °C water (Namdari *et al.*, 1996). Since most of the data points fell on or near a straight line of the plot, each predicted value was very close to its empirical counterpart. Based on the results of these studies, the parameters (fixed value) were: Vmaxc (4.7), Km (81.9), Klc (139.8), Kac (0.5), H (3.35), and RC (2.57).

3.2.3. Scaling of OTC Elimination Rate Constants. When the PBPK model was implemented with a different body weight or water temperature from that of the 15 °C data

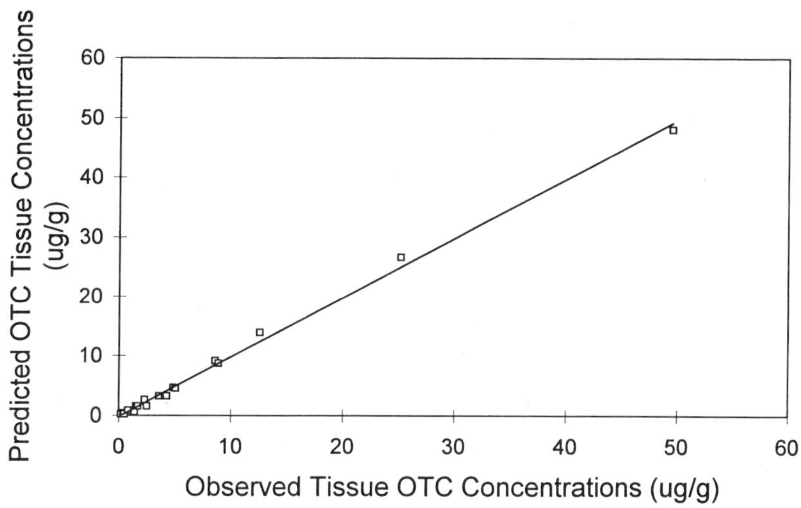

Figure 2. The "best-fit" model predictions (line) *versus* the observed (blocks) *in vivo* data.

set (Namdari *et al.*, 1996), Kl, Vmax, and Ka were scaled from the following allometric equations:

$$V\max = V\max c(BW^{0.7})(1+0.1(T-9.0)) \tag{7}$$

$$Kl = Klc(BW^{-0.3})(1+0.1(T-9.0)) \tag{8}$$

$$Ka = Kac(BW^{-0.3}) \tag{9}$$

where Vmaxc, Klc, and Kac are the scaling constants for Vmax, Kl, and Ka, respectively. These allometric equations had been adapted from a mammalian study (Travis, 1987) and were used successfully to scale the hepatic and renal clearances of pyrene from trout (Law *et al.*, 1991). The allometric equations assumed a 10% change in the Vmax and Kl values per degree of water temperature because Ellis *et al.*, (1978) reported that each degree change in water temperature generally resulted in a 10% change in the metabolic rate of fish.

3.3. Monte Carlo Analysis

Monte Carlo simulations were conducted with the aid of a Fortran subroutine to generate the Latin hypercube sampling (Iman and Shortencarier, 1984). In the Monte Carlo method, a probability distribution for each of the input parameters was randomly sampled. After selecting the parameter values, the set of differential and algebraic equations describing the movement of OTC through the salmon was solved numerically with the assistance of a Fortran-based software package, Advanced Continuous Simulation Language (ACSL, 1991). This process was repeated a large number of times until a probability distribution for the population was created. It took the computer less than 4 min to complete 400 Monte Carlo simulations. Monte Carlo simulations also could be implemented by translating the PBPK model into Excel's Visual Basic (1993) and using the Crystal Ball (1993) as a source to generate the distributions (Law and Achenbach, 1995).

However, it took a Pentium based PC with a 90 mhz procesor at least 3 hours to complete 400 Monte Carlo simulations for the PBPK model. In view of the long computational time, this procedure was used only in the preliminary studies.

Four independent runs of Monte Carlo simulations (400, 500, 600 and 800 iterations) were performed based on the experimental conditions of the 9 °C data set (Namdari *et al.*, 1996) to test if the PBPK model retained its fidelity with a large-scale probabilistic data set. The impact of a different number of iterations on the convergence and stability of the model outputs was evaluated by comparing the summary statistics of these runs. The summary statistics between the 400 and 500 iterations indicated substantial differences. However, differences among summary statistics after 500 and 600 iterations were of less significance. The summary statistics of the 600 and 800 iterations were very similar with the exception of the skewness, kurtosis and coefficient of variation. Independent runs, each with 600 Monte Carlo simulations, also were conducted. All of the estimated means, standard deviations, variances and 90th and 95th percentiles agreed within 1% (data not shown). Therefore, about 500 or 600 Monte Carlo iterations were required to maintain convergence, stability and fidelity of the output distributions in the model.

3.4. Experimental Data

Although there was a wealth of published data on the uptake and elimination of OTC by freshwater trout (Salte and Liestφl, 1983; Silven *et al.*, 1968; Herman *et al.*, 1969; Ljungberg *et al.*, 1969; McCracken *et al.*, 1976; Norlander *et al.*, 1987; Björklund and Bylund, 1990; Jacobsen, 1989), little was known of the tissue distribution of OTC in seawater salmon. A previous study from our laboratory (Namdari *et al.*, 1996) contained well-documented OTC kinetic data in the chinook salmon and was used to develop and validate the population based physiological model (Figure 1).

Tissue : blood partition coefficients were estimated by *in vitro* and *in vivo* tissue binding methods: (a) *In vitro* tissue binding was performed with dialysis bags prepared from a molecularporous membrane tubing (Spectrum Medical Industries, Inc. Los Angeles, CA). Briefly, about 0.5 ml of a tissue homogenate (25% w/v) or whole blood (25% v/v) in 0.1 M Tris-HCl buffer (pH 7.4) was put into a dialysis bag and dialyzed against the same buffer containing 0.4 mM OTC for 5 days at 4°C. Both the tissue homogenate and the buffer outside the dialysis bag were assayed for OTC by HPLC according to the procedure of Namdari *et al.*, (1996). *In vitro* tissue : blood binding were calculated as described by Lin *et al.*, (1982), (b) *In vivo* tissue binding experiments were conducted by putting 3 trout (412 g) in a 160-l fiberglass tank containing recirculating, aerated seawater at 10°C. The seawater in the tank was changed weekly. The trout were treated daily with 100 mg/kg OTC medicated feed for a month. The fish were sacrificed by a sharp blow on the head. Samples of major organs and blood were removed from the fish and analyzed for OTC residues according to the procedure of Namdari *et al.*, (1996). *In vivo* tissue : blood binding were calculated according to Dedrick *et al.*, (1973).

3.5. Sensitivity Analysis

The Spearman rank correlation constants were used to assess the sensitivity of the model parameters. The Spearman rank correlation constants were computed by assessing the one-to-one correlations between the input parameters and the area under the curve for muscle (AUCM). The AUCM was a measure of the amount of OTC accumulated by the muscle and was calculated by integrating the time course of OTC concentrations in the muscle numerically. Spearman correlation constants provided a meaningful measure of the degree to which the input parameter changed together with the AUCM. Therefore, if a parameter and the AUCM had a high Spearman correlation constant, it meant that the parameter had a significant impact on the muscle OTC concentration and the derived WP. Six hundred Monte Carlo simulations were conducted for the sensitivity analysis based on the 9 °C data set of Namdari *et al.*, (1996).

3.6. WP Determination

3.6.1. The PBPK Model Approach. The population based physiological model (Figure 1) was used to predict the WP of OTC in the chinook salmon. WP was defined as the time when the OTC residue concentration in muscle was at or below 0.1 ppm for the 99th percentile population with 95% certainty. Because the tolerance limit of OTC in the US was set at 2 ppm, the model also was used to predict a WP for salmon with a 2 ppm or less muscle OTC residue. The WP were determined graphically from the time intercepts between the 0.1 ppm (or 2 ppm) OTC residue level and the 95% upper confidence limit of the PBPK model simulation. The 95% confidence intervals were determined from the outputs of Monte Carlo simulation using Crystal Ball (1993).

3.6.2. The 90% Upper Prediction Limit Method. The WP of OTC in chinook salmon also was determined according to the statistical approach reported by Salte and Liestφl (1983). Briefly, the terminal elimination phases of the 9 °C and 15 °C data sets (Namdari *et al.*, 1996) were fitted separately by linear regression. The 90% upper prediction limit of the regression line was then calculated with the Statistical Analysis System (SAS) package (1996). The WP of OTC was defined as the time intercepts between the detection limit of the OTC assay (0.1 ppm) and the 90% upper prediction limit of the regression line.

4. RESULTS

4.1. Model Outputs

Monte Carlo simulation (600 iterations) on the modified PBPK model (Figure 1) was conducted for chinook salmon based on the experimental conditions of the 9 °C data set (Namdari *et al.*, 1996). Figure 3 shows the probability or frequency distribution of OTC concentrations in the muscle of salmon at day 1 post-dosing. The probability distribution displayed in Figure 3 shows all the model-predicted values within ±2.6 STD of the population mean, or approximately 99% of the 600 simulations. It also shows the probability of salmon tissue containing OTC at specified concentrations. Arrowheads under the x-axis (Figure 3) show the upper limit and lower limit of the 95% confidence interval of the simulated results. The results of the probability distribution also can be summarized in a table of statistics or as percentiles.

4.2. Model Validation

The population based physiological model was validated with the 9 °C data set of Namdari *et al.*, (1996) because the deterministic PBPK model (Law, 1992) had been validated by data sets with different dose, species and administration route. Figures 4, 5 and 6 show the model-predicted 95% confidence intervals of OTC residue levels in the liver,

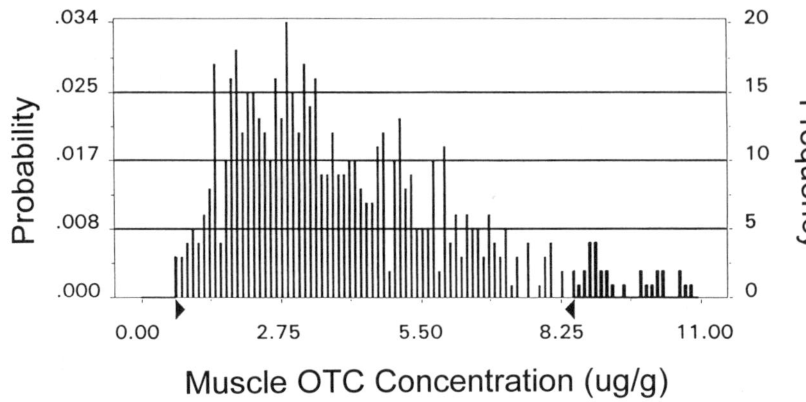

Figure 3. Model outputs of Monte Carlo simulations.

kidney and muscle of chinook salmon, respectively. The mean values of model predictions and empirical results (Namdari *et al.*, 1996) also were included in these figures for comparison. However, the STD of the empirical data were omitted from these figures for clarity; they had been reported previously (Namdari *et al.*, 1996). The empirical and the simulated OTC residues in the liver, kidney and muscle were found to decrease with increasing withdrawal time. The simulated 95% confidence intervals converged at or near the 0.1 ppm OTC residue level but still bracketed the empirical results.

4.3. Graphic Determination of WP

Figure 7 shows the graphic method used to determine a PBPK model-derived WP of OTC. Withdrawal period was determined from the time intercepts between the model

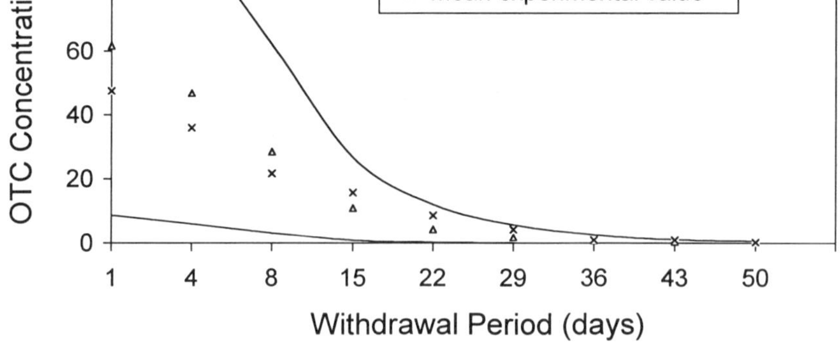

Figure 4. Model predicted 95% confidence intervals and observed OTC concentrations in the liver.

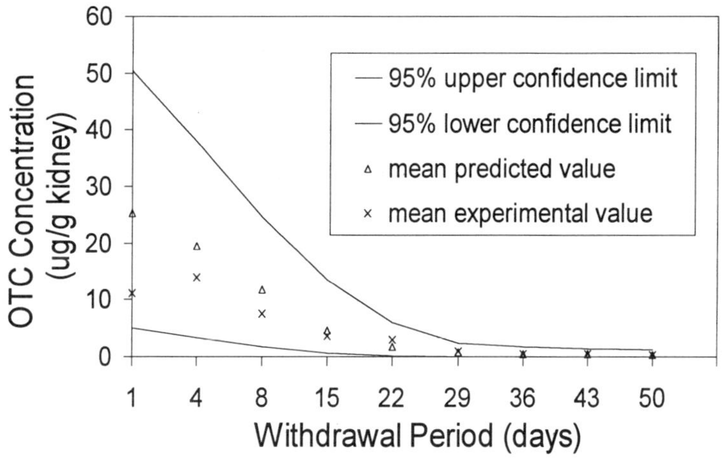

Figure 5. Model predicted 95% confidence intervals and observed OTC concentrations in the kidney.

predicted 95% upper confidence limit and the 0.1 ppm OTC residue level. A WP of 60 days was determined from this plot.

4.4. Model-Predicted WP at Different Water Temperatures

Table 2 summarizes the WP estimated by the PBPK model and the empirical methods at different water temperatures. The PBPK model-predicted WP (column 2) were based on the Canadian maximum muscle residue level of 0.1 ppm OTC and the experimental conditions of the 9 °C data set (Namdari *et al.*, 1996). The WP increased with decreasing water temperature, but was not related linearly to the water temperature. Column 3 shows the WP determined by the 90% upper prediction limit method (Salte and Liestφl, 1983). They were very similar to those predicted by the PBPK model at 9 °C

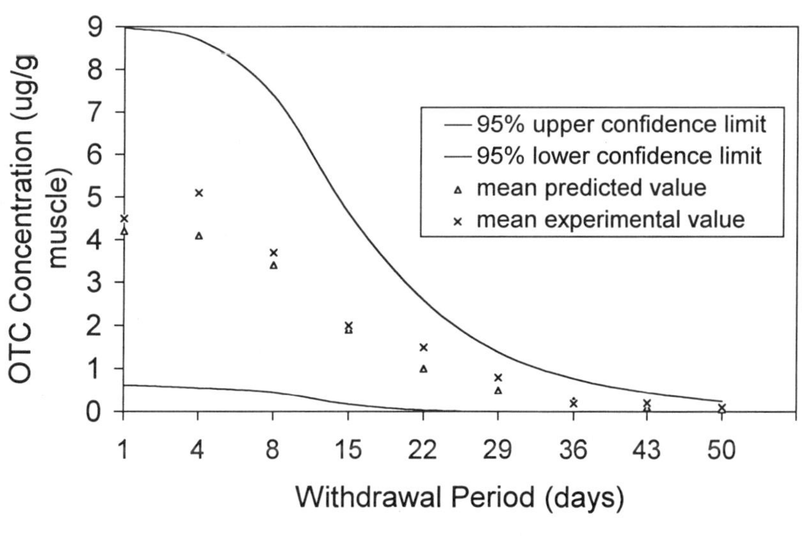

Figure 6. Model predicted 95% confidence intervals and observed OTC concentrations in the muscle.

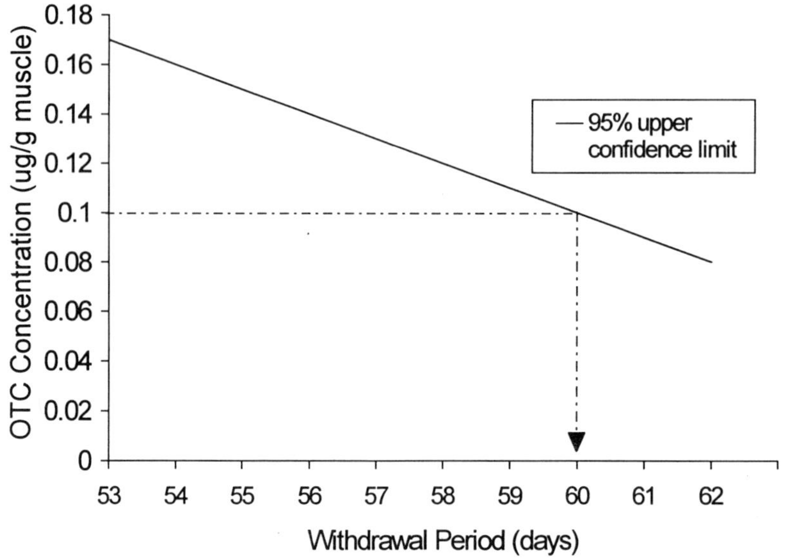

Figure 7. WP prediction using the model simulated 95% upper confidence limit.

Table 2. A comparison of the WP derived by the PBPK model and the empirical methods at different water temperatures.

Water temperature range (°C)	PBPK model (0.1 ppm MRL)[a,b]	Salte and Liestφl's method[a,c]	Canadian guidelines[a,d]	PBPK model (2.0 ppm tolerance limit)[a,c]
4–9	74		80	
4–6	81			
6–8	67			
8–10	60	62		27
10–12	55			24
12–14	52			21
14–16	49	49		16
10–16	51		40	

[a]Post-dosing day WP.
[b]Based on the time intercepts of 0.1 ppm OTC concentration (Canadian maximum residue level) and the 95% upper confidence limit; simulation was performed using the experimental conditions of the 9°C data set but with the appropriate water temperature range.
[c]Calculated according to the statistical approach of Salte and Liestφl.
[d]Forty days when the water temperature is at or above 10°C and 80 days at a lower water temperature range.
[e]Based on the time intercepts of 2.0 ppm OTC concentration (U.S. tolerance limit) and the 95% upper confidence limit; simulation was performed using the experimental conditions of the 9°C data set but with the appropriate water temperature range.

and 15 °C water. However, there were only 2 WP calculated by the empirical approach because two sets of pharmacokinetic data were available (Namdari *et al.*, 1996). Column 4 lists the recommended WP of OTC in Canada. The last column shows the PBPK model-predicted WP based on the experimental conditions of the 9 °C data set (Namdari *et al.*, 1996) but using the US tolerance limit of 2 ppm. A maximum WP of 27 days was predicted for chinook salmon in 8–10 °C seawater.

4.5. Model Parameter Sensitivity

The sensitivity analysis of model parameters was conducted with 19 of the model parameters because Vmax, Km, Kl, Ka, H, and RC were excluded from the analysis. The sensitivity of the 19 parameters was found to be highly dependent on the exposure conditions for which the simulation was performed, the dose surrogate and the time point being considered (data not shown). Table 3 depicts the Spearman correlation table for 9 of the parameters with AUCM at 60 days post-dosing based on the 9 °C data set of Namdari *et al.*, (1996). The sensitivity analysis showed that the fraction of dose absorbed was the parameter most strongly affecting model predictions. Liver volume and muscle : blood partition coefficient were of secondary importance with a correlation constant of about 0.4. Body weight and T were of even less importance for the model predictions. Since the correlation constants of the remaining parameters were less than 0.1 in absolute value, there was no amplification of variability from these parameters to the model outputs.

5. DISCUSSION

To my knowledge, this is the first application of the population based physiological model to assess a WP of OTC for the chinook salmon. The model-predicted WP are 49 days and 60 days for the 15 °C and 9 °C studies (Namdari *et al.*, 1996), respectively (Table 2). These WP are very close to the 49 days and 62 days calculated by the 90% upper prediction limit method (Salte and Liestφl, 1983) using the same data. These results show that the PBPK model is an accurate and reliable tool of WP determination. There are several advantages of using the PBPK model to determine the WP of OTC: (a) the model can be used for quantitative extrapolation well beyond the experimental conditions of the depletion study because it is based largely on the anatomy and physiology of the chinook salmon. In contrast, the WP of the empirical method cannot be extrapolated to other fish species, or even to the same fish species differing in exposure conditions because it is derived from an "average" fish in a "typical" exposure environment, (b) the PBPK model helps to make key pharmacokinetic parameters more explicit in the

Table 3. Spearman rank correlation constants of model parameters.

Model parameters	AUCM[a]
Fraction of dose absorbed (F)	0.725
Liver weight (VLC)	−0.414
Muscle : blood partition coefficient (RM)	0.397
Body weight (BW)	−0.262
Water temperature (T)	−0.104
Kidney weight (VKC)	−0.032
Gill : blood partition coefficient (RGL)	0.024
Gut weight (VGTC)	−0.020
Blood flow in gut (QGTC)	−0.011

[a]Spearman correlation constants with <0.01 are not included.

prediction of WP, and (c) by coupling the PBPK model with the Monte Carlo analysis, the importance of a pharmacokinetic parameter on the WP can be quantified.

The current Canadian WP for OTC in farmed fish (Agriculture Canada, 1984) is 40 days when the water temperature is at or above 10 °C and 80 days at lower water temperatures (Table 2). The model-predicted WP agree with the Canadian WP when the water temperature is <10 °C, but disagree with the Canadian WP if the water temperature is >10°C (Table 3). The current US WP for OTC in farmed fish is 21 days when the fish are dosed between 55–82.5 mg/kg OTC for 10 days; OTC is not allowed to be used for the salmonids when the water temperature is below 9 °C (G. Stehly, personal communication). The model-predicted WP agree with the US WP when the water temperature is >12 °C, but disagree with the U.S. WP if the water temperature is <12 °C. It should be noted that the model-predicted WP in Table 2 are calculated using only OTC residues in the muscle whereas the tolerance limit in the U.S. is defined as the OTC residues in the muscle and the adhering skin.

In general, there are two types of uncertainty in a PBPK model: the uncertainty that results from inter-individual variation and the uncertainty that results from a lack of knowledge in the model. Inter-individual variation is the difference in exposure that occurs between one fish and another. This can be measured and reported. In contrast, uncertainty due to a lack of knowledge in the model parameters is difficult or impossible to measure. In the present study, variability is presumed to result mainly from inter-individual variation such as body weight, tissue volumes, blood flows, tissue : blood partition coefficients and water temperature. Variability resulting from analytical procedure, other environmental factors and day-to-day difference within the same fish is ignored.

It is difficult, if not impossible, to assign probability distributions to the model parameters since this requires an evaluation of uncertainty from different sources. Moreover, many parameters with assigned probability distributions are themselves not well characterized. Because there is considerable uncertainty involved in the assignment of probability distributions to the model parameters, the assigned probability distributions are very tentative at this time.

The muscle is a membrane-limited compartment. This is based on the following lines of evidence: (a) preliminary studies conducted with a flow-limited muscle compartment invariably result in model-predicted OTC residues higher than those observed experimentally. In other words, the transfer of OTC from blood into the muscle cells is slower than would be predicted on the basis of published blood flow to the muscle due perhaps to limited membrane permeability or maldistribution (shunt) of blood flow; (b) Randall (1970) has shown that the white muscle of trout, which constitutes the bulk of the body, is poorly vascularized and is used in burst activity of short duration. These findings are consistent with a membrane-limited compartment for the muscle; (c) the kinetic data used to support a membrane-limited compartment in the muscle have been obtained with laboratory salmon (Namdari *et al.*, 1996) which may differ to sea-pen cultured salmon in muscular physiology. Indeed, recent studies with sea-pen cultured salmon (Namdari *et al.*, 1998; Brocklebank *et al.*, 1997) show that the PBPK model underpredicts OTC residues in the muscle although the model is able to describe closely the kinetics of OTC in all other tissues (unpublished observation). If the mass transfer coefficient of muscle (H) in the model were increased arbitrarily by 10 fold, simulated tissue concentrations including those of the muscle would agree closely with the experimental data. Apparently, the muscle cells of sea-pen cultured salmon are more penetrable by OTC or are better perfused by the blood than those of the laboratory salmon. Because sea-pen cultured salmon are kept in coastal seawaters and are physically more active than the laboratory salmon, it is not surprising that sea-pen cultured salmon have higher blood flows and OTC residue levels in the muscle than laboratory salmon.

Spearman's rank correlation constants have been used to quantitatively determine the input parameters which have the greatest effect on the WP. These constants are calculated by examining the one-to-one correlations of the input parameters and the AUCM. The sensitivity analysis shows that the pharmacokinetics of OTC in the chinook salmon are mainly controlled by 5 parameters: the fraction of dose absorbed, liver volume, muscle: blood partition coefficient, fish weight and water temperature (Table 3). The other 14 parameter distributions are insensitive to the model-predicted WP because their Spearman's correlation constants are less than 0.1 in absolute value. Therefore, the exact values of these parameters are unimportant since they do not have a significant impact on the model simulation. However, only the major features exhibited by this sensitivity analysis are likely to be valid because only a part of the covariance structure of the parameters (e.g., the dependence on body weight and water temperature) has been implemented. Also, the parameters with high correlations may be due to covariance between parameters and not necessarily due to a high sensitivity of the model. It is also possible that the 6 parameters (Vmax, Km, Kl, Ka, H, and RC) which are excluded from the sensitivity analysis, are more important than the 19 parameters with assigned probability distributions in affecting the model predictions. Indeed, both Ka and Kl also have been shown to be sensitive parameters of the deterministic model (Law, 1992).

Although there is considerable uncertainty in the assignment of probability distribution and correlation to the parameter, the population based physiological model is a useful tool for predicting the WP of OTC in the chinook salmon. The PBPK model is especially useful to determine a WP for "off-label" use of OTC since the model can incorporate treatment specific information such as the fish weight, dose regime, bioavailability and water temperature in the WP determination.

6. REFERENCES

Abedini, S., Namdari, R., and Law, F.C.P., 1998, Comparative pharmacokinetics and bioavailability of oxytetracycline in rainbow trout and chinook salmon. *Aquacult.* 162:23.

Achenbach, T., Abedini, S., Cox, W., and Law, F.C.P., 1998, Development of a physiologically based pharmacokinetic (PBPK) model for oxytetracycline in cattle. *Toxicol. Sci.* 42(1-S):140.

ACSL., 1991, *Advanced Continuous Simulation Language*. Mitchell and Gauthier Associates, Concord, MA.

Agriculture Canada, 1984, *Compendium of Medicating Ingredients Brochures*. Feeds and Fertilizer Division, Canada Department of Agriculture. Ottawa, ONT.

Ahmad, T.S., and Matty, A.J., 1989, The effects of feeding antibiotics on growth and body composition on carp (*Cyprinus carpio*). *Aquacult.* 77:211.

Baggot, J.D., 1977, *Principles of Drug Disposition in Domestic Animals*, W.B. Saunders, Philadelphia, PA.

Barron, M.G., Tarr, B.D., and Hayton, W.L., 1987, Temperature-dependence of cardiac output and regional blood flow in rainbow trout, *Salmo gairdneri* Richardson. *J. Fish Biol.* 31:735.

Björklund, H., and Bylund, G., 1990, Temperature-related absorption and excretion of oxytetracycline in rainbow trout (*Salmo gairdneri.* R.). *Aquacult.* 84:363.

Björklund, H., and Bylund, G., 1991, Comparative pharmacokinetics and bioavailability of oxolinic acid and oxytetracycline in rainbow trout (*Oncorhynchus mykiss*). *Xenobiotica.* 21:1511.

Brocklebank, J.R., Namdari, R., and Law, F.C.P., 1997, An oxytetracycline residue depletion study to assess the physiologically based pharmacokinetic (PBPK) model in farmed Atlantic salmon. *Can. Vet. J.* 38:645.

Brown, L.A., 1992, Constraints on chemotherapy: a manufacturer's view, in: *Chemotherapy in Aquaculture: from Theory to Reality*, C. Michel and D.J. Alderman, eds., Office International Des Epizooties, Paris, France.

Cravedi, J.P., Choubert, G., and Delous, G., 1987, Digestibility of chloramphenicol, oxolinic acid, and oxytetracycline in rainbow trout and influence of these antibiotics on lipid digestibility. *Aquacult.* 60:133.

Crystal Ball., 1993, *Crystal Ball Forecasting and Risk Analysis for Spreadsheet Users. Version 3.0*, Decisioneering, Inc., Denver, CO.

Daxboeck, C., 1981, *A Study of the Cardiovascular System of the Rainbow Trout (Salmo gairdneri) at Rest and During Swimming Exercise*. Ph.D. Thesis, University of British Columbia. Vancouver, B.C.

Dedrick, R.L., Zaharko, D.S., and Lutz, R.J., 1973, Transport and binding of methotrexate *in vivo*. *J. Pharm. Sci.* 62:882.

Elema, M.O., Hoff, K.A., and Kristensen, H.G., 1996, Bioavailability of oxytetracycline from medicated feed to Atlantic salmon (*Salmo salar* L.). *Aquacult.* 144:7.

Ellis, A.E., Roberts, R.J., and Tytler, P., The anatomy and physiology of teleosts, in: *Fish Physiology*. R. J. Roberts, ed., Baillière Tindall, London.

Excel., 1993, *Microsoft Excel Spreadsheet with Business Graphic and Database*. Version 5.0, Microsoft Corporation, Redmond, WA.

FDA, 1994, General Principles for Evaluating the Safety of Compounds used in Food-Producing Animals. Center for Veterinary Medicine, Food and Drug Administration, Public Health Services, U.S. Department of Health and Human Services. Washington, D.C.

Gibaldi, M., and Perrier, D., 1975, *Pharmacokinetics*, Marcel Dekker, New York, NY.

Gingerich, W.H., Pityer, R.A., and Rach, J., 1987, Estimates of plasma, packed cell and total blood volume in tissues of the rainbow trout (*Salmo gairdneri*). *Comp. Biochem. Physiol.* 87A(2):251.

Herman, R.L., Collins, D., and Bullock, G., 1969, Oxytetracycline residues in different tissues of trout. U.S. Fish and Wildl. Serv. Tech. Pap. 37:1.

Iman, R.L., and Shortencarier, M.J., 1984, *A Fortran 77 Program and User's Guide for Generation of Latin Hypercube and Random Samples for Use with Computer Models*, National Technical Information Service, Springfield, VA.

Jacobsen, M.D., 1989, Withdrawal times of freshwater rainbow trout, *Salmo gairdneri* Richardson, after treatment with oxolinic acid, oxytetracycline and trimethoprim. *J. Fish. Dis.* 12:29.

Kalbfleisc, J.G., 1985, *Probability and Statistical Inference*, Springer-Verlag, New York, NY.

Law, F.C.P., Abedini, S., and Kennedy, C.J., 1991, A biologically based toxicokinetic model for pyrene in rainbow trout. *Toxicol. Appl. Pharmacol.* 110:390.

Law, F.C.P., 1992, A physiologically based pharmacokinetic model of oxytetracycline for salmonids, in: *New Frontiers in Pharmacology*, P. Conlon, ed., Proceedings of the 8th Biennial Symposium of the American Academy of Veterinary Pharmacology and Therapeutics and the American College of Veterinary Clinical Pharmacology, Guelph, Ontario.

Law, F.C.P., and Achenbach, T.E., 1995, A Monte Carlo approach for the physiologically-based toxicokinetic model of waterborne pyrene in trout: spreadsheet implementation. Proceedings of the Second Society of Environmental Toxicology and Chemistry World Congress. Vancouver, B.C.

Lin, J.H., Sugiyama, Y., Awazu, S., and Hanano, M., 1982, *In vitro* and *in vivo* evaluation of the tissue-to-blood partition coefficient for physiological pharmacokinetic models. *J. Pharmacokinet. Biopharm.* 10:637.

Ljungberg, O., Silven, L., and Johansson, N., 1969, Residues of sulphamerazine and terramycin in artificially fed rainbow trout. Proceedings of the 5th Symposium of the World Association of Veterinary Hygiene. World Association of Veterinary Hygiene, Berlin, Germany.

McCracken, A., Fidgeon, S., O'Brien, J.J., and Anderson, D., 1976, An investigation of antibiotic and drug residues in fish. *J. Appl. Bacteriol.* 40:61.

Namdari, R., Abedini, S., and Law, F.C.P., 1996, Tissue distribution and elimination of oxytetracycline in seawater chinook and coho salmon following medicated- feed treatment. *Aquacult.* 144:27.

Namdari, R., Abedini, S., Albright, L., and Law, F.C.P., 1998, Tissue distribution and elimination of oxytetracycline in sea-pen cultured chinook salmon, *Oncorhynchus tshawytscha*, and Atlantic salmon, *salmo salar*, following medicated-feed treatment. *J. Appl. Aquacult.* 8:39.

Namdari, R., Abedini, S., and Law, F.C.P., 1999, A comparative tissue distribution study of oxytetracycline in rainbow trout *Oncorhynchus mykiss* (Walbawn) and chinook salmon *Oncorhynchus tshawytscha* (Walbawn). *Aquacult. Res.* (in press).

Nordlander, I., Johnsson, H., and Österdahl, B., 1987, Oxytetracycline residues in rainbow trout analyses by a rapid HPLC method. *Food Addit. Contam.* 4:291

Nouws, J.F.M., Grondel, J.L., Boon, J.H., and van Ginneken, V.J.T., 1992, Pharmacokinetics of antimicrobials in some fresh water fish species, in: *Chemotherapy in Aquaculture: from Theory to Reality*, C. Michel and D.J. Alderman, eds., Office International Des Epizooties, Paris, France.

Randall, D.J., 1970, The circulatory system. in: *Fish Physiology*, W.S. Hoar and D.J. Randall, eds., Academic Press, New York, NY.

Ritchel, W.A., 1980, *Handbook of Basic Pharmacokinetics*, Drug Intelligence Publications Inc., Hamilton, IL.

Rogstad, A., Hormazabal, V., Ellingsen, O.F., and Rasmussen, K.E., 1991, Pharmacokinetic study of oxytetracycline in fish. I. Absorption, distribution, and accumulation on rainbow trout in freshwater. *Aquacult.* 96:219.

Rubenstein, R.Y., 1981, *Simulation and the Monte Carlo Method*, John Wiley & Son, New York, NY.

Salte, R., and Liestφl, K., 1983, Drug withdrawal from farmed fish. Depletion of oxytetracycline, sulfadiazine and trimethoprim from muscular tissue of rainbow trout (*Salmo gairdneri*). *Acta Vet. Scand.* 24:418.

SAS., 1996, *The SAS System for Window Release 6.12*, SAS Institute Inc., Cary, NC.

Schnick, R.A., 1992, An overview of the regulatory aspects of chemotherapy in aquaculture, in: *Chemotherapy in Aquaculture: from Theory to Reality*, C. Michel and D.J. Alderman, eds., Office International Des Epizooties, Paris, France.

Silven, L., Johansson, N., and Ljungberg, O., 1968, Terramycin and sulphamerazine residues in muscle tissue and internal organs of rainbow trout after oral drug administration. Bull. Off. Int. Epiz. 69:1465.

Stevens, E.D., 1968, The effects of exercise on the distribution of blood to various organs in rainbow trout. *Comp. Biochem. Physiol.* 25:615.

Thorarensen, H., 1994, *Gastrointestinal Blood Flow in Chinook Salmon (Oncorhynchus Tshawytscha)*. Ph.D. Thesis. Simon Fraser University. Burnaby, B.C

Travis, C.C., 1987, Interspecies and dose-route extrapolation. in: *Drinking Water and Health. Pharmacokinetics in Risk Assessment*. National Academy Press, Washington, DC.

9

COMPARATIVE PHARMACOKINETICS OF SARAFLOXACIN IN RAINBOW TROUT AND CHANNEL CATFISH

Guy R. Stehly, Jeffery R. Meinertz, and William H. Gingerich

U.S. Geological Survey
Biological Resources Division
Upper Midwest Environmental Sciences Center
2630 Fanta Reed Road
La Crosse, Wisconsin 54603

1. ABSTRACT

Sarafloxacin is a broad spectrum fluoroquinolone potentially effective in controlling a number of bacterial infections in fish. Of the aquaculture therapeutants that are under consideration for approvals in the International Association of Fish and Wildlife Agencies proposal, sarafloxacin was chosen as the best example of an orally administered drug that would be readily metabolized by fish, and therefore a model compound to study the crop grouping concept. The crop grouping concept hypothesizes that species of fish can be grouped based on defining characteristics such as activity level, phylogeny, and temperature preference. As an initial test of the crop grouping concept, the similarities and differences in the pharmacokinetics of sarafloxacin after a single intraarterial dose of 5 mg/kg were tested in two surrogate fish species (rainbow trout and channel catfish) that differ greatly in group defining characteristics. The mean apparent volume of distribution of individual trout was approximately 10 L/kg and 4 L/kg for catfish while total body clearance was 0.05 L/kg/h for trout and 0.17 L/kg/h for catfish. Rainbow trout had an extended terminal elimination with a half-life of approximately 158 h when compared to 18 h for channel catfish.

2. INTRODUCTION

Public concerns about human food safety, human health, and environmental impacts have resulted in stricter interpretation and enforcement of regulations by the U.S. Food and Drug Administration. These actions have curtailed the availability and use of

Xenobiotics in Fish, edited by *Smith et al.*
Kluwer Academic / Plenum Publishers, New York, 1999.

drugs essential to maintain fish health on hatcheries (Geyer, 1993). Lack of approved drugs and chemicals dramatically reduces the effectiveness and increases the cost of fish production. Drug and chemical manufacturers are reluctant to undertake efforts to gain approval of aquaculture drugs because the market potential for these products is less than the annual sales target for research investment. Only three therapeutants and one anesthetic are currently approved and available to hatchery managers in the United States.

The greatest need in aquaculture is for drugs that control infectious diseases because outbreaks can cause major or complete loss of production on a hatchery. The concept of crop grouping hypothesizes that species of fish can be grouped based on defining characteristics such as phylogeny, temperature preference, and activity level. Fish species within a group may have similar abilities to absorb, eliminate and metabolize xenobiotics or their abilities in these areas could be normalized based on some measurable factor. If this were true, complete regulatory testing could be conducted on one or several surrogate species with limited testing on other species within the crop grouping. Costs for expensive studies, such as depletion of the marker compound, could be reduced by testing only a few species while resulting in broad approvals. It is unfeasible to conduct these studies on the large number of species (>50) cultured in public aquaculture (Gingerich *et al.*, 1998).

Approvals of aquaculture therapeutants is currently being pursued by a cooperative effort between the federal government and state natural resource agencies through the International Association of Fish and Wildlife Agencies (IAFWA). Two federal research facilities are generating the bulk of the required data for approvals of eight high priority aquaculture therapeutants and anesthetics as well as generating data to support the concept of crop grouping. A compound that is metabolized to at least a moderate degree would be desirable to test the crop grouping concept. This allows investigation on species differences in metabolite profile and degree of metabolism. Sarafloxacin hydrochloride (HCl) (Abbott Laboratories, North Chicago, IL) is a broad spectrum fluoroquinolone potentially effective in controlling a number of bacterial infections in fish. Of the aquaculture therapeutants under consideration for approvals in the International Fish and Wildlife Agencies proposal, sarafloxacin was chosen as the best example of an orally administered drug that would be readily metabolized by fish (Schnick *et al.*, unpublished). The metabolism of sarafloxacin in fish has not been reported. Structurally similar fluoroquinolones are readily metabolized by humans; these functional groups are commonly metabolized in fish for other compounds (Lech and Bend, 1980; Sorgel *et al.*, 1989).

Pharmacokinetic data, including residue depletion from edible tissue and metabolic profile on several fish species are needed as an initial test of the concept of crop grouping. The adaptation of analytical methods for sarafloxacin HCl in fish edible tissue and plasma to other fish species and subsequent plasma pharmacokinetics were initially needed to obtain comparative data. The objectives of this study were to; (1) develop and validate an analytical method for sarafloxacin in the plasma of rainbow trout and channel catfish and (2) to develop comparative pharmacokinetic information for sarafloxacin in rainbow trout and channel catfish.

3. MATERIALS AND METHODS

3.1. Test Chemical

Sarafloxacin HCl [6-fluoro-1-(4-fluorophenyl)-1,4-dihydro-4-oxo-7-(1-piperazinyl)-3-quinolinecarboxylic acid hydrochloride] (Figure 1) was obtained from Abbott Laboratories (North Chicago, IL). Sarafloxacin HCl chemical purity was assessed by reversed

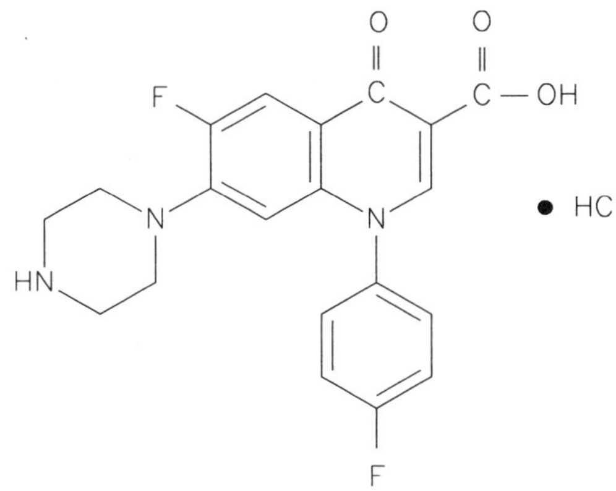

Figure 1. Structure of sarafloxacin HCl.

phase (C_{18}) high performance liquid chromatography (HPLC) with a gradient mobile phase of 0.002 M phosphoric acid as the aqueous phase and 3:2 acetonitrile:methanol for the organic phase. Peaks obtained with the UV absorbance trace by a diode array detector, at eight wavelengths from 220 nm to 375 nm, and with a fluorescence detector with excitation at 280 nm and emission above 389 nm for standards and solvent blanks were compared. Purity was determined to be >99%.

Sarafloxacin hydrochloride stock solutions in HPLC mobile phase were used to spike fish plasma and for analytical standards. Stoppered stock solutions were stable for at least two weeks at room temperature.

3.2. Test Animals

Rainbow trout (*Oncorhynchus mykiss*, Erwin strain) were obtained as eyed eggs from Ennis National Fish Hatchery (Ennis, MT). Eggs were hatched in incubation trays and the fish held and reared to the appropriate size (400 to 750 g) in fiberglass tanks or concrete raceways supplied with flowing well water (about 12 °C) at the Upper Midwest Environmental Sciences Center (UMESC). Trout were acclimated (≥1 month) at 12 °C ± 2 °C before dosing was initiated. Channel catfish (*Ictalurus punctatus*) were obtained from Senecaville State Fish Hatchery (Senecaville, OH) as fry and reared as a single lot to the appropriate size (200 to 500 g) in fiberglass tanks, stainless steel tanks, or outdoor ponds at UMESC. Catfish were acclimated (≥1 month) at 21 °C ± 2 °C at least 1 month before experiments were initiated. Fish were used in all studies without regard to gender. Because all fish received the same experimental treatment, randomization was not required.

3.3. Diet

Sterling Silver Cup trout feed of the appropriate size was obtained from Murray Elevators (Murray, UT). Fish were offered the diet daily during weekdays at a rate appropriate to maintain growth, approximately 1 to 2% body weight/day (Haskell, 1959, Huner and Dupree, 1984). Fish were not fed the day before cannulation and during the dosing-blood sampling procedures.

3.4. Analytical Methods

The HPLC method for sarafloxacin in rainbow trout and catfish plasma used a simple sample cleanup procedure and fluorescence detection. Three mL Empore columns

(3M Industrial and Consumer Sector, St Paul, MN) were conditioned with 5 mL methanol followed by 5 mL of 0.002 M o-phosphoric acid. With the vacuum off, the columns were loaded sequentially with 250 μL of 0.1 M o-phosphoric acid, the plasma sample (250 μL), and 250 μL water. The columns were aspirated then washed with 100 μL water and 100 μL of 0.1 M o-phosphoric acid. The sarafloxacin was eluted into volumetric glassware with 5 × 100 μL of 7:3 (v:v) methanol:0.5 M o-phosphoric acid. Filtered samples were injected on an HPLC for quantitation. The following equipment and conditions were used to quantify sarafloxacin in the fish plasma; A Hewlett-Packard 1090 Series II HPLC (San Fernando, CA), YMC-PAC ODS-A 250 × 4.6 mm ID column (Wilmington, NC) with a YMC S5-120A guard column, a mobile phase of 18% acetonitrile, 82% 85 mM $NaHPO_4/H_3PO_4$ and 20 mM triethylamine at 1.0 mL/min, fluorescence detector set at a wavelength of 280 nm with a 389 nm cutoff filter, and column heater at 36 °C.

Accuracy of the method was determined in fortified samples of fish plasma with sarafloxacin in the HPLC mobile phase to produce concentrations of approximately 10, 100, 1,000 and 5,000 ng/mL. The percent recovery from each sample indicated the accuracy of the method. The precision of the method was reported as the relative standard deviation of the spiked samples at each concentration. Limits of detection and quantitation were determined from the variability of replicate plasma samples spiked with 10 ng/mL of sarafloxacin (Keith *et al.*, 1983).

Based on an earlier determination, sarafloxacin in the extract from plasma was stable at room temperature in sealed HPLC vials for at least 8 d for rainbow trout and 10 d for channel catfish.

3.5. Cannulation, Dosing, and Sampling

Rainbow trout were cannulated similar to a method previously described (Beyenbach *et al.*, 1975). Briefly, a length of narrow bore (PE-50, OD 0.965 mm, Intramedic; Sparks, MD) polyethylene tubing was inserted into the dorsal aorta of an anesthetized fish and sutured into place. Channel catfish were cannulated in a similar manner with placement of the cannula as previously described (Stehly and Plakas, 1993). After cannulation, each fish was allowed to recover individually in flow through aquaria used in the dosing experiments for at least 48 h before dosing.

Dose delivery and recovery of residual sarafloxacin were determined in the apparatus used for dosing and blood sampling. The apparatus consisted of a 1 mL syringe attached to a three way valve that was connected to the PE-50 cannulation tubing. The delivery efficiency was assessed by an injection of the dosing and flush solution with a syringe pump into volumetric glassware. A dosing solution at approximately 5 mg/mL sarafloxacin was made by dissolving sarafloxacin hydrochloride in 0.1 M NaOH adjusted to pH 10.5 with HCl (Martinsen *et al.*, 1993a), and administered at a dosage of 5 mg/kg. Dosage volume was based on the weight of the individual fish measured during cannulation; e.g., a 500 g fish received 0.5 mL of the dosing solution. The dosing solution was delivered through the cannula followed by purging the cannula with 250 μL of 0.1 M NaOH, adjusted to pH 10.5 with HCl, replacement of a 60 cm section of the cannula, and an additional 250 μL flush of 0.1 M NaOH at pH 10.5. Volumes of dosing and flush solutions were minimized to avoid possible reaction of the fish to the basic solution. Fish did not display any unusual reaction to administration of the dose. It was determined that the delivery efficiency of the syringe pump with the apparatus used was 96.3% with a negligible recovery of residual sarafloxacin in the first and subsequent blood samples taken after dosing. The replacement of a 60 cm section of cannula was required to minimize the carryover of residual sarafloxacin in the first plasma sample.

Blood samples (0.5 mL) were placed in 0.65 mL microcentrifuge tubes and centrifuged at approximately 10,000 RCF for 10 minutes to obtain plasma. Plasma was drawn off with a micropipet and immediately processed for HPLC analysis. Extracts of sarafloxacin from the plasma were stored at room temperature in sealed vials until analysis by HPLC. A blood sample was taken from each fish before dosing to serve as a baseline value for the subsequent analyses.

3.6. Determination of the Plasma Pharmacokinetics of Sarafloxacin in Rainbow Trout and Channel Catfish after i.a. Administration

Optimal times for sampling of blood after dosing were determined by preliminary experiments. Three cannulated fish of each species were dosed at 5 mg/kg with sarafloxacin and serial blood samples taken from 0 to 48 h. Based on these preliminary studies, blood sampling times for rainbow trout were at 1, 2, 4, 7, 10, 24, 48, and 96 h and at 1, 2, 4, 8, 24, 48 and 96 h for channel catfish. Eight rainbow trout and ten channel catfish were used for determination of pharmacokinetic parameters.

Preliminary analysis of the concentration of sarafloxacin in plasma vs time data for individual fish, indicated that the catfish data fit best to a two compartment model. The data for individual trout however, had mixed results of best fit to two or three compartment models. The pharmacokinetic parameters total body clearance, apparent volume of distribution, and mean residence time were therefore determined by noncompartmental analysis from calculation of the area under the curve and area under the moment curve for the sarafloxacin in the plasma vs time curves for individual fish (Barron *et al.*, 1990). A terminal elimination half-life was calculated for each individual fish by log linear regression of the plasma sarafloxacin concentrations from 10 h to 96 h for rainbow trout and 8 h to 96 h for channel catfish.

4. RESULTS

The precision and accuracy of the HPLC method for sarafloxacin in channel catfish and rainbow trout plasma are summarized in Table 1. Recoveries ranged from 90.6 to 116% for rainbow trout and channel catfish; relative standard deviations ranged from 1.4 to 7.2%. The limit of detection for sarafloxacin in rainbow trout plasma was 1.8 ng/mL with a limit of quantitation of 6.1 ng/mL. For channel catfish plasma, the limit of detection was 1.6 ng/g with a limit of quantitation of 5.8 ng/g.

The mean and standard deviation of the sarafloxacin concentrations in plasma at each sampling time are shown in Figure 2 for rainbow trout and Figure 3 for channel catfish. The mean and standard deviation of the total body clearances, apparent volumes of distribution, mean residence times, and half-lives determined on individual fish by noncompartmental analyses are shown in Table 2.

5. DISCUSSION

The HPLC analytical method for sarafloxacin in fish plasma was simple and sensitive, with relatively high recoveries and a low variation between repetitions. Because

Table 1. Accuracy and precision of the HPLC determination of sarafloxacin in channel catfish and rainbow trout plasma. Recovery was based on the amount determined from the standard curve and the nominal amount of the fortification.

Nominal spike concentration (ng/g)	Recovery %	RSD %
Rainbow trout		
10	116	6.4
100	90.6	1.4
1,000	99.4	4.3
5,000	91.1	5.9
Channel catfish		
10	92.3	7.2
100	97.3	3.6
1,000	95.6	2.0
5,000	94.5	4.2

Table 2. Mean values for pharmacokinetic parameters (standard deviation) determined for individual rainbow trout (n = 8) and channel catfish (n = 10).

	Total body clearance (L/h/kg)	Apparent volume of distribution (L/kg)	MRT (h)	Half-life (h)
Rainbow trout	0.046 (0.014)	9.67 (2.98)	221.5 (90.81)	158.0 (69.42)
Channel catfish	0.171 (0.019)	3.95 (0.42)	23.34 (3.01)	17.90 (2.27)

sarafloxacin in the extracts from fish plasma was stable in sealed vials at room temperature for at least 8 d, processing was completed immediately and HPLC analysis initiated after all samples were collected for an individual fish.

Rainbow trout and channel catfish, two species proposed as surrogates for the crop grouping concept, differ markedly in the pharmacokinetics of sarafloxacin. In general

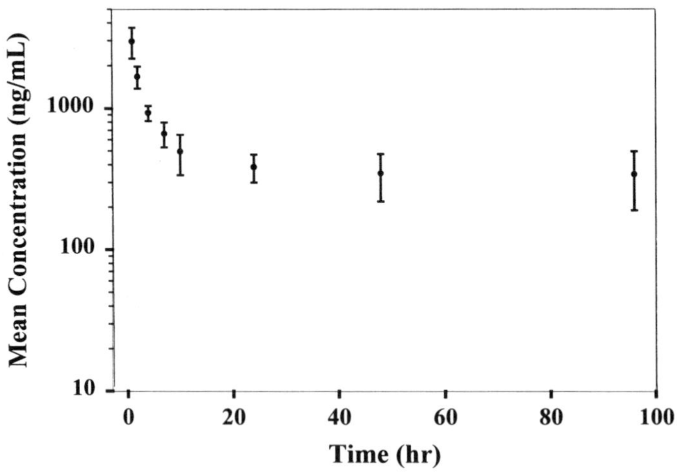

Figure 2. Mean concentrations of sarafloxacin in plasma vs time plot for rainbow trout after an intraarterial dose at 5 mg/kg. Error bars are the standard deviation of eight fish.

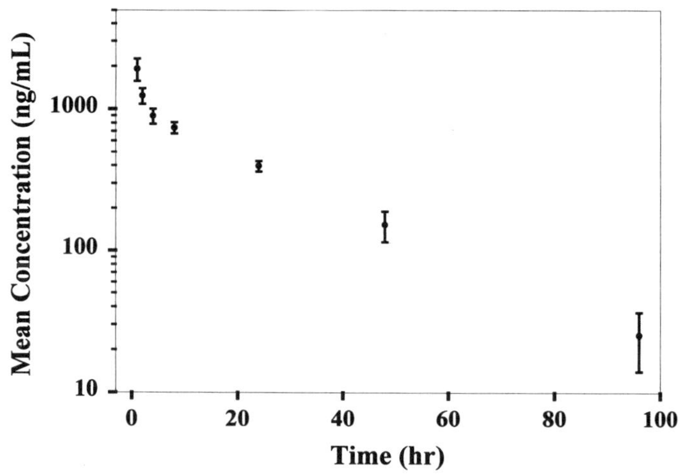

Figure 3. Mean concentrations of sarafloxacin in plasma vs time plot for channel catfish after an intraarterial dose at 5 mg/kg. Error bars are the standard deviation of ten fish.

the catfish had a distributive phase that was significantly diminished in magnitude when compared to the rainbow trout and the terminal elimination phase for the catfish was much steeper than the rainbow trout. The mean value for the apparent volume of distribution for the trout was approximately 2.4X the value for catfish, while the total body clearance in the trout was 27% of the value in the catfish. The mean residence time and terminal half-life of sarafloxacin were much greater in the trout when compared to the catfish.

Several studies have been conducted on the pharmacokinetics of sarafloxacin in Atlantic salmon (Martinsen *et al.*, 1993a, Martinsen *et al.*, 1993b, Martinsen *et al.*, 1994, Martinsen and Horsberg 1995). The results of Martinsen's studies indicated a wide range of values for pharmacokinetic parameters, notably the apparent volume of distribution and the terminal half-life. Martinsen indicated that cannulation of fish appeared to increase the apparent volume of distribution and increase the terminal half-life, although size of the fish and acclimation temperature may have had some effect in these studies (Marinsen *et al.*, 1993b). In cannulated Atlantic salmon acclimated at 12 °C (932 g), the following pharmacokinetic parameters were estimated: total body clearance, 0.22 L/kg/h; apparent volume of distribution, 13.6 L/kg; and terminal half-life, 45.3 h. In comparison to the present study, rainbow trout have a smaller total body clearance and an extended half-life, but a similar apparent volume of distribution. In comparison, catfish have similar values for total body clearance, apparent volume of distribution, and half-life compared to the Atlantic salmon. In other words, rainbow trout and Atlantic salmon differ mainly in the elimination of sarafloxacin while catfish differ mainly in its capacity to accumulate sarafloxacin. Atlantic salmon most likely would be represented in the same crop group as rainbow trout.

Because our experiments were conducted with fish of similar size and at acclimation temperatures slightly below the preferred temperature of each species, the differences in pharmacokinetics of sarafloxacin were thought to be due to differences in the defining characteristics of these surrogate species (i.e., effect of phylogeny, activity level and temperature preference on anatomy, physiology and physiochemical interaction of sarafloxacin with tissues). Although values for these pharmacokinetic parameters would be expected to vary with a number of internal and external variables, extrapolation of the pharmacokinetics of sarafloxacin conducted on the trout and catfish, based on plasma pharmacokinetic studies, is limited. Incorporation of the information from these studies with data required for the development of physiologically based pharmacokinetic models would allow a biological basis for the extrapolation of the pharmacokinetics of sarafloxacin for different conditions (e.g., temperature, dose, size) and allow an analysis of the factors that have the most affect on the pharmacokinetics of sarafloxacin. Therefore, although differences were observed in the pharmacokinetics of sarafloxacin in these two surrogate species, the present studies do not prove or disprove, the crop grouping concept.

In summary, channel catfish and rainbow trout, two proposed surrogate species for the crop grouping concept, differ greatly in the pharmacokinetics of sarafloxacin; the apparent volume of distribution and terminal half-life are greater and total body clearance less in rainbow trout when compared to channel catfish.

6. ACKNOWLEDGMENTS

We wish to acknowledge J.M. Anderson, M.J. Kinsella, C.T. Myers, P.A. Rudd and G.A. Steinke for technical assistance and the suggestions of the anonymous reviewers.

7. REFERENCES

Barron, M.G., Stehly, G.R., and Hayton, W.L., 1990, Pharmacokinetic modeling in aquatic animals. I. Models and concepts. *Aquat. Toxicol.* 17:187–212.

Beyenbach, K.W., and Kirschner, L.B., 1975, Kidney and urinary bladder function of the rainbow trout in Mg^{++} and Na^{+} excretion. *Am. J. Physiol.* 229:89–393.

Geyer, R.E., 1993, FDA enforcement policy for aquaculture. *Salmonid* 17(1):4–10.

Gingerich, W.H., Stehly, G.R., Clark, K.J., and Hayton, W.L., 1998. Crop grouping: A proposal for public aquaculture. Vet. Human Toxicol. 40:24–31.

Haskell, D.C., 1959, Trout growth in hatcheries. *N. Y. Fish Game J.* 6:204–236.

Huner, J.V., and Dupree, H.K., 1984. Chapter 4. Methods and economics of channel catfish production, and techniques for the culture of flathead catfish and other catfishes: *Report to the Fish Farmers (3rd): Status of Warmwater Fish Farming and Progress in Fish Farming Research.*, U.S. Department of Interior, Fish and Wildlife Service, Washington DC, pgs 44–82.

Keith, L.H., Crummett, W., Degan Jr., J., Libby, R.A., Taylor, J.K., and Wentler, G., 1983, Principals of environmental analysis. *Anal. Chem.* 55:2210–2218.

Lech, J.J., and Bend, J.R., 1980, Relationship between biotransformation and the toxicity and fate of xenobiotic chemicals in fish. *Environm. Health. Perspec.* 34:115–131.

Martinsen, B., Horsberg, T.E., Sohlberg, S., and Burke, M., 1993a, Single dose kinetic study of sarafloxacin after intravenous and oral administration of different formulations to Atlantic salmon (*Salmo salar*) held in sea water at 8.5 °C. *Aquaculture* 118:37–47.

Martinsen, B., Sohlberg, S., Horsberg, T.E., and Burke, M., 1993b, Single dose kinetic study of sarafloxacin after intravascular and oral administration to cannulated Atlantic salmon (*Salmo salar*) held in seawater at 12 °C. *Aquaculture* 118:49–52.

Martinsen, B., Sohlberg, S., Horsberg, T.E., and Burke, M., 1994, Multiple-dose pharmacokinetic and depletion studies of sarafloxacin in Atlantic salmon, *Salmo salar* L. *J. Fish Dis.* 17:111–121.

Martinsen, B., and Horsberg, T.E., 1995, Comparative single-dose pharmacokinetics of four quinolones, oxolinic acid, flumequine, sarafloxacin, and enrofloxacin, in the Atlantic salmon (*Salmo salar*) held in seawater at 10 °C. *Antimicrob. Agents Chemother.* 39:1059–1064.

Schnick, R.A., Gingerich, W.H., Dawson, V.K., Stehly, G.R., and Griffin, B., unpublished, "Approval of drugs for public fish production" Proposal submitted to the International Association of Fish and Wildlife Agencies by the National Fisheries Research Center-La Crosse, La Crosse, WI and Fish Farming Experimental Laboratory, Stuttgart, AR on March 4, 1994.

Sorgel, F., Jaehde, U., Naber, K., and Stephan, U., 1989, Pharmacokinetic disposition of quinolones in human fluids and tissues. *Clin. Pharmacokinet.* 16(suppl. 1):5–24.

Stehly, G.R., and Plakas, S.M., 1993, Pharmacokinetics, tissue distribution, and metabolism of nitrofurantoin in the channel catfish *(Ictalurus punctatus)*. *Aquaculture* 113:1–10.

10

PREDICTING THE TOXICOKINETICS OF TRIFLURALIN IN RAINBOW TROUT USING CLEARANCE-VOLUME PHARMACOKINETIC MODELS

I. R. Schultz[1] and W. L. Hayton[2]

[1]Battelle PNNL
Molecular Biosciences Division
P.O. Box 999-P7-56
Richland, WA 99352
[2]College of Pharmacy
Ohio State University
500 W. 12th, Columbus, OH 43210

1. ABSTRACT

Trifluralin (TF) is a lipophilic, pre-emergent herbicide widely used in agriculture and known to bioconcentrate in fish. We have characterized the accumulation of TF in rainbow trout under a variety of experimental conditions. Our approach has been to use static water exposure systems and intravascular dosing in combination with clearance-volume pharmacokinetic (CV-PK) models to obtain quantitative estimates of uptake clearance, apparent volume of distribution and elimination due to xenobiotic metabolism. This paper will briefly discuss pertinent physicochemical data for TF and review the toxicokinetics of TF in rainbow trout. Emphasis will be placed on physiological interpretations of TF model parameters and practical aspects of modeling TF toxicokinetics with CV-PK models.

2. INTRODUCTION

Trifluralin (TF) is a dinitroaniline type, pre-emergent herbicide that has been widely used since 1961 to control grass and broadleaf weeds. Additional applications for TF have included control of larval mycosis in penaied shrimp culture (Williams *et al.*, 1986) and human parasitic infections caused by trypanosomatid protozoans (Chan *et al.*, 1993).

Table 1. Trifluralin Physicochemical and Toxicological Values.

Parameter	Value	Ref
Water Solubility (25°C)	0.3–0.6 mg/L	a, b
Log K_{ow}	5.0–5.3	a, c–d
Bioconcentration Factor		
Fathead minnow	1,060	e
Guppy	3,250–5,080	f
Sauger	5,800	g
Golden Redhorse	2,800	g
Acute Toxicity-Fish (96 hr LC_{50})		
Trout (12°C)	0.041–0.042 mg/L	h, i
Fathead minnow (18°C)	0.105 mg/L	h
Bluegill Sunfish (24°C)	0.047 mg/L	j
Acute Toxicity-Mammals (LD_{50})		
Wistar Rat	1,930–2,270 mg/kg	k
Beagle Dog	>10,000 mg/kg	k
No-Observed-Effect Level—Rat	41 mg/kg/day	k

a) Briggs, 1981; b) Wauchope, *et al.*, 1992; c) Ali, 1984; d) Brown and Flagg, 1981; e) Macek, *et al.*, 1976; f) Yockim, *et al.*, 1980; g) Spacie and Hamelink, 1979; h) Johnson and Finley, 1980; i) Macek, *et al.*, 1969; j) Cope, 1965; k) Ebert, *et al.*, 1992.

Recent annual agriculture use of TF exceeded 12,000 metric tons in the U.S.A. (Grover *et al.*, 1997).

The importance and widespread use of TF in agriculture has spurred a large number of studies examining the physicochemical and toxicological properties of TF. Important observations from these studies are that TF is lipophilic, sparingly soluble in water and has very low or negligible acute and chronic toxicity in mammals (Ebert *et al.*, 1992). In contrast to the high tolerance of TF by mammals, TF has been demonstrated to have high acute toxicity to some aquatic organisms including rainbow trout (*Onchorynchus mykiss*), and to cause vertebral deformities in fish during chronic exposures (Couch *et al.*, 1979). Trifluralin is also known to extensively accumulate in fish with field and laboratory studies indicating the bioconcentration factor (BCF) for TF in fish ranges between 1000–5800 (Macek *et al.*, 1976; Spacie and Hamelink, 1979). A summary of pertinent physicochemical and toxicological values for TF are shown in Table 1.

Although TF accumulates in fish, extensive biotransformations can occur and a large fraction (>30%) of an administered dose may be converted to metabolites in fish and mammals (Emerson and Anderson, 1966; Reinbold and Metcalf, 1976). The metabolism of TF appears to be complex, and a variety of specific metabolites have been identified in fish and rodents (Erkog and Menzer, 1985; Nelson *et al.*, 1977; Reinbold and Metcalf, 1976). Common metabolites formed by fish and rodents are summarized in Figure 1. The primary initial metabolic transformation of TF appears to be oxidation of one N-propyl side group and subsequent N-dealkylation. After N-dealkylation, several transformations can occur, involving further dealkylation, nitroreduction and conjugation with amino acids (Erkog and Menzer, 1985; Figure 1).

The combination of environmental persistence and complex metabolism makes TF a useful xenobiotic for studying the influence of environmental and physiological factors on the accumulation of pesticides in fish. In this paper, we review our studies of TF toxicokinetics in rainbow trout, with emphasis on the role of body size and lipid content as determinants of the extent of accumulation. Also, we present additional data to

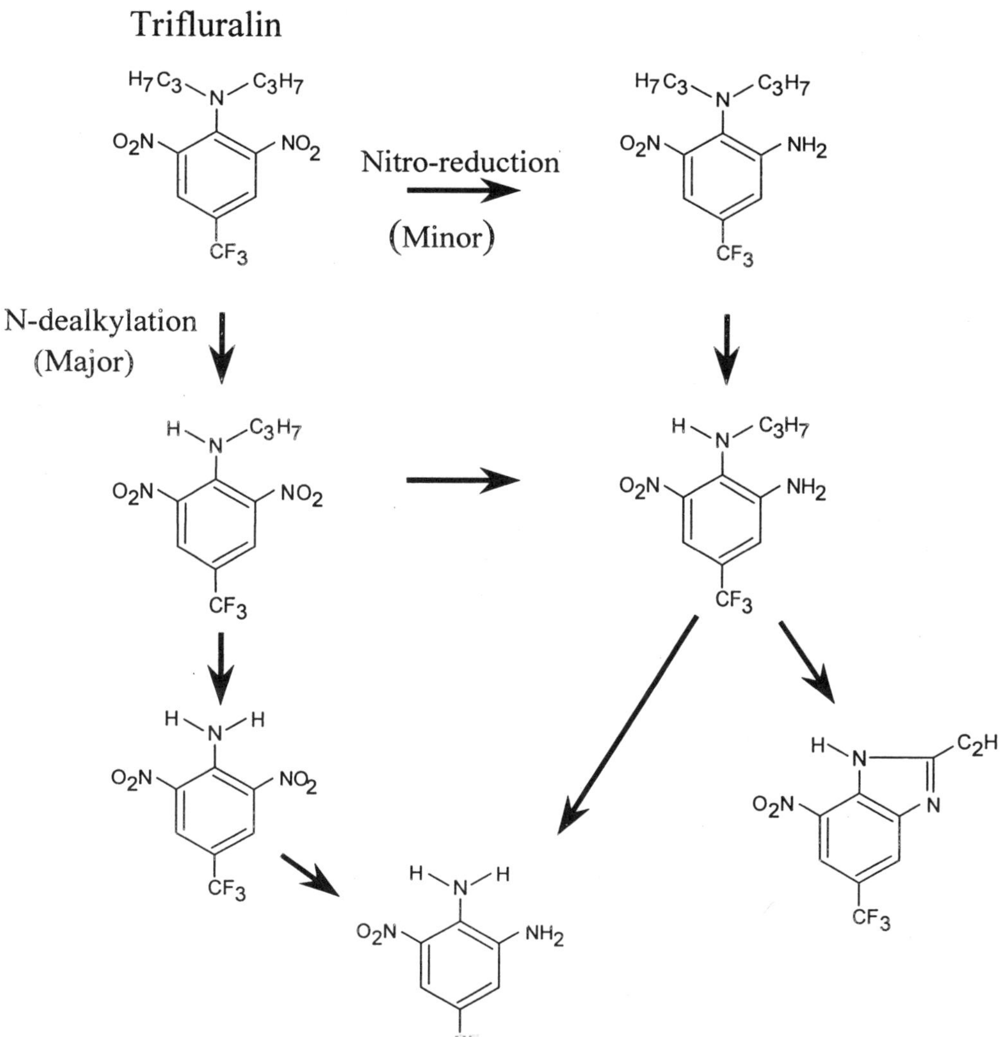

Figure 1. Metabolic pathways of trifluralin. The predominant initial biotransformation is N-dealkylation. Adapted from Reinbold and Metcalf (1976) and Erkog and Menzer (1985).

demonstrate the applicability of predicting TF accumulation using simple approaches to estimating TF pharmacokinetic model parameters in trout.

3. EXPERIMENTAL DESIGN: DATA COLLECTION

We used a combination of intravascular administration and static water exposures to TF to collect data sets which would allow quantitative interpretations of absorption, distribution and biotransformation of TF. The approach used for intravascular injection of TF in free swimming trout is similar to that described in previous papers in this series (Chapters 5, 6), with serial sampling of blood plasma continuing until 300–350 hrs. Static

Table 2. Water quality conditions for rainbow trout (5.1 g; $n = 2$) maintained in four liters of water.

Parameter	0 hr	96 hr
pH	7.0	6.9
Ammonia (mg/L)	<0.04	0.65
Nitrate (mg/L)	1.50	1.4
Alkalinity (mg/L)	25.8	32.0
O_2 (% saturation)	100	95
Phosphorus (mg/L)	0.31	0.6
Tot. Org. Car. (mg/L)	2.84	5.30

Note: Oxygen was slowly bubbled (0.5 ml/min) into the beakers.

water exposures relied on the use of ^{14}C-labeled TF and serial sampling of individually exposed trout. At selected time points, trout (typically $n = 4$/time point), were euthanized and the exposure water and fish carcass analyzed for TF content. Blood plasma may also be collected and analyzed for TF. The total amount of metabolites formed during the exposure was calculated from the difference between the total quantity of ^{14}C in the exposure water and fish carcass and that comprised by TF. This approach to water exposures offered certain advantages with TF because the exposure water concentration quickly declined to meet the rising whole animal concentration, which permitted characterization of the pharmacokinetics over a much shorter time period compared with continuous water flow exposures. The use of ^{14}C-labeled TF permitted determination of mass balance and allowed accurate characterization of the amount of metabolites formed, which would otherwise be difficult, given the complexity of TF metabolic pathways. The main disadvantages of this exposure system were related to the potential loss of chemical from the exposure system due to degradation, adsorption or volatilization and the possible decline of water quality to a degree which compromised the health of the fish. Loss of chemical can be accommodated by performing careful preliminary experiments without fish (Schultz and Hayton, 1993). Also, we have found that trout can be successfully maintained for 96 hrs in static systems if small amounts of oxygen (e.g., <1 ml/min) are continuously bubbled into the exposure tank using micro gas dispersion tubes and by adjustment of the loading density of fish for the particular size of trout in use. An example of the water quality conditions before and after a 96 hr static exposure is shown in Table 2 for a five g trout maintained in a four liter beaker at 12°C.

We modified our static water exposure protocol for use with larger sized trout which had been fitted with a dorsal aortic cannula. When using cannulated trout, the exposure water and blood plasma are serially sampled from individual fish. A disadvantage of this protocol is that without measurement of whole body levels of TF, characterization of distribution and biotransformation is not possible and this limits the kinetic analysis to characterization of uptake (Schultz and Hayton, 1994). The experimental designs utilized for water exposures of trout to TF are summarized in Figure 2 a,b.

4. PHARMACOKINETIC ANALYSIS: CLEARANCE-VOLUME MODELS

The TF and metabolite concentration-time data collected during water exposures were analyzed using a 2-compartment clearance-volume pharmacokinetic (CV-PK)

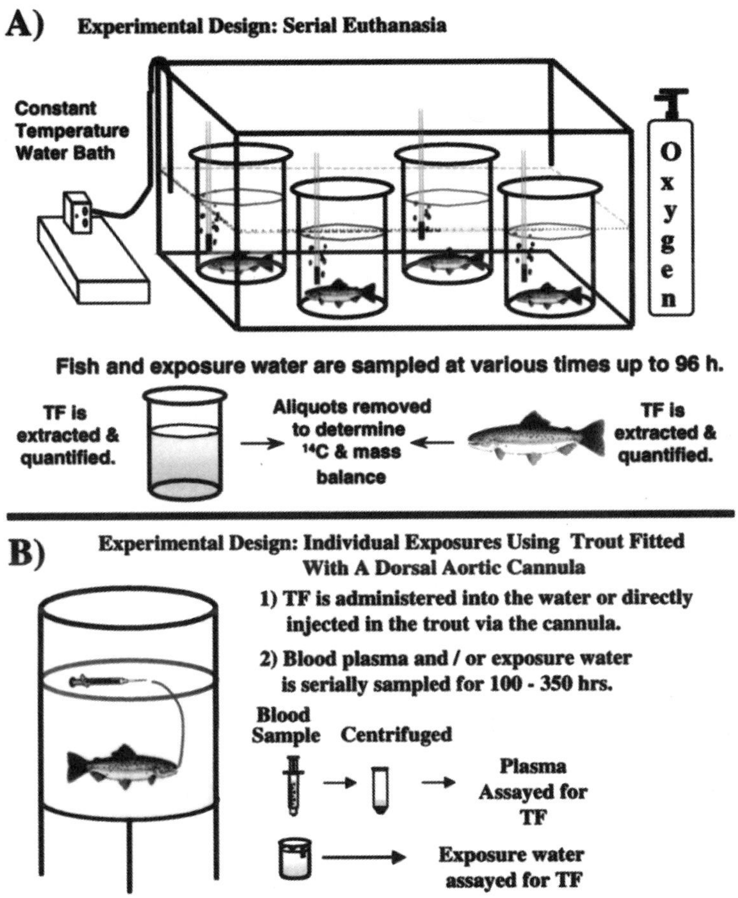

Figure 2. Diagrammatic representations of the experimental protocols for water exposure of TF to rainbow trout. A) Serial euthanasia of smaller sized trout (<100 g). Trout were placed in a solution of [^{14}C-ring labeled]TF (2 ng/ml) and sampled at selected times up to 96 h. After the exposure, the concentration of unchanged TF in the water and fish carcass was determined and the quantity of TF metabolites formed calculated as the difference between the amount of total radioactivity and the amount of unchanged TF in the fish and exposure water. B) Repetitive sampling of larger trout fitted with a dorsal aortic cannula.

model similar to the models discussed in Chapters 5 and 6. This model was modified to include parameters for cardiac output and a plasma : water distribution coefficient, which allowed modeling of the arterial plasma concentration (Schultz and Hayton, 1993). The observed concentration-time profiles for TF and metabolites were then simultaneously fit to the model based equations to obtain estimates of the kinetic parameters. An overview of the steps involved in obtaining pharmacokinetic parameters for TF is depicted in Figure 3.

The plasma concentration-time profiles measured after intravascular administration of TF appeared triexponential and a three compartment CV-PK model was used to analyze the data. This model was discussed by Stehly *et al.* (Chapter 6), and was used to obtain estimates of the apparent volume of distribution at steady-state (V_{ss}), total body clearance (Cl_b) and elimination half-life ($t_{1/2}$). Throughout the remainder of this paper, the V_{ss} will be used in conjunction with extravascular distribution because >90% of the

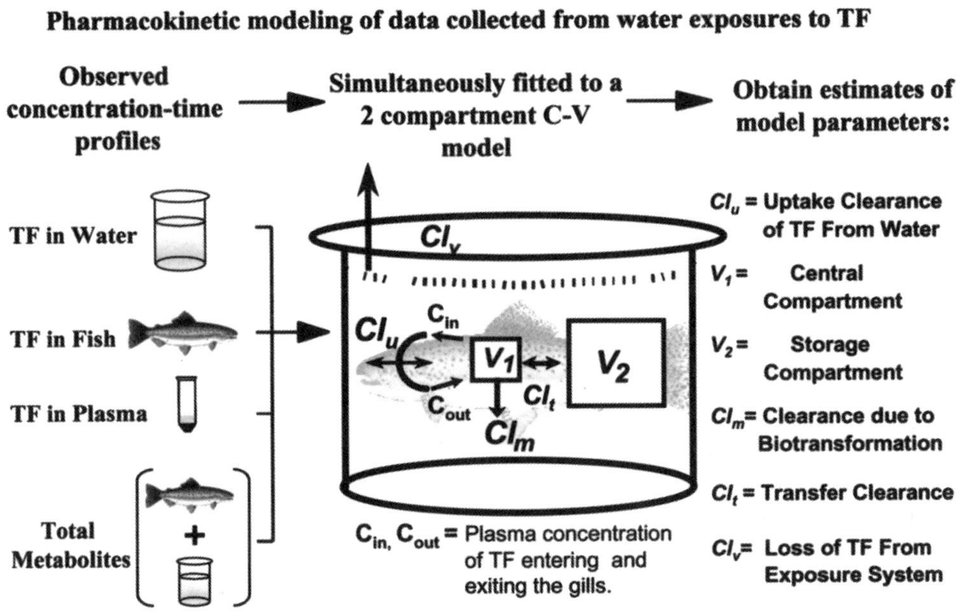

Figure 3. Overview of steps involved in determining TF toxicokinetic parameters after water exposures. A schematic representation of the clearance-volume (C-V) pharmacokinetic model is shown. See Schultz and Hayton (1993) for a mathematical description of the model.

V_{ss} for TF in trout is comprised by a peripheral storage compartment (Schultz and Hayton, 1993; 1997).

5. EXPERIMENTAL RESULTS: WATER EXPOSURES

Our purpose in experiments using water exposures of TF was to characterize the accumulation kinetics in a wide range of trout sizes under consistent environmental conditions. This was achieved by exposing five different body size classes of trout, acclimated to 12 °C, to TF. The three smallest size classes (0.18 to 83.6 g) were exposed to TF using the serial euthanasia design and subsequent estimation of kinetic parameters based on whole body values of TF. Larger trout (92.0 to 3395 g) were fitted with an aortic cannula and plasma and exposure water repetitively sampled. The 83.6 and 92.0 g size groups provided an overlap in body sizes between experimental designs.

The observed and model predicted concentration-time profiles for TF in the exposure water, whole fish, and total metabolites for the 0.18, 4.07, and 83.6 g body weight size classes are shown in Figure 4. Absorption of TF from water was rapid as evidenced by both the rapid decline in water concentrations and increases in body levels during the initial 24 h of exposure (Figure 4, top row). The metabolite-profile appeared linear until approximately 75% of the dose had been absorbed which occurred between 24 and 48 hr. The most apparent differences in TF kinetics among the size classes was the extent of metabolism and the pattern of TF accumulation in the body (Figure 4, middle and bottom rows). The two smallest trout sizes (0.18 and 4.07 g) converted a greater fraction of the dose to metabolites and the whole body levels of TF peaked earlier (~50 hr) compared to the 83.6 g trout.

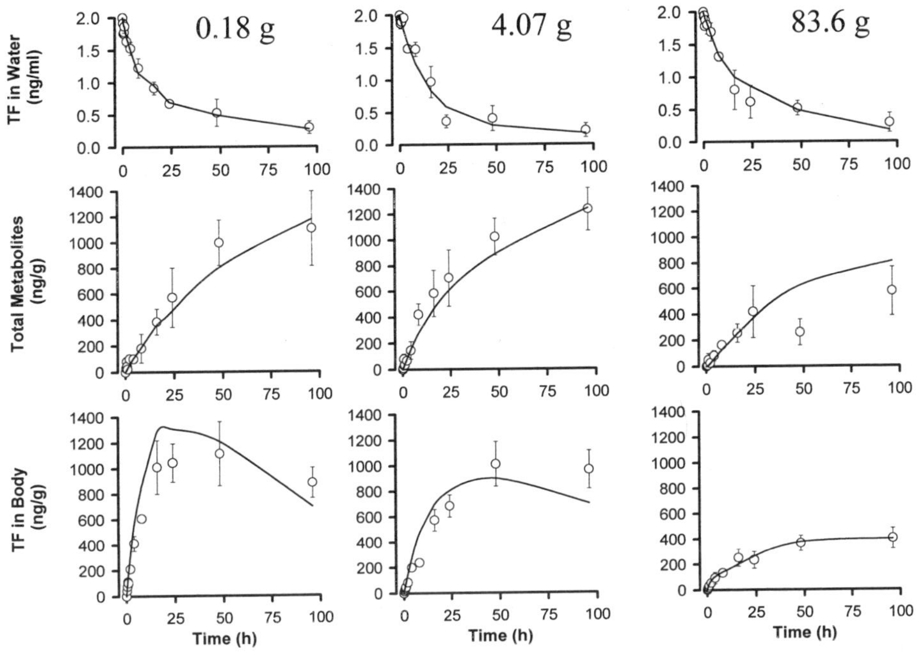

Figure 4. Trifluralin kinetic profiles for the exposure water concentration, total metabolites formed and the whole body TF concentration. The initial exposure water concentration of TF was 2 ng/ml. Rainbow trout with average body weights of 0.18, 4.07 and 83.6 g are shown. Symbols represent the mean ± SD from a minimum of four fish and the solid lines show the model predicted values. Data is from Schultz and Hayton (1993, 1994).

Results from water exposures of cannulated trout to TF indicated that plasma concentrations rapidly increased during the initial 12 hr of exposure and were up to 1 to 2 orders of magnitude greater than exposure water concentrations of TF. A representative graph of the plasma and water concentrations of TF from a single 3,395 g trout is shown in Figure 5. The large disparity between plasma and water concentrations of TF suggested TF was binding to plasma proteins, which increased the solubility of TF in plasma. To investigate this in more detail, we measured the plasma binding of TF using blood plasma obtained from donor trout of various body sizes. Plasma binding was determined using equilibrium dialysis, a technique which allows separation of bound and unbound fractions of low molecular weight chemicals across a semi-permeable membrane. Results of these experiments, shown in Figure 6, demonstrated that >99% of TF in plasma was bound to macromolecules. Trifluralin binding appeared to be increased in plasma obtained from smaller sized trout (Figure 6). These observations suggest that uptake of TF at the gill can be characterized as water flow-limited (Hayton and Barron, 1987).

The similarity between the observed and model predicted values shown in Figure 4 indicated that the two compartment model adequately characterized the accumulation kinetics of TF in trout. The parameter estimates obtained from the non-linear least squares fitting of model based equations to the observed data are shown in Tables 3 and 4. Comparison of the parameters obtained from the different size classes for which whole body data was available, indicated that the storage capacity or steady-state apparent volume of distribution (V_{ss}) for TF increased with body size while metabolic clearance (Cl_m), did not appear to vary systematically among the size classes. However, this result

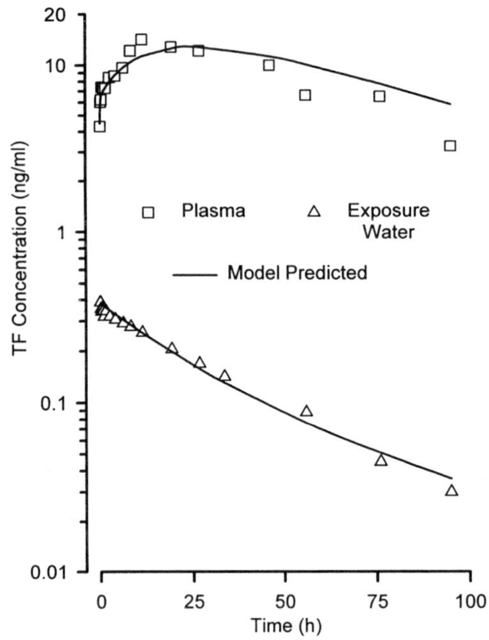

Figure 5. Trifluralin concentration-time profiles determined from a 3,395 g trout fitted with a intra-arterial cannula and exposed to TF at an initial concentration of 0.6 ng/ml. Data is from Schultz and Hayton (1994).

indicates metabolic clearance assumes greater importance as an elimination pathway for TF in larger trout. This can be best appreciated by comparing the uptake:metabolic clearance ratios from Tables 3 and 4. This ratio is 13.1 for 0.18 g trout and decreases to 2.4 in the 83.6 g trout.

Uptake clearance values reported in Table 4 were quite large and corresponded to the size of the ventilation volume in trout (Boddington *et al.*, 1979; Mckim *et al.*, 1986). This observation was a further indication that the uptake of TF was controlled by the flow of water across the gills. An allometric analysis of the Cl_u values (Figure 7) showed that uptake of TF scaled to a body weight exponent of 0.66. The allometry of ventilation volume and oxygen consumption in fish is known to scale to comparable values (Yamamoto 1991, 1992), which suggests that changes in the uptake of TF due to body size influences can be predicted from differences in ventilation volume or perhaps, resting oxygen consumption rates.

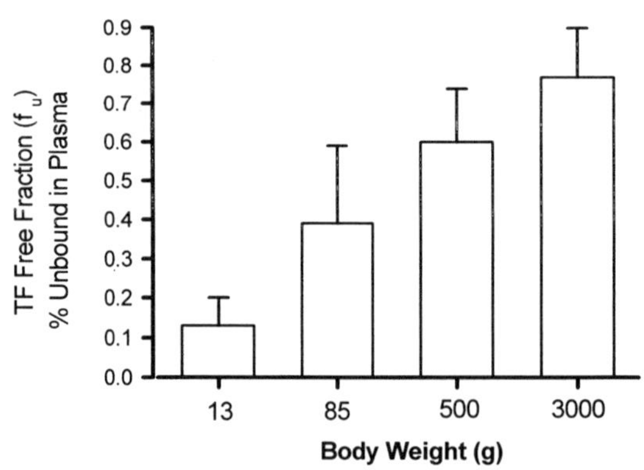

Figure 6. Binding of TF to blood plasma. Values shown are the mean ± SD of three determinations made using plasma obtained from trout whose average body weight was 13 g (samples pooled from $n = 20$ trout), 85 g (samples pooled from $n = 5$ trout), 400 g ($n = 4$) and 3,000 g ($n = 2$). Data from Schultz and Hayton (1993, 1994).

Table 3. Model parameters describing TF clearance due to metabolism and the apparent volume of distribution at steady-state in trout of different body sizes (values are from Schultz and Hayton, 1994).

	Model Parameter	
Fish weight (g)	Metabolism (Cl_m) $ml\,hr^{-1}\,g^{-1}$	Distribution (V_{ss}) $ml\,g^{-1}$
0.18 ± 0.03	22.2 ± 6.0	2,067 ± 716
4.07 ± 0.73	36.0 ± 14.0	2,564 ± 984
83.6 ± 19.6	14.0 ± 2.3	3,240 ± 2,036

Table 4. Uptake clearance of TF in trout of different body sizes. (values are from Schultz and Hayton, 1994).

Fish weight (g)	Uptake (Cl_u) $ml\,hr^{-1}\,g^{-1}$
0.18 ± 0.03	290 ± 211
4.07 ± 0.73	145 ± 89
83.6 ± 19.6	33.3 ± 3.9
92 ± 34	42.0 ± 8.8
451 ± 64	25.0 ± 8.8
3,395 ± 707	10.8 ± 1.7

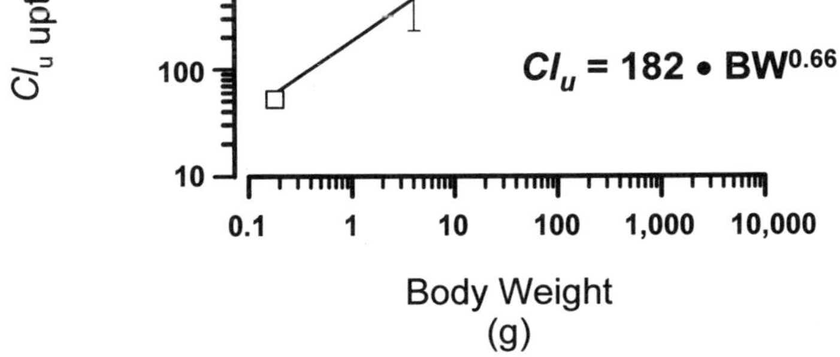

Figure 7. Uptake clearance of TF as a function of body weight in trout. The allometric equation describing uptake is shown on the figure. Data from Schultz and Hayton (1994).

6. EXPERIMENTAL RESULTS: INTRAVASCULAR ADMINISTRATION

Our objective in experiments studying the elimination of TF in trout after intravascular administration was to asses the influence of body fat on the extent of extravascular distribution and elimination of TF (Schultz and Hayton, 1997). In these studies, we used large, hatchery reared rainbow trout and adult summer-run steelhead trout. Steelhead are an anadromous form of rainbow trout that spend up to 3 years in saltwater before returning to freshwater to spawn (Leider *et al.*, 1986). Upon re-entering freshwater steelhead do not feed, which eventually causes exceptionally low lipid levels in non-reproductive tissues. This causes large, effective differences between steelhead and hatchery trout in lipid content of storage tissues and permitted an analysis of the influence of body fat on the peripheral distribution of TF.

We injected TF into free swimming trout via an intraarterial cannula and determined the plasma concentration-time profiles of TF. In both groups, the plasma concentration of TF declined in a tri-exponential manner (Figure 8). Pharmacokinetic analysis of the profiles indicated the V_{ss} was smaller in the steelhead while Cl_b (which equals the sum of all elimination pathways) was similar (Table 5). These results indicated that the primary difference in the pharmacokinetics of TF between hatchery trout and steelhead was in the extent of the peripheral distribution. The elimination half-life ($t_{1/2\beta}$) was also significantly shorter in steelhead, which was attributed to the reduced distribution space that TF must be cleared from. The smaller V_{ss} in the steelhead as compared with the hatchery trout, appeared to be due to the lower lipid content of the steelhead (Table 5). A plot of V_{ss} versus lipid content suggested that the body lipid content influenced the degree of extravascular distribution of TF in trout (Figure 9).

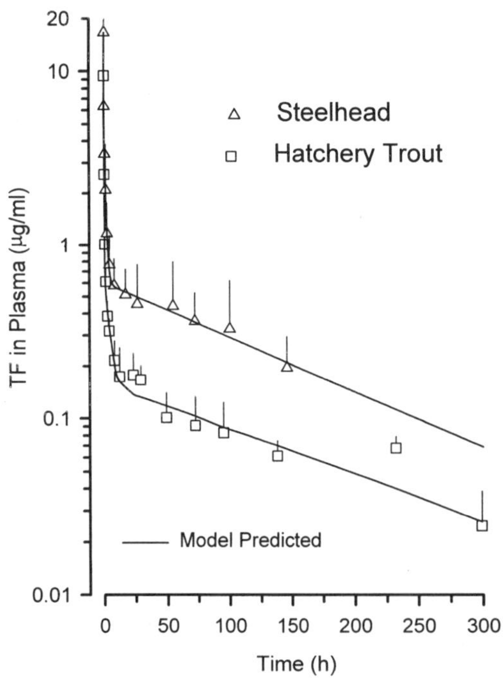

Figure 8. Plasma concentration-time profiles after intraarterial injection of 1 mg/kg TF to hatchery and steelhead trout fitted with a dorsal aortic cannula. Symbols represent experimentally determined values (mean ± SD, n = 5–7 hatchery trout; n = 5–6 steelhead) and the line is the least squares fit to the observed data using a three compartment clearance-volume model. Error bars not shown fit within the data point. Data is from Schultz and Hayton (1997).

Table 5. Selected model parameters and lipid content of steelhead and hatchery trout that received a 1.0 mg/kg intra-arterial dose of TF (values are from Schultz and Hayton, 1997). Mean values (± SD) are shown.

Item	Steelhead[a]	Hatchery[b]
Sample size	7	6
V_{ss} (ml g^{-1})	1.33 ± 0.5[c]	5.73 ± 1.7
Cl_b (ml h^{-1} g^{-1})	0.024 ± 0.01	0.045 ± 0.025
$t_{1/2}$ (h)	61.5 ± 22.3[d]	140 ± 50
Lipid (% BW)	1.2 ± 0.8[c]	8.0 ± 1.2

Asterisks indicate values significantly different from Hatchery trout: $^*p < 0.05$; $^{**}p < 0.001$.

[a]Mean body weight ± SD = 3123 ± 625 g.

[b]Mean body weight (± SD) = 3392 ± 557 g.

[c]Significantly different from Hatchery trout (p < 0.001).

[d]Significantly different from Hatchery trout (p < 0.05).

Although steelhead and hatchery trout are taxonomically identical, there were undoubtedly a number of physiological differences between them, some of which may contribute to the differences in the V_{ss} of TF. This is clear in Figure 9, where it is apparent that within either group the correlation was weak between the lipid content of the individual fish and the corresponding values for V_{ss}. Because TF is lipophilic and the one clear physiological difference between steelhead and the hatchery fish was body fat, we examined how the affinities of TF for adipose and lean tissues would have to vary to account for the differences in V_{ss} values between the two groups. We performed a linear regression of the V_{ss} versus fractional body lipid content values shown in Figure 9 and derived an equation to describe the contribution of lean (V_L) and adipose (V_A) tissue to the V_{ss} (Figure 9). This analysis indicated that the distribution of TF into lean tissues

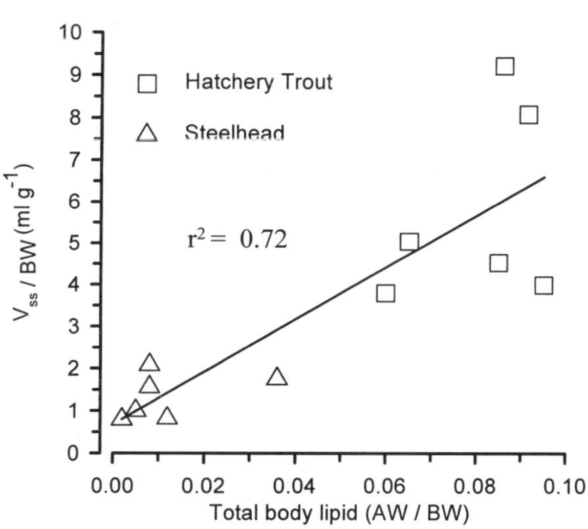

V_A, V_L = apparent volume of adipose and lean tissues.

AW, BW = adipose and body weight.

$$\frac{Vss}{BW} = VL + (VA - VL) \bullet \frac{Aw}{BW}$$

$$V_L = 0.68 \text{ ml g}^{-1}$$

$$V_A = 64 \text{ ml g}^{-1}$$

$$\frac{VA}{VL} = 94$$

Figure 9. The relationship between the apparent volume of distribution at steady-state (V_{ss}) and total body lipid for steelhead and hatchery trout. Shown on the right is calculation of the apparent volumes of lean and adipose tissues based on linear regression analysis of the observed data. Data is from Schultz and Hayton (1997).

approximated the total body water volume and was consistent with a lack of intracellular accumulation of TF in lean tissue; i.e., TF binding in lean tissue was similar to that in plasma. In contrast, TF was predicted to concentrate in body fat and the ratio of TF distribution between adipose and lean tissues was 94, which represents the relative affinity of TF for the lipid compared with the lean portion of the fish (i.e. fat : lean partition coefficient).

7. PREDICTION OF TF BIOCONCENTRATION

An important aspect in aquatic risk assessments of pesticides is exposure determination and estimation of internal dose. Pharmacokinetic models provide a means to extrapolate dose estimates across changing body sizes, fish species and environmental conditions through the appropriate adjustment of model parameters. Procedures for adjusting pharmacokinetic parameters which rely upon existing physiological data in addition to simple, *in-vitro* methods would be particularly valuable. We have recently tested experimental approaches for extrapolating the uptake, apparent distribution and biotransformation of TF across multiple fish species (Schultz and Hayton, 1999). Our experimental work with TF in trout provided a data base to test these procedures for the prediction of body size influences on the bioconcentration of TF.

The BCF can be estimated from model parameters using the following equation $Cl_u \cdot V_{ss}/(Cl_u + Cl_m)$. Therefore, scaling relationships developed for these pharmacokinetic parameters would be the most useful. The equation shown in Figure 9 provides a means

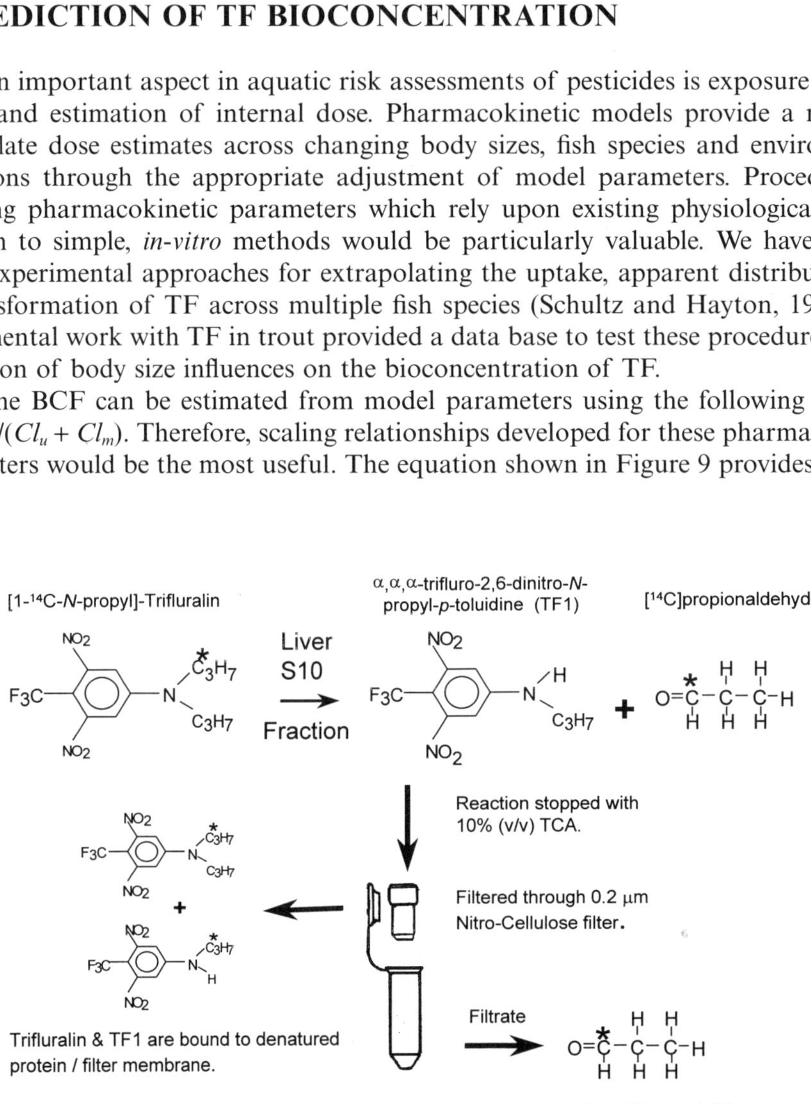

Figure 10. The determination of TF metabolism in liver homogenates. The N-depropylation of [1-^{14}C-*N*-propyl]trifluralin produced [1-^{14}C]propionaldehyde and α,α,α-trifluro-2,6-dinitro-*N*-[1-^{14}C]propyl-*p*-toluidine (TF1). Separation of [1-^{14}C]propionaldehyde from TF and TF1 was achieved by stopping the reaction with trichloroacetic acid and filtering the denatured protein mixture through a 0.22-μ cellulose acetate filter. TF and TF1 become bound to the denatured protein/filter complex while the [1-^{14}C]propionaldehyde was collected in the filtrate.

to predict the V_{ss} of TF based on lipid content. We now focus on physiological attributes which might be determinants of uptake and biotransformation. Because the uptake of TF is controlled by the ventilation volume, changes in this physiological parameter should predict changes in uptake clearance with body size. However, measurement of ventilation volume is impractical in small fish, and we chose to use the resting routine oxygen consumption as a surrogate. To estimate metabolic clearance, we developed an *in-vitro* method for measuring the N-depropylase activity of TF in liver homogenates (Schultz and Hayton, 1993). The biochemical basis of the N-depropylase assay is summarized in Figure 10. This activity was used to characterize the initial rate of TF metabolism, which allowed calculation of intrinsic metabolic clearance from the ratio of V_{max}/K_m (Rane *et al.*, 1977). We measured the initial rate of TF metabolism using starting concentrations of TF ranging from 0.8 to 16 μM. We then calculated the V_{max} and K_m from a non-linear, least squares fit of the Michaelis-Menten equation to observed values for TF metabolism in liver S10. The intrinsic metabolic clearance measured in liver S10 was scaled to whole fish values based on the relative mass of the liver and the binding of TF in blood, which we have assumed has a free fraction of 0.005. This adjustment allowed direct comparisons with parameter estimates for metabolic clearance obtained from *in-vivo* experiments, which were referenced to the exposure water and therefore based on the unbound concentration of TF. This approach to scale-up of intrinsic metabolic clearance assumed liver metabolism of TF can be described by the well-stirred or venous-equilibrium model (see Wilkinson (1987) for a review of liver clearance models).

Results from these experiments (Figures 11 and 12, Table 6) indicate that scaling of pharmacokinetic model parameters for body size is possible. We observed a significant, linear relationship between resting oxygen consumption and uptake clearance ($r^2 = 0.95$;

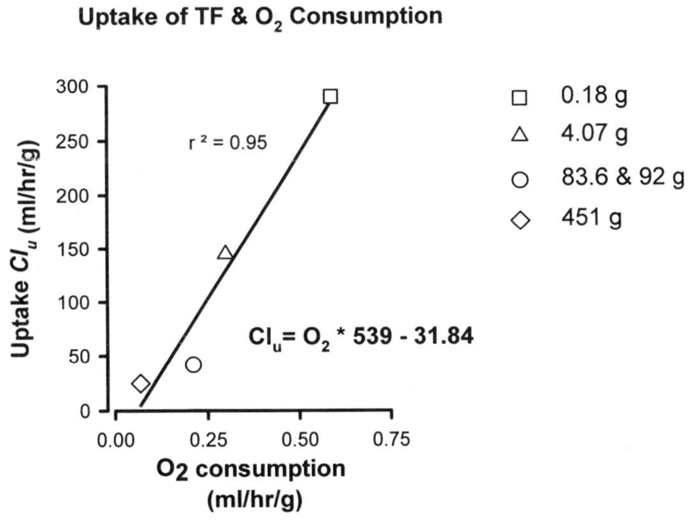

Figure 11. The relationship between the uptake of TF and the resting oxygen consumption in trout of different body sizes. Oxygen consumption measurements were performed (12°C; $n = 8$–20 size group) in sealed volumetric flasks (0.5–6.0 l) and the change in O_2 concentration measured after 0.5–3 hr. The oxygen consumption rate used for the 451 g trout was the mean value reported in Mckim *et al.*, (1986) for trout with an average body weight of 723 g.

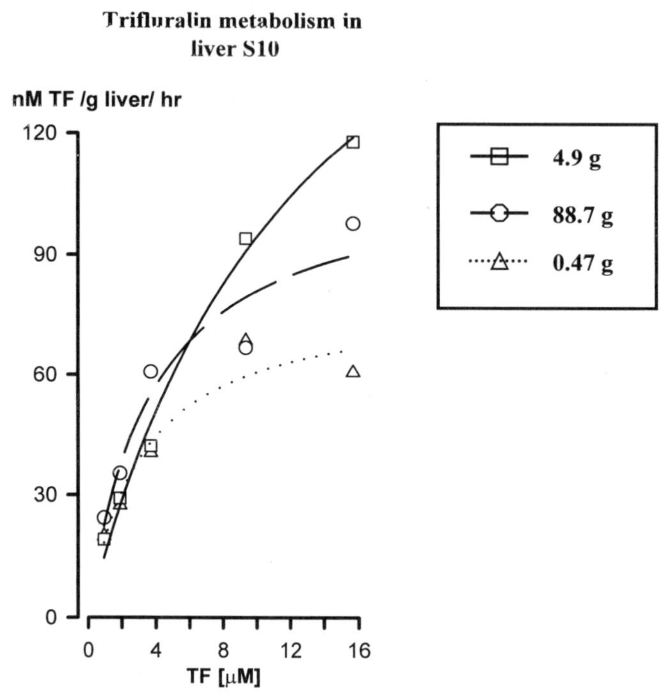

Figure 12. Trifluralin metabolism in liver S10 fractions prepared from trout of various body weights. Liver S10 fractions were incubated at 12°C with [1-^{14}C-*N*-propyl]trifluralin for 20 min and the formation of [^{14}C]-propionaldehyde determined to quantify initial TF metabolism. The lines represent the nonlinear least squares fit of the data to the Michaelis-Menten equation which was used to calculate the V_{max}/K_m ratios shown in Table 6.

Figure 11). The use of oxygen consumption as an indirect measure of uptake clearance offers several advantages due to its ease of measurement and the extensive literature values pertaining to the allometry of oxygen consumption in fish. This data base could be used to estimate the body size differences in the uptake of pesticides with ventilation-rate-limited uptake.

Comparison of the *in-vitro* derived metabolic clearance values to those determined from *in-vivo* experiments indicated it is possible to predict whole animal biotransformation rates from *in-vitro* data (Table 6). However, the scaled values tended to underestimate clearance in trout which formed metabolites more rapidly *in-vivo* (0.47 & 4.9 g sizes) and overestimate clearance in trout which metabolized TF at a lower rate *in-vivo* (87.7 g size; see Figure 4). These discrepancies were likely caused by assumptions made for blood binding of TF (which has not been experimentally determined) and omission of hepatic

Table 6. Scale-up of in-vitro metabolic clearance values.

Fish Wt. (mean ± SD) (g)	n^a	Relative Liver Mass (g-liver/g-fish)	Intrinsic Metabolic Clearance (ml hr^{-1} g liver^{-1})	Adjusted^{b,c} Metabolic Clearance (ml hr^{-1} g-fish^{-1})
0.47 ± 0.01	47	0.010	7.41	14.8 (66.7%)
3.0 ± 0.8	15	0.0152	8.85	26.9 (74.7%)
87.7 ± 5.0	4	0.0108	13.4	28.9 (206%)

[a]Number of fish from which pooled liver tissue was obtained.
[b]Adjusted metabolic clearance was the intrinsic clearance divided by f_u(0.005) multiplied by the relative mass of the liver.
[c]Values in parenthesis represent the percentage of *in-vivo* determined values for trout of similar sizes shown in Table 3.

Table 7. Bioconcentration factors calculated directly from model parameters or from scaled parameters based on the relationships shown in Figures 9, 10 and Table 6.

	Bioconcentration Factor[a]	
FW (g)	Model Predicted	Scaled
0.18	1,920	1,430
4.07	2,050	2,047
83.6	2,280	1,007

[a]BCF = $Cl_u * V_{ss}/(Cl_u + Cl_m)$.

blood flow from the model, which limited our calculations to intrinsic as opposed to total metabolic clearance.

The calculation of a BCF for TF in trout of different body sizes was made using model parameters (Tables 3 and 4) and by scaling. The scaled BCF values used a V_{ss} that was calculated from the relationship shown in Figure 9 and using lipid fractions of 0.019, 0.038 and 0.074 for the 0.18, 4.07 and 83.6 g sizes respectively (Schultz and Hayton, 1994). These estimates were converted to water referenced values using a plasma f_u of 0.0013 (0.18 & 4.07 g trout) and 0.0039 (83.6 g trout) for TF (Figure 6). The model predicted and scaled BCF estimates are shown in Table 7 and appear to be similar. These BCF estimates are also comparable to values reported for other fish species (Table 1). This similarity supports the use of the scaling approaches described in this paper for extrapolation of pharmacokinetic parameters across fish of varying body sizes. This offers advantages to existing methods based on empirical relationships between lipophilicity and BCF, which may overestimate the BCF of pesticides that are extensively metabolized by fish.

8. REFERENCES

Ali, S.M. 1984, Dissipation kinetics and adsorption-desorption thermodynamics of trifluralin in soil. *Diss. Abstr. Int.*, 44:3745–b.

Boddington, M.J., Mackenzie, B.A., and DeFreitas, A.S., 1979, Respirometer to measure the uptake efficiency of waterborne contaminants in fish. *Ecotoxicol. Environ. Safety* 3:383–393.

Briggs, G.G., 1981, Theoretical and experimental relationships between soil adsorption, octanol-water partition coefficients, water solubilites, bioconcentration factors, and the parachor. *J. Agric. Food Chem.* 29:1050–1059.

Brown, D.S., and Flagg, E.W., 1981, Empirical predictions of organic pollutant sorption in natural sediments. *J. Environ. Qual.* 10:382–386.

Chan, M.M., Grogl, M., Chen, C.C., Bienen, E. J., and Fong, D., 1993, Herbicides to curb human parasitic infections: in vitro and in vivo effects of trifluralin on the trypanosomatid protozoans. *Proc. Natl. Acad. Sci.* 90:5657–5661.

Cope, O.B., 1965, Some responses of fresh-water fish to herbicides. *Proc. South. Weed. Sci.* 18:439–445.

Couch, J.A., Winstead, J.T., and Goodman, L.R., 1979, Vertebral dysplasia in young fish exposed to the herbicide trifluralin. *J. Fish Dis.* 7:157–163.

Ebert, E., Leist, K.H., Hack, R., and Ehling, G., 1992, Toxicology and hazard potential of trifluralin. *Food Chem. Toxicol.* 30:1031–1044.

Emmerson, J.L., and Anderson, R.C., 1966, Metabolism of trifluralin in the rat and dog. *Toxicol. Appl. Pharmacol.* 9:84–97.

Erkog, E.U., and Menzer, R. E., 1985, Metabolism of trifluralin in rats. *J. Agric. Food Chem.* 33:1061–1070.

Grover, R., Wolt, J.D., Cessna, A.J., and Schiefer, H.B. 1997, Environmental fate of trifluralin. *Rev. Environ. Contam. Toxicol.* 153:1–64.

Hayton, W.L., and Barron, M. G., 1990, Rate limiting barriers to xenobiotic uptake by the gill. *Environ. Toxicol. Chem.* 9:151–157.

Johnson, D.W., and Finley, M.T., 1980, Handbook of acute toxicity of chemicals to fish and aquatic invertebrates: summaries of toxicity tests conducted at Columbia National Fisheries Research Laboratory, 1965–78. *U.S. Dept. of the Interior, Fish and Wildlife Service* Publ. 137.

Koyama, J., 1996, Vertebral deformity susceptibilities of marine fishes exposed to herbicide. *Bull. Environ. Contam. Toxicol.* 56:655–662.

Leider, S.A., Chilcote, M.W., and J.J. Loch. (1986). Comparative life history characteristics of hatchery and wild steelhead trout (*Salmo gairdneri*) of summer and winter races in the Kalama River, Washington. *Can. J. Fish. Aquat. Sci.* 43:1398–1409.

Macek, K.J., Hutchinson, C., and Cope, O.B., 1969, The effect of temperature on the susceptibility of bluegills and rainbow trout to selected pesticides. *Bull. Environ. Contam. Toxicol.* 4:174–183.

Macek, K.J., Lindberg, M.A., Sauter, S., Buxton, K.S., and Costa, P.A., 1976, Toxicity of four pesticides to water fleas and fathead minnows. EPA-600/3-76-099.

McKim, J.M., Schmeider, P.K., and Erickson, R. J., 1986, Toxicokinetic modeling of ^{14}C-pentachlorophenol in the rainbow trout (*Salmo gairdneri*). *Aquatic Toxicol.* 9:59–80.

Nelson, J.O., Kearny, P.C., Plimmer, J.R., and Menzer, R.E., 1977, Metabolism of trifluralin, profluralin, and fluchloralin by rat liver microsomes. *Pestic. Biochem. Physiol.* 7:73–82.

Rane, A., Wilkinson, G.R., and Shand, D.G., 1977, Prediction of hepatic extraction ratio from *in-vitro* measurement of intrinsic clearance. *J. Pharmacol. Exp. Ther.* 200:420–424.

Reinbold, K.A., and Metcalf, R.L., 1976, Effects of the synergist piperonyl butoxide on metabolism of pesticides in green sunfish. *Pestic. Biochem. Physiol.* 6:401–412.

Schultz, I.R., and Hayton, W.L., 1993, The toxicokinetics of trifluralin in rainbow trout. *Aquatic Toxicol.* 26:287–306.

Schultz, I.R., and Hayton, W.L., 1994, Body size and the toxicokinetics of trifluralin in rainbow trout. *Toxicol. Appl. Pharmacol.* 129:138.

Schultz, I.R., and Hayton, W.L., 1997, Influence of Body Fat on Trifluralin Toxicokinetics in Rainbow Trout (*Oncorhynchus mykiss*). *Environ. Toxicol. Chem.* 16:997–1001.

Schultz, I.R., and Hayton, W.L., 1999, Interspeices saling of the bioaccumulaiton of lipophilic xenobiotics in fish: an example using trifluralin. *Environ. Toxicol. and Chem.* 18:1440–1449.

Spacie, A., and Hamelink, J.L., 1979, Dynamics of trifluralin accumulation in river fishes. *Environ. Sci. Technol.* 13:817–822.

Stehly, G.R., and Hayton, W.L., 1989, Disposition of pentachlorophenol in rainbow trout (Salmo gairdneri): effect of inhibition of metabolism. *Aquatic Toxicol.* 14:131–148.

Wauchope, R.D., Buttler, T.M., Hornsby, A.G., Augustine-Beckers, P.W.M., and Burt, J.P., 1992, The SCS/ARS/CES pesticide properties data base for environmental decision making. *Rev. Environ. Contam. Toxicol.* 123:1–155.

Williams, R.R., Bell, T.A., and Lightner, D.V., 1986, Degradation of trifluralin in seawater when used to control larval mycosis in penaeid shrimp culture. *World Aquacult. Soc.* 17:8–12.

Wilkinson, G.R., 1987, Clearance approaches in pharmacology. *Pharmacol. Rev.* 39:1–47.

Yamamoto, K., 1991, Relationship of respiration to body weight in the carp *Cyprinus carpio* under resting and normoxic conditions. *Comp. Biochem. Physiol.* 100A:113–116.

Yamamoto, K., 1992, Relationship of respiration to body weight in the tilapia *Oreochromis niloticus* under resting and normoxic conditions. *Comp. Biochem. Physiol.* 103A:81–83.

Yockim, W.D., Isensee, A.R., and Walker, E.A., 1980, Behavior of trifluralin in aquatic model ecosystems. *Bull. Environ. Contam. Toixcol.* 24:134–141.

11

DISPOSITION AND METABOLISM OF MALACHITE GREEN AND OTHER THERAPEUTIC DYES IN FISH

Steven M. Plakas[1], Daniel R. Doerge[2], and Sherri B. Turnipseed[3]

[1]Gulf Coast Seafood Laboratory
U.S. Food and Drug Administration
P.O. Box 158
Dauphin Island, AL 36528
[2]National Center for Toxicological Research
U.S. Food and Drug Administration
Jefferson, AR 72079
[3]Animal Drugs Research Center
U.S. Food and Drug Administration
P.O. Box 25087
Denver, CO 80225

1. ABSTRACT

The triphenylmethane dye malachite green has been used in aquaculture worldwide for treatment of parasitic and fungal infections in fish and shellfish. Other dyes (i.e., acriflavine, methylene blue) exhibit similar antimicrobial properties and are widely marketed for treatment of external infections in ornamental fish. None of these compounds are regulated for use in food fish production in the United States. Concern about the health risks associated with drug residues in aquaculture products prompted our studies of the uptake, metabolism, and elimination of selected dyes in the channel catfish. Malachite green was the primary focus because of its toxicological properties and long history of use in aquaculture. Analytical methods were developed for the determination of parent compound and metabolites in fish tissues. Distribution and persistence of residues in the tissues, including edible flesh, were characterized. This chapter summarizes current knowledge of the disposition and metabolism of malachite green and other therapeutic dyes in aquatic species.

Xenobiotics in Fish, edited by *Smith et al.*
Kluwer Academic / Plenum Publishers, New York, 1999.

2. INTRODUCTION

Antibacterial activities and other therapeutic properties of dyes were first reported in the early 1900s. Subsequently, formulations from the acridine, azo, triphenylmethane, and thiazine dye classes came into widespread use in human and veterinary medicine, principally as topical antiseptics and disinfectants. For example, acridine dyes were widely used in the World Wars as wound disinfectants. Most therapeutic applications of the dyes in humans and in terrestrial food animals have been discontinued because of safety considerations and the development of more effective drugs.

In aquatic species, certain dyes remain in use as an economical and effective means for treatment of parasitic and fungal infections. Malachite green has a long history of use as an antifungal agent in fish and fish eggs. Acriflavine and methylene blue are also used to treat and prevent fish diseases. Although none of these drugs are approved for food fish production in the United States, they are readily available and widely marketed for ornamental (non-food) fish applications. The potential exists for their improper or illegal use in domestic and foreign aquaculture.

We examined the pharmacokinetics and metabolism of malachite green in response to food safety concerns about residues in the food supply. Target analytes were identified and methods were developed for monitoring residues of malachite green and other dyes in aquaculture products. This chapter reviews available information on the disposition and metabolism of therapeutic dyes in fish. Therapeutic applications in fish and toxicological properties of these compounds are briefly reviewed.

2.1. Structures

Malachite green is a triphenylmethane dye prepared commercially as the chloride or oxalate salt of its dye cation (Fig. 1). The oxalate salt (zinc-free) form has been recommended for fishery use. In aqueous solutions, the dye cation (chromatic form) combines with available hydroxyl ions to produce the non-ionized and colorless carbinol base (pKa, 6.9; Fig. 1). Formation of the carbinol compound is slow, requiring up to 5 hr to attain equilibrium under alkaline conditions (Lanzing, 1965; Alderman, 1985; Plakas *et al.*, 1996). The carbinol base is much more lipophilic compared with the dye cation and is believed to be the form of the drug which enters cells.

Acriflavine is an acridine dye available commercially as a mixture of acriflavine and proflavine (Fig. 1). Typically, a ratio of 2:1 (acriflavine:proflavine) is encountered in commercial products. Proflavine can be obtained commercially in purified form. Acriflavine has been purified from the commercial product for use in pharmacokinetic and metabolism studies (Yu, 1996; Plakas *et al.*, 1997).

Methylene blue belongs to the phenothiazine class of dyes (Fig. 1). Because of the nature of the thiazine structure, methylene blue is completely ionized in its aquatic applications.

2.2. Therapeutic Applications in Aquatic Species

The first reported use of malachite green was to control fungal infections in hatchery fish (Foster and Woodbury, 1936). Since that time, malachite green has been used routinely in aquaculture as an antifungal and antiparasitic agent, and as a general hatchery disinfectant. Few known chemicals rival its effectiveness in treating external infections of fish and fish eggs, particularly those caused by fungi of the genus *Saprolegnia*

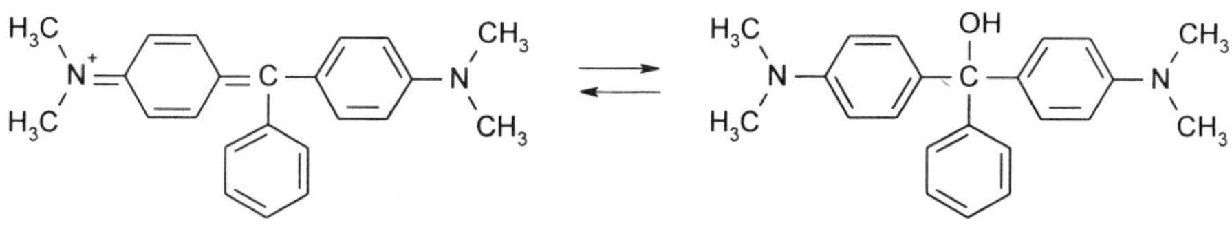

Malachite green (chromatic form) Malachite green (carbinol form)

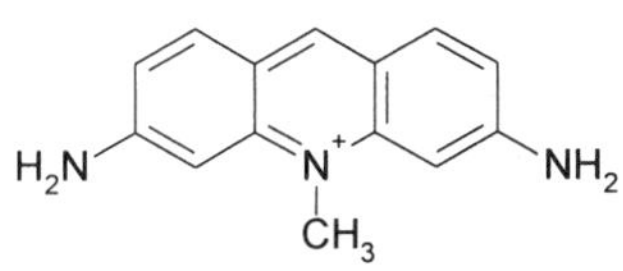

Acriflavine

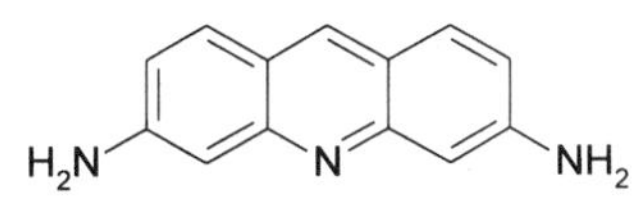

Proflavine

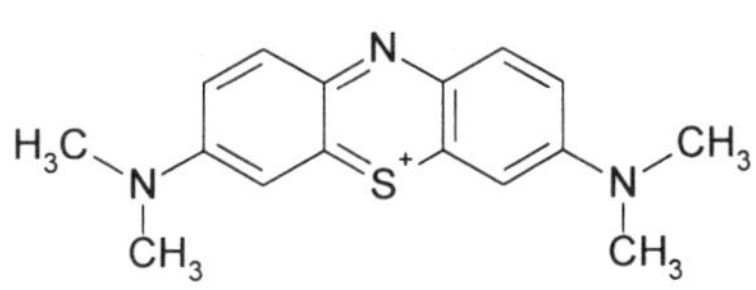

Methylene blue

Figure 1. Structures of the therapeutic dyes malachite green (chromatic and carbinol forms), acriflavine, proflavine, and methylene blue. Dye cations are shown without corresponding anions.

and by the protozoan parasite *Ichthyophthirius multifiliis* (Meyer and Schnick, 1989; Bailey and Jeffrey, 1989; Wahli *et al.*, 1993). Malachite green is also effective against systemic protozoan infections (Clifton-Hadley and Alderman, 1987; Alderman and Clifton-Hadley, 1988). Bath or dip treatments are the most common means of administration of malachite green to fish. However, treatment conditions (e.g., drug concentration, duration of exposure) have varied widely, particularly with respect to life stage of the animal.

Acriflavine is used to treat fungal and parasitic infections in fish (Richards, 1977; Gratzek, 1983; Stuart, 1983; Kabata, 1985). Its antimicrobial activities vary compared with malachite green (Martin, 1968; Seenappa and Manohar, 1983; Paperna, 1984). Acriflavine also exhibits antibacterial and antiviral activities (Fijan and Voorhees, 1969; Elliott and Amend, 1978; Carson *et al.*, 1993; Schmidtke and Carson, 1995). Hines and Watts (1995) reported sex-manipulating activity of acriflavine in fish. In the ornamental fish industry, acriflavine is often recommended for prevention and treatment of parasitic velvet disease (caused by *Oodinium sp.*), general disinfection of fish eggs, and wound antisepsis. Acriflavine also reduces transport mortality in fish (Gratzek, 1983). Acriflavine is typically administered as a bath treatment at varying concentrations depending on exposure duration (usually 5–10 ppm for several hours or days).

Methylene blue is used as an antifungal or antiparasitic agent in fish and fish eggs, and to reduce post-shipment mortality of ornamental fish (Richards, 1977; Stuart, 1983; Gratzek, 1983). It has been proposed as an alternative to malachite green for treatment of fungal infections in fish; however, Bailey and Jeffrey (1989) found it unsuitable for use as an aquatic fungicide. Methylene blue is effective in treating *Ichthyophthirius multifiliis* infections in fish (Antychowicz, 1977) and in destroying myxosporidian spores (Seenappa and Manohar, 1983). It is also effective in treating nitrite intoxication (methemoglobinemia) in trout (Wedemeyer and Yasutake, 1978) and is marketed for treatment of nitrite and cyanide poisoning in ornamental fish. Typical dosages for bath treatment of external infections range from 1 to 5 ppm for one or more hours but, as with other dyes, have varied considerably.

2.3. Toxicological Properties

Culp and Beland (1996) recently reviewed the toxicological properties of malachite green, including its toxicity to aquatic species. Malachite green is a respiratory enzyme poison (Werth and Boiteux, 1968a), a teratogen (Werth, 1958; Meyer and Jorgenson, 1983), and a tumor enhancer (Fernandes *et al.*, 1991) in mammals. It can bind to or intercalate in DNA and is mutagenic in the Ames test after metabolic activation (Clemmensen *et al.*, 1984). Malachite green and its metabolites are structurally related to known carcinogenic triphenylmethane dyes (e.g., gentian violet), and may have similar toxicities (Doerge *et al.*, 1998a). In 1993, malachite green was one of several chemicals chosen for rodent carcinogenicity testing under the National Toxicology Program. It was selected based on structural considerations and the potential for worker and consumer exposures through its applications in the dye and aquaculture industries (Culp and Beland, 1996). These studies are scheduled for completion around the year 2000. Culp and Beland (1996) summarized acute and sub-acute toxicity data in aquatic species. Malachite green toxicity in fish is influenced by environmental conditions of exposure (Lanzing, 1965; Bills *et al.*, 1977, 1993).

Acriflavine and proflavine were once widely used in human medicine as topical antiseptics, but use has now been largely discontinued because of toxicological considerations. Human contact sensitivity to proflavine associated with its widespread use as an antiseptic is well documented (Goh, 1989a,b; Lim *et al.*, 1992). Allergic reactions in humans may be severe (Goh, 1986). Contact sensitivity was documented in a fishery worker after repeated application of acriflavine as an antiparasitic agent in aquarium fish (Goh, 1987). Acriflavine and proflavine are DNA-intercalating agents and mutagens (Bruce and Heddle, 1979; Speck and Rosenkranz, 1980; Rogers and Back, 1982; Neidle *et al.*, 1988), but have not been fully evaluated for carcinogenic activity (IARC, 1977; 1980). Santamaria *et al.* (1979) demonstrated photocarcinogenic activity of proflavine in mice. Acriflavine can suppress the immune response (Samuelson *et al.*, 1965) and the expression of xenobiotic-metabolizing enzymes (Kim *et al.*, 1998) in mammals. In fish, the acute toxicity of acriflavine has been examined (Willford, 1967; Wellborn, 1971; Okafor and Taege, 1996). It is much less toxic compared with malachite green.

Methylene blue intercalates in DNA and is a frameshift mutagen (Imray and MacPhee, 1975; MacPhee and Imray, 1978; Speit and Vogel, 1979; Chung *et al.*, 1981). Methylene blue and acridine dyes are photosensitizers and their light-induced DNA damage has been characterized in cellular systems (Epe *et al.*, 1993). Barbosa and Peters (1971) reviewed its toxicity in living organisms. In fish, acute toxicity of methylene blue is comparable to acriflavine (Willford, 1967).

2.4. Analytical Methodology

Pfenning (1998) and Roybal *et al.* (1996) recently reviewed analytical methods for determination of dye residues in animal tissues and fluids. Most of the available methods were developed for terrestrial animal tissues and food products. This section briefly reviews those methods which are relevant to fish.

Initially, methods for determining malachite green residues in fish tissues were limited to measurements of the parent compound on the basis of its visible light absorption. Subsequently, liquid chromatographic methods that enable the simultaneous analysis of the chromatic parent compound and its colorless leuco metabolite were introduced (Allen and Meinertz, 1991; Allen *et al.*, 1992; Allen *et al.*, 1994). The latter methods incorporated a post-column reactor containing lead oxide in which leucomalachite green, after separation on the analytical column, was oxidized back to the chromatic compound for visible light detection. However, when applied to fish tissues, some interconversion of analytes occurred (Allen *et al.*, 1994). Modifications by Roybal *et al.* (1995) enabled extraction and analysis of both compounds in fish muscle without interconversion. Plakas *et al.* (1995) optimized extraction efficiencies for these analytes in muscle and plasma of catfish for application in pharmacokinetic and metabolism studies. Hajee and Haagsma (1995) reported a post-column oxidation method for analysis of malachite green and its metabolites in eel plasma. Rushing and Thompson (1997) described a procedure for the simultaneous analysis of malachite green, gentian violet, and their leuco metabolites in fish muscle. Others have used electrochemical cells for detection of chromatic and leuco compounds (Sagar *et al.*, 1994) or for post-column oxidation of leucomalachite green prior to its visible light detection (Swarbrick *et al.*, 1997; Rushing and Hansen, 1997). Turnipseed *et al.* (1995a,b) developed gas and liquid chromatographic/mass spectrometric methods for confirmation of malachite green and leucomalachite green in catfish muscle. More recently, Doerge *et al.* (1998b) developed an isotope dilution liquid chromatographic/mass spectrometric method for quantification and structural confirmation of malachite green and its metabolites in fish. In the above methods, extraction and analysis of malachite green under low pH conditions (e.g., pH 4) ensured complete conversion of its carbinol base to dye cation (chromatic) form.

Relatively few methods are available for determination of acriflavine and proflavine residues in fish. Plakas *et al.* (1997) developed a liquid chromatographic method for the simultaneous determination of unchanged acriflavine and proflavine in fish muscle. Methods have also been developed for the isolation and identification of their metabolites in fish tissues (Yu, 1996; Yu *et al.*, 1997; Plakas *et al.*, 1998).

There are few published methods for determination of methylene blue residues in fish. As with other dyes, analytical methods initially focused on measurements of the parent compound based upon its visible light absorbance. Nakagawa *et al.* (1984) developed a spectrophotometric assay for determination of methylene blue in fish tissues. Kasuga *et al.* (1991) developed a liquid chromatographic method for simultaneous determination of methylene blue and malachite green in trout tissues, however analyte recoveries were low. Turnipseed *et al.* (1997) reported high recovery and sensitivity in their procedure for quantification of methylene blue in catfish muscle. No methods have been developed or evaluated for characterization of metabolites of methylene blue in fish. In contrast, liquid chromatographic methods have been described for parent compound and metabolites in terrestrial animals, specifically for residues in milk (Roybal *et al.*, 1992; Munns *et al.*, 1992).

3. DISPOSITION AND METABOLISM

3.1. Malachite Green

Malachite green has been used since the 1930s, however information on its absorption and elimination in fish was not available until the 1980s (Poe and Wilson, 1983; Nakagawa *et al.*, 1984; Allen and Hunn, 1986; Kietzmann *et al.*, 1990). In earlier studies, quantitative information was limited to uptake and elimination data for the parent drug or total residues, with little insight into metabolism.

Metabolism of malachite green to its colorless, reduced derivative leucomalachite green has long been observed in mammals. Early studies in rats concluded that reduction of malachite green was a detoxification pathway (Werth and Boiteux, 1968b). Evidence of absorption and metabolism of malachite green in fish was first reported by Poe and Wilson (1983). These authors noted that in tissues from malachite green-treated fish, particularly in visceral adipose tissue, a green color developed on exposed tissue surfaces during prolonged frozen storage. They deduced that the color change was due to oxidation of leucomalachite green back to the parent compound. Subsequently, the extensive metabolism of malachite green to leucomalachite green in fish was demonstrated by liquid chromatographic techniques and its effect in prolonging residue elimination was recognized (Bauer *et al.*, 1988; Allen *et al.*, 1994). As analytical methodology improved, more detailed descriptions of the disposition and metabolism of malachite green in fish appeared in the literature.

Malachite green is rapidly absorbed and metabolized in fish and fish eggs during bath treatment. In short-term bath exposures (≤1 hr), residue concentrations in fish tissues readily exceeded water concentrations (Alderman and Clifton-Hadley, 1993; Plakas *et al.*, 1996). In channel catfish exposed to 0.8 ppm malachite green for 1 hr, plasma concentrations of parent compound and leuco metabolite were 3.3 and 1.9 ppm, respectively; muscle concentrations were 1.2 and 1.5 ppm, respectively (Plakas *et al.*, 1996). After a 24-hr bath exposure to malachite green, nearly 80% of the initial amount of drug in the water was recovered in experimental rainbow trout; in muscle, accumulated residues were composed mostly (90%) of the leuco metabolite (Bauer *et al.*, 1988). Residues were accumulated in rainbow trout eggs during repeated treatment, primarily as the leuco metabolite (Meinertz *et al.*, 1995). Residues were also incorporated in eggs of malachite green-treated adult fish (Allen, 1990). In addition to topical applications, malachite green has been administered orally in fish. Alderman and Clifton-Hadley (1993) were unable to detect residues in tissues after oral dosing in rainbow trout. However, Schmahl (1992) demonstrated efficacy of orally-administered malachite green against parasitic infections in fish skin, indicating bioavailability by this route.

Few studies have characterized the pharmacokinetics of malachite green in fish. Alderman and Clifton-Hadley (1993) measured tissue concentrations of malachite green in rainbow trout during and after waterborne exposure (1.6 ppm for 40 min) at two temperatures. Uptake and elimination of the parent compound in all tissues, except serum, were more rapid at the higher temperature. Serum data suggested saturable uptake with peak concentrations (C_{max}) of 13.0 and 13.5 ppm at 8 and 16 °C, respectively. Plakas *et al.* (1996) provided the first detailed study of the pharmacokinetics of malachite green in channel catfish. After intravascular dosing (0.8 mg/kg), plasma concentrations of parent compound exhibited a rapid, triphasic decline (Fig. 2), with a terminal elimination half-life of 6.2 hr. Volume of distribution at steady-state (1.77 L/kg) indicated extensive extravascular distribution of malachite green. Other pharmacokinetic values are pre-

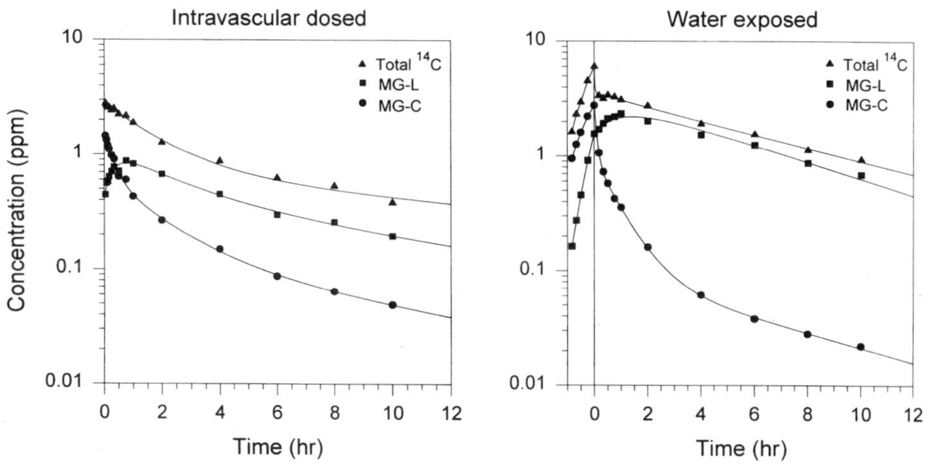

Figure 2. Elimination of parent malachite green (MG-C), leucomalachite green (MG-L), and total radioactive residues (total ^{14}C) in plasma of channel catfish after intravascular dosing (0.8 mg/kg) or waterborne exposure (0.8 ppm for 1 hr) to ^{14}C-labeled malachite green. Solid line for MG-C represents the compartmental model fit of the mean observed data. For waterborne exposure, the vertical line at 0 hr represents the end of the exposure period. From: Plakas *et al.* (1996), with permission (NRC Canada).

sented in Table 1. The major metabolite leucomalachite green was rapidly accumulated in plasma after intravascular dosing (C_{max}, 0.88 ppm at 0.75 hr), and was eliminated more slowly compared with the parent compound (Fig. 2). During waterborne exposure (0.8 ppm for 1 hr), plasma concentrations of parent compound rapidly exceeded the water concentration, with a C_{max} of 2.8 ppm (Fig. 2). After dosing, parent compound exhibited a triphasic decline with a terminal elimination half-life of 4.7 hr. Peak concentrations of the leuco metabolite (C_{max}, 2.4 ppm) in plasma occurred at 1 hr after bath treatment. These data demonstrate a major role of metabolism in clearance of malachite green.

Malachite green and its metabolites are widely distributed in the tissues of fish. Residue concentrations are generally highest in excretory tissues and lowest in muscle or plasma (Kasuga *et al.*, 1992; Alderman and Clifton-Hadley, 1993; Plakas *et al.*, 1996). After bath treatment with ^{14}C-labeled malachite green (0.8 ppm for 1 hr), total residue concentrations in catfish tissues ranged from 3.2 ppm in muscle to 33 ppm in liver (Plakas *et al.*, 1996). Concentrations in visceral adipose tissue increased with time after dosing (from 1.5 ppm at 0 hr to 26.7 ppm at 336 hr), suggesting redistribution and storage of the leuco metabolite in this tissue after its formation elsewhere. Lipid content of the tissues

Table 1. Pharmacokinetic values for the dyes malachite green, proflavine, and acriflavine after intravascular dosing in channel catfish.

Parameter[a]	Malachite green[b]	Proflavine[c]	Acriflavine[c]
$t_{1/2}$ (hr)	6.24	8.71	11.4
Cl_b (L/h/kg)	0.328	4.58	1.68
V_{ss} (L/kg)	1.77	47.0	21.2

[a]Parameter abbreviations: $t_{1/2}$, terminal elimination half-life; Cl_b, total body clearance; V_{ss}, volume of distribution at steady-state. [b]From: Plakas *et al.* (1996). [c]From: Plakas *et al.* (1998).

is apparently an important variable influencing distribution and elimination of malachite green and its metabolites. Elimination half-lives of leucomalachite green were positively correlated with lipid content in muscle of rainbow trout (Bauer *et al.*, 1988).

Leucomalachite green is eliminated more slowly compared with malachite green in fish. Plakas *et al.* (1996) examined residue concentrations in muscle and plasma of channel catfish after waterborne exposure to malachite green (0.8 ppm for 1 hr). In plasma, parent compound declined below measurable levels (<0.005 ppm) within 1–2 days after dosing while the leuco metabolite was quantifiable for up to 4 weeks; in muscle, parent and leuco compounds persisted for up to 2 and 6 weeks, respectively (Fig. 3). Calculated half-lives for malachite green and leucomalachite green in catfish muscle were 2.8 and 10 days, respectively (Plakas *et al.*, 1996). Allen and Hunn (1986) reported a half-life of 1.5 days for malachite green in muscle of rainbow trout. Bauer *et al.* (1988) reported half-lives ranging from 10 to 40 days for leucomalachite green in trout muscle, depending on tissue lipid content. In rainbow trout fry dosed with ^{14}C-labeled malachite green, the whole body half-life of total radioactive residues (composed mostly of leucomalachite green) was 13 days, when corrected for growth dilution (Meinertz *et al.*, 1995).

Environmental factors, particularly pH and temperature, influence uptake and elimination of malachite green residues in fish. Peak concentrations and rates of elimination of malachite green in rainbow trout tissues generally increased with water temperature (Alderman and Clifton-Hadley, 1993). Residue concentrations in tissues of channel catfish increased dramatically with pH of the exposure water (Plakas *et al.*, 1996). In the latter study, the sums of the malachite green and leucomalachite green concentrations were 5-fold higher in muscle and 8-fold higher in plasma after exposure of fish at pH 8 compared with pH 6. The lipophilic carbinol base of malachite green appears to be more readily absorbed than its dye cation.

Leucomalachite green is a major metabolite of malachite green in fish. Several minor metabolites have also been observed in chromatograms of tissue extracts (Klein *et al.*, 1991; Allen *et al.*, 1994; Hajee and Haagsma, 1995; Meinertz *et al.*, 1995; Plakas

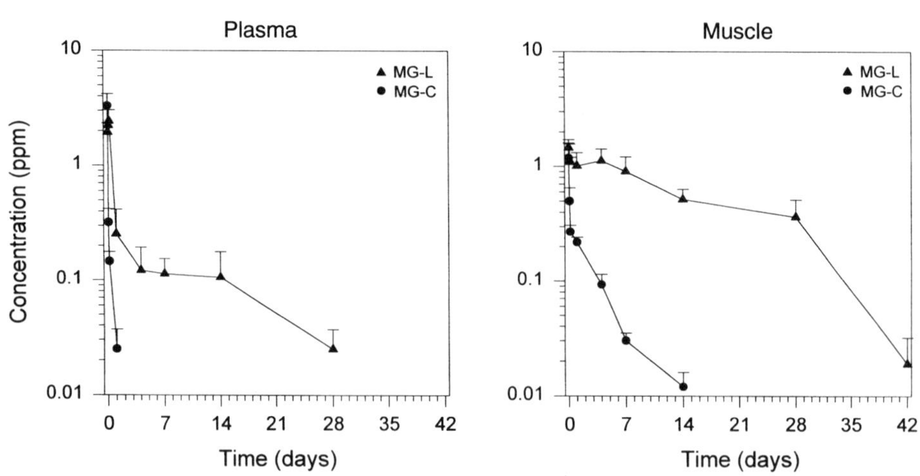

Figure 3. Elimination of parent malachite green (MG-C) and leucomalachite green (MG-L) in plasma and muscle of catfish after waterborne exposure (0.8 ppm for 1 hr). From: Plakas *et al.* (1996), with permission (NRC Canada).

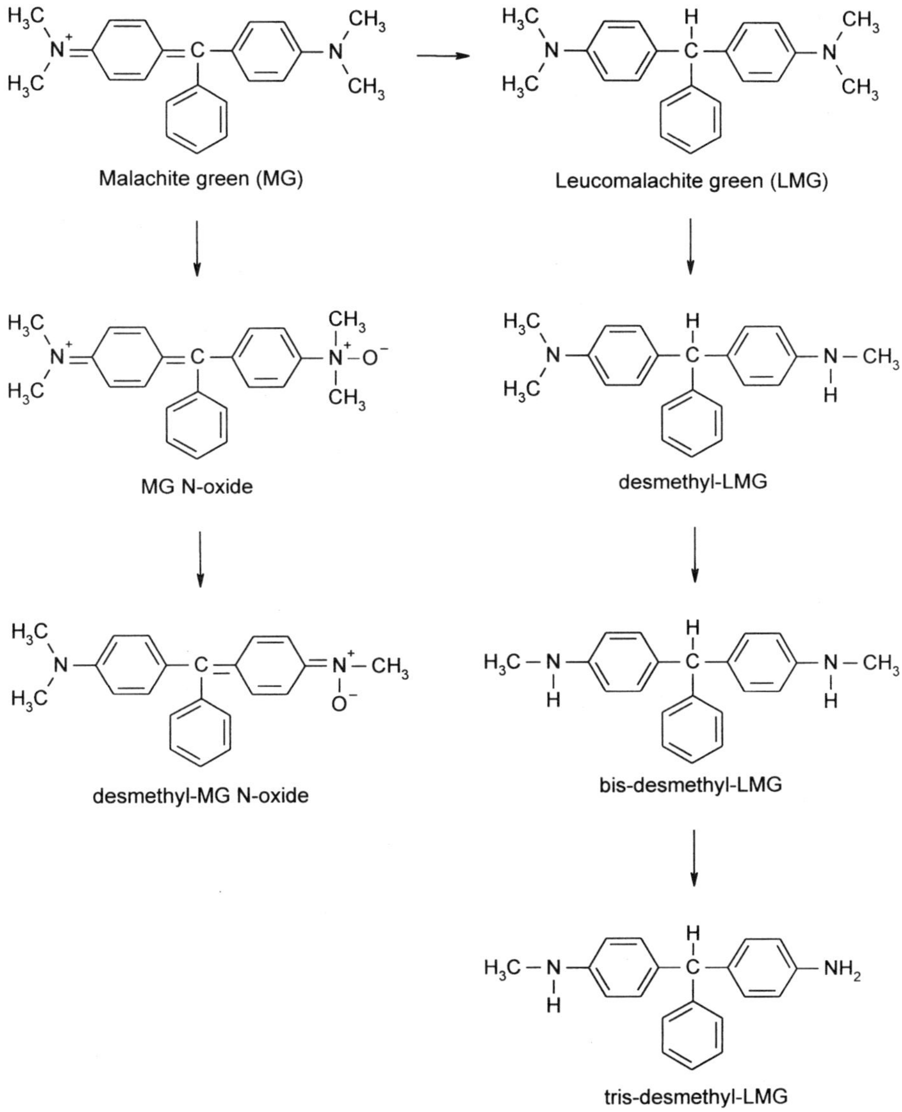

Figure 4. Structures of malachite green metabolites in fish muscle identified by liquid chromatography/mass spectrometry (Doerge *et al.*, 1998b).

et al., 1996). These metabolites were believed to be demethylated derivatives of leucomalachite green. Recently, Doerge *et al.* (1998b) confirmed the sequential demethylation of leucomalachite green in fish by using liquid chromatography/mass spectrometry. The structures of the demethylated metabolites identified in fish muscle are shown in Fig. 4. An oxidation product of the parent compound (N-oxide) was also isolated. The finding of the primary arylamine metabolite of malachite green in fish muscle is a potentially significant food safety issue. This compound is structurally related to known carcinogens, such as pararosaniline. In addition to direct consumption, these metabolites could also be formed in the human body through consumption of malachite green-treated fish.

Human intestinal microflora are capable of metabolizing malachite green to leucomalachite green (Henderson *et al.*, 1997). The presence of a minor, irreversibly-bound metabolite(s) of malachite green was also determined in fish tissues (Law *et al.*, 1993), which would escape detection by existing methodology.

3.2. Acriflavine

Proflavine and acriflavine are poorly absorbed in fish during waterborne exposure. Plakas *et al.* (1997) examined residue concentrations in muscle of channel catfish after waterborne exposure to commercial acriflavine (10 ppm for 4 hr). Parent proflavine and acriflavine concentrations in muscle were 0.014 and 0.010 ppm, respectively, immediately after dosing. Similar results were obtained in catfish dosed individually with purified acriflavine and proflavine; muscle concentrations were <1% of the exposure water concentration (Plakas *et al.*, 1998). Measurement of total radioactive residues after waterborne exposure to ^{14}C-proflavine confirmed its relatively poor absorption and distribution in catfish muscle (Plakas *et al.*, 1998). In rainbow trout, acriflavine and proflavine were also poorly absorbed and distributed in muscle during waterborne exposure (Yu, 1996).

The pharmacokinetics of proflavine and acriflavine have been characterized in two species of fish. Plakas *et al.* (1998) examined the pharmacokinetics of proflavine and acriflavine individually after intravascular dosing (1 mg/kg) in channel catfish. Plasma concentrations of proflavine and acriflavine were best described by two- and three-compartment pharmacokinetic models, respectively (Fig. 5). Terminal elimination half-lives for proflavine and acriflavine were 8.7 and 11.4 hr, respectively. Volumes of distribution were very large, indicating extensive extravascular distribution of these drugs (Table 1). In rainbow trout, terminal elimination half-lives of proflavine and acriflavine in plasma were 25.9 and 38.3 hr, respectively (Yu, 1996). Volumes of distribution of proflavine and acriflavine in trout were 28.2 and 31.2 L/kg, respectively, and total body clearances were 1.05 and 1.08 L/hr/kg, respectively. Large volumes of distribution appear characteristic of the dyes.

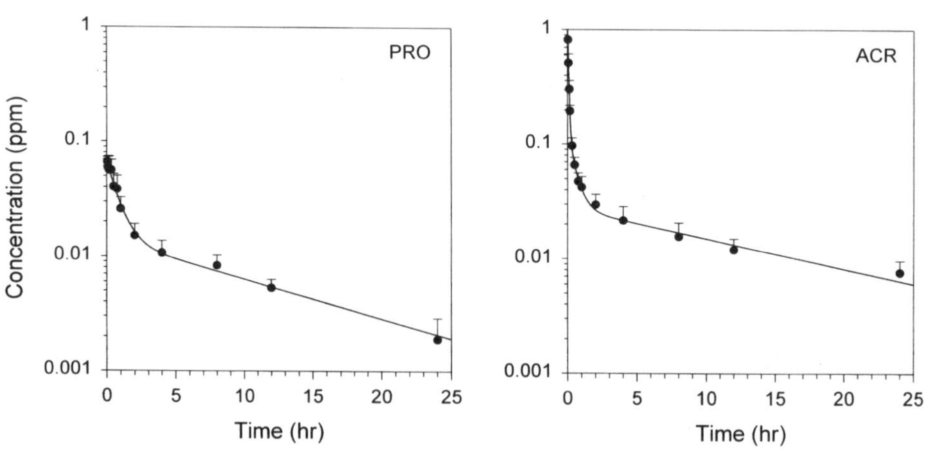

Figure 5. Elimination of parent proflavine (PRO) and acriflavine (ACR) in plasma of catfish after intravascular dosing (1 mg/kg). Solid lines represent the compartmental model fit of the mean observed data. From: Plakas *et al.* (1998), with permission (Taylor & Francis Ltd.).

Proflavine and acriflavine are widely distributed in fish tissues. After intravascular dosing with ^{14}C-labeled proflavine or acriflavine, total residue concentrations were highest in liver and trunk kidney and were lowest in plasma of channel catfish (Plakas *et al.*, 1998). A similar pattern of distribution was found after waterborne exposure; in catfish dosed with ^{14}C-proflavine (10 ppm for 4 hr), total residue concentrations were highest in the liver, head kidney, trunk kidney, and spleen (ranging from 1.7–3.9 ppm) and were lowest in muscle, adipose tissue, and plasma (0.07–0.08 ppm). In rainbow trout, total residue concentrations were also relatively high in the excretory organs and very low in muscle and plasma (Yu, 1996).

Elimination of proflavine and acriflavine residues from fish tissues is generally slow. Plakas *et al.* (1998) examined elimination of total radioactive residues in tissues of channel catfish after waterborne exposure to ^{14}C-proflavine. At 336 hr after dosing, total residue concentrations in liver, kidney, and spleen of catfish were 5–8% of their initial (0 hr elimination) values; in skin, residue concentrations were relatively unchanged, indicating high affinity or binding to tissue components. Plakas *et al.* (1998) also examined residue concentrations after waterborne exposure of channel catfish to unlabeled proflavine or acriflavine. In muscle, concentrations of the parent compounds (the major residues present) increased slightly from 0 to 12 hr after dosing, and then slowly declined (Fig. 6); elimination half-lives of proflavine and acriflavine were 1.5 and 5.3 days, respectively. The initial increase in residue concentrations in muscle after transfer of fish to clean water may have resulted from redistribution of residues from other tissues, or possibly migration from skin. In rainbow trout, proflavine and acriflavine residues were also slowly eliminated from the tissues, particularly kidney and skin (Yu, 1996). Persistence of residues in fish skin, if consumed, is a food safety concern.

Metabolism of proflavine and acriflavine has been characterized in channel catfish (Plakas *et al.*, 1998) and rainbow trout (Yu *et al.*, 1997). Administration of drugs by intravascular route was necessary to achieve adequate residue levels for metabolite identification. Yu *et al.* (1997) isolated and identified three conjugates of proflavine in rainbow trout; 3-*N*-glucuronosyl proflavine, 3-*N*-acetylproflavine, and 3-*N*-glucuronosyl-6-*N*-acetylproflavine (Fig. 7). The glucuronosyl conjugate of proflavine was the only metabo-

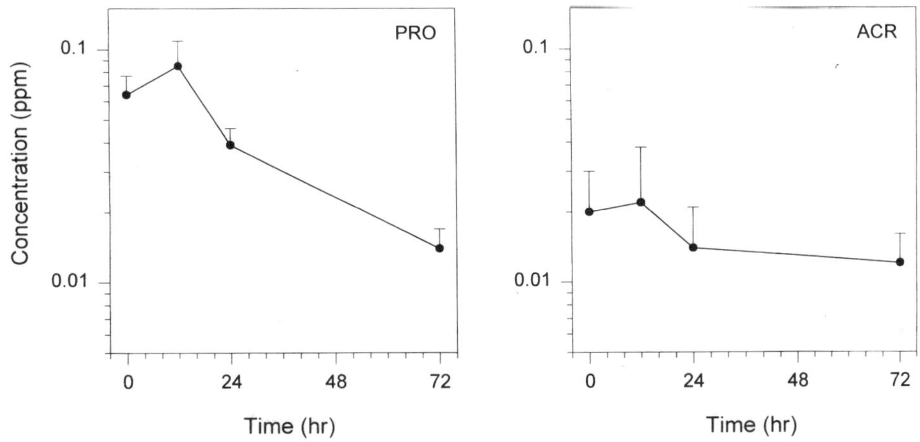

Figure 6. Elimination of parent proflavine (PRO) and acriflavine (ACR) from muscle of catfish after waterborne exposure (10 ppm for 4 hr). From: Plakas *et al.* (1998), with permission (Taylor & Francis Ltd.).

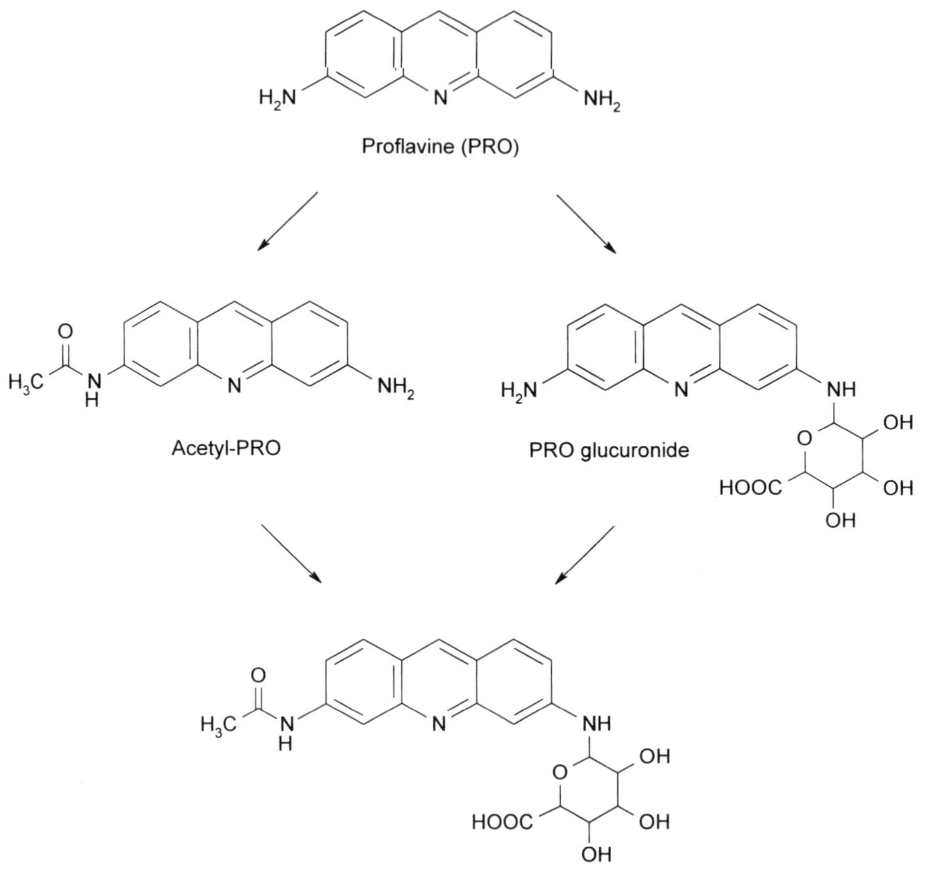

Figure 7. Structures of proflavine metabolites in fish tissues identified by liquid chromatograph/mass spectrometry (Yu, 1996; Plakas *et al.*, 1998).

lite in plasma while all three metabolites were found in liver and bile. In acriflavine-dosed fish, at least three metabolites were observed in chromatograms of tissue extracts, but their structures were not confirmed (Yu, 1996). In channel catfish, the metabolic profile of proflavine was qualitatively the same as in trout (Plakas *et al.*, 1998). At 24 hr after intravascular dosing, >95% of total residues in catfish muscle were in the form of the parent drug. In kidney, residues were composed mostly of acetylproflavine (65%) and unchanged drug (35%). In liver, the major metabolite was proflavine glucuronide (50%), with lesser amounts of acetylproflavine (10%) and acetylproflavine glucuronide (15%). In catfish dosed with purified acriflavine under the same conditions, parent acriflavine comprised >90% of total residues in all tissues examined. There was no interconversion of proflavine and acriflavine in catfish.

3.3. Methylene Blue

Methylene blue is poorly absorbed in fish during waterborne exposure. In eels exposed to 3 ppm for 1–3 hr, dye concentrations in the liver were 0.9–1.2 ppm; residues in muscle were undetectable (Nakagawa *et al.*, 1984). Turnipseed *et al.* (1997) examined methylene blue residues in muscle of channel catfish after waterborne exposure (1 ppm for 1 hr). Parent dye concentrations were quantifiable (~0.01 ppm) in only 2 of 4 fish examined. A higher dosage (5 ppm for 1 hr) yielded similar results; muscle concentrations were <1% of the exposure water concentration. Plakas (unpublished) estimated the elimination half-life of methylene blue in catfish muscle as ~9 hr. Tonogai *et al.* (1982) reported accumulation of methylene blue in gill tissue of carp during waterborne exposure and suggested its role in depressing gill function.

Measurement of methylene blue in animal tissues by existing methodology probably represents the sum of parent compound and its leuco metabolite. Leucomethylene blue is readily oxidized under ambient conditions and is unavailable commercially. Turnipseed *et al.* (1997) produced the leuco compound *in vitro* by ascorbic acid treatment of methylene blue, but it rapidly reverted to the dye cation upon standing or injection into a liquid chromatographic system.

DiSanto and Wagner (1972a) devised a method for measuring leucomethylene blue in human urine indirectly as the parent dye after its release from a stabilized complex by acid treatment. They recovered a large percentage (74%) of an orally administered dose of methylene blue in the urine, mostly (78%) in the form of leuco metabolite (DiSanto and Wagner, 1972b). The nature of the acid-labile complex in urine was not elucidated. Attempts to recover leucomethylene blue from animal tissues and milk were unsuccessful (DiSanto and Wagner, 1972a; Munns *et al.*, 1992; Turnipseed *et al.*, 1997). Since other dyes (e.g., malachite green, gentian violet) are extensively metabolized to their leuco forms, metabolism of methylene blue to leucomethylene blue may be anticipated in fish and other animals. Its presence in tissues, if measured as the parent compound, would compromise pharmacokinetic modeling of methylene blue.

Demethylated metabolites of methylene blue were tentatively identified in milk of methylene blue-treated cattle (Roybal *et al.*, 1992; Munns *et al.*, 1992). Metabolite structures are shown in Fig. 8. Leuco forms of these demethylated derivatives may also exist in animal tissues or fluids, but have not been isolated. In channel catfish, small amounts of demethylated metabolites, particularly azure B, were isolated from muscle (Turnipseed *et al.*, 1997). However, improved extraction and separation procedures are needed to fully characterize the metabolic profile of methylene blue in fish.

4. SUMMARY

The therapeutic applications of dyes, particularly malachite green, are well known in fish. However, their use in aquaculture is highly restricted or banned because of toxicological considerations. Disposition and metabolism studies of the dyes provide valuable information in assessing potential hazards and in developing analytical methods for their control. Malachite green is rapidly absorbed and metabolized in fish. The major metabolite, leucomalachite green, is slowly eliminated from tissues, including the edible flesh. Demethylated metabolites of leucomalachite green, including a primary arylamine, have been isolated from fish muscle. Leucomalachite green and its demethylated derivatives structurally resemble known arylamine carcinogens. Leucomalachite green is an appropriate target analyte for monitoring the use of malachite green in food fish. Acriflavine and methylene blue are poorly absorbed in fish and may be less of a problem in regards to residues in aquaculture products.

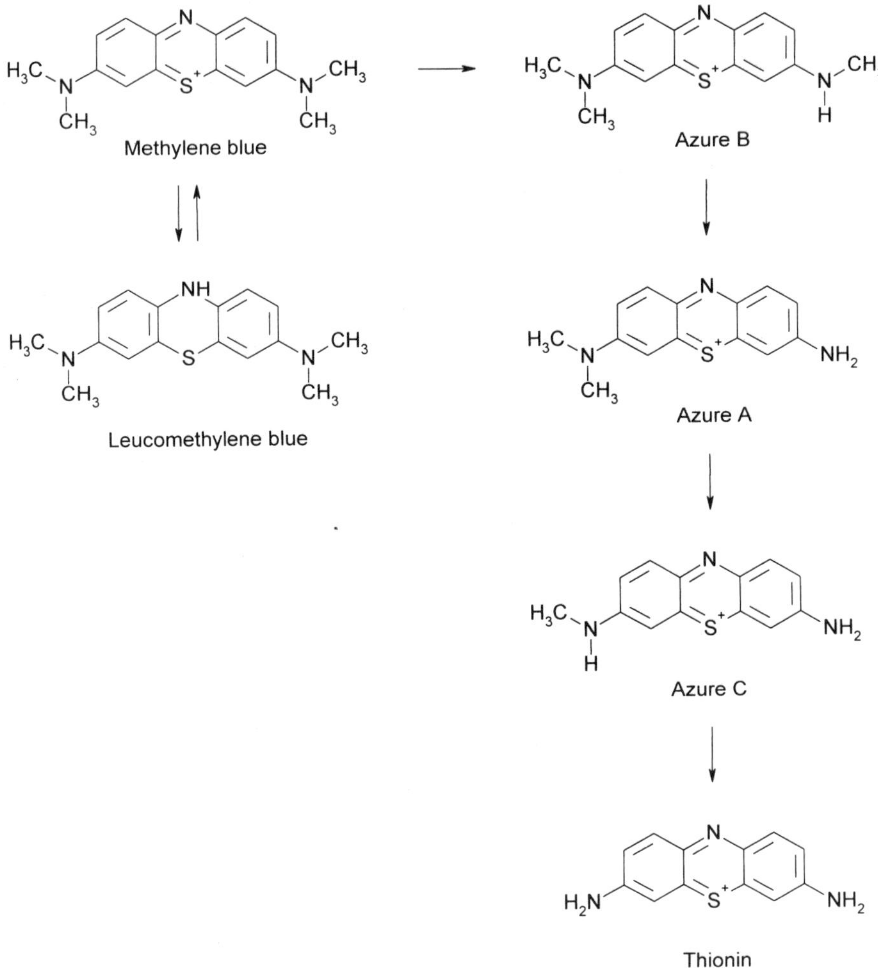

Figure 8. Probable metabolic scheme for methylene blue in animals. Leuco forms of demethylated metabolites are also presumed to exist (Roybal *et al.*, 1992).

5. ACKNOWLEDGEMENTS

We thank K.R. El Said and F.A. Bencsath (FDA) for reviewing this manuscript.

6. REFERENCES

Alderman, D.J., 1985, Malachite green: a review, *J. Fish Dis.* 8:289–298.

Alderman, D.J., and Clifton-Hadley, R.S., 1988, Malachite green therapy of proliferative kidney disease in rainbow trout: field trials, *Vet. Rec.* 122:103–106.

Alderman, D.J., and Clifton-Hadley, R.S., 1993, Malachite green: a pharmacokinetic study in rainbow trout, *Oncorhynchus mykiss* (Walbaum), *J. Fish Dis.* 16:297–311.

Allen, J.L., 1990, Residues of malachite green in muscle, eggs, and fry from treated Atlantic salmon and chinook salmon, U.S. Fish and Wildlife Service, *Investigations in Fish Control* 101, 4 pp.

Allen, J.L., and Hunn, J.B., 1986, Fate and distribution studies of some drugs used in aquaculture, *Vet. Human Toxicol.* 28:21–24.

Allen, J.L., and Meinertz, J.R., 1991, Post-column reaction for simultaneous analysis of chromatic and leuco forms of malachite green and crystal violet by high-performance liquid chromatography with photometric detection, *J. Chromatogr.* 536:217–222.

Allen, J.L., Meinertz, J.R., and Gofus, J.E., 1992, Determination of malachite green and its leuco form in water, *J. AOAC Int.* 75:646–649.

Allen, J.L., Gofus, J.E., and Meinertz, J.R., 1994, Detemination of malachite green residues in the eggs, fry, and adult muscle tissue of rainbow trout (*Oncorhynchus mykiss*), *J. AOAC Int.* 77:553–557.

Antychowicz, J., 1977, The influence of some chemical compounds upon the development of *Ichthyophthirius multifiliis*, *Bull. Vet. Inst. Pulaway*, 21:35–43.

Bailey, T.A., and Jeffrey, S.M., 1989, Evaluation of 215 candidate fungicides for use in fish culture, U.S. Fish and Wildlife Service, *Investigations in Fish Control 99*, 9 pp.

Barbosa, P., and Peters, T.M., 1971, The effects of vital dyes on living organisms with special reference to methylene blue and neutral red, *Histochem. J.* 3:71–93.

Bauer, K., Dangschat, H., Knoeppler, H.-O., and Neudegger, J., 1988, Uptake and excretion of malachite green by rainbow trout, *Archiv Lebensmittelhyg.* 39:97–102.

Bills, T.D., Marking, L.L., and Chandler, J.H., Jr., 1977, Malachite green: its toxicity to aquatic organisms, persistence and removal with activated carbon, U.S. Fish and Wildlife Service, *Investigations in Fish Control* 75, 6 pp.

Bills, T.D., Marking, L.L., and Howe, G.E., 1993, Sensitivity of juvenile striped bass to chemicals used in aquaculture, U.S. Fish and Wildlife Service, *Resource Publication* 192, 11 pp.

Bruce, W.R., and Heddle, J.A., 1979, The mutagenic activity of 61 agents as determined by the micronucleus, *Salmonella*, and sperm abnormality assays, *Can. J. Genet. Cytol.* 21:319–334.

Carson, J., Schmidtke, L.M., and Munday, B.L., 1993, *Cytophaga johnsonae*: a putative skin pathogen of juvenile farmed barramundi, *Lates calcarifer* Bloch, *J. Fish Dis.* 16:209–218.

Chung, K.-T., Fulk, G.E., and Andrews, A.W., 1981, Mutagenicity testing of some commonly used dyes, *Appl. Env. Microbiol.* 42:641–648.

Clemmensen, S., Jensen, J.C., Jensen, N.J., Meyer, O., Olsen, P., and Wurtzen, G., 1984, Toxicological studies on malachite green: a triphenylmethane dye, *Archiv. Toxicol.* 56:43–45.

Clifton-Hadley, R.S., and Alderman, D.J., 1987, The effects of malachite green upon proliferative kidney disease, *J. Fish Dis.* 10:101–107.

Culp, S.J., and Beland, F.A., 1996, Malachite green: a toxicological review, *J. Amer. Coll. Toxicol.* 15:219–238.

DiSanto, A.R., and Wagner, J.G., 1972a, Pharmacokinetics of highly ionized drugs I: methylene blue—whole blood, urine, and tissue assays, *J. Pharm. Sci.* 61:598–602.

DiSanto, A.R., and Wagner, J.G., 1972b, Pharmacokinetics of highly ionized drugs II: methylene blue—absorption, metabolism, and excretion in man and dog after oral administration, *J. Pharm. Sci.* 61:1086–1090.

Doerge, D.R., Chang, H.C., Divi, R.L., and Churchwell, M.I., 1998a, Mechanism for inhibition of thyroid peroxidase by leuco-malachite green, *Chem. Res. Toxicol.* 11:1098–1104.

Doerge, D.R., Churchwell, M.I., Gehring, T.A., Pu, Y.M., and Plakas, S.M., 1998b, Analysis of malachite green and metabolites in fish using liquid chromatography atmospheric pressure chemical ionization mass spectrometry. *Rapid. Commun. Mass Spectrom.* 12:1625–1634.

Elliott, D.G., and Amend, D.F., 1978, Efficacy of certain disinfectants against infectious pancreatic necrosis virus, *J. Fish Biol.* 12:277–286.

Epe, B., Pflaum, M., and Boiteux, S., 1993, DNA damage induced by photosensitizers in cellular and cell-free systems, *Mut. Res.* 299:135–145.

Fernandes, C., Lalitha, V.S., and Rao, K.V.K., 1991, Enhancing effect of malachite green on the development of hepatic pre-neoplastic lesions induced by *N*-nitrosodiethylamine in rats, *Carcinogenesis* 12:839–845.

Fijan, N.N., and Voorhees, P.R., 1969, Drug senstivity of *Chondrococcus columnaris*, *Veterinarski Arhiv, Zagreb.* 39:259–267.

Foster, F.J., and Woodbury, L., 1936, The use of malachite green as a fish fungicide and antiseptic, *Prog. Fish-Cult.* 18:7–9.

Goh, C.L., 1986, Contact sensitivity to proflavine, *Int. J. Dermatol.* 25:449–451.

Goh, C.L., 1987, Occupational dermatitis from proflavine, *Contact Dermatitis* 17:256.

Goh, C.L., 1989a, Contact senstivity to topical medicaments, *Int. J. Dermatol.* 28:25–28.

Goh, C.L., 1989b, Contact sensitivity to topical antimicrobials. (I). Epidemiology in Singapore, *Contact Dermatitis* 21:46–48.

Gratzek, J.B., 1983, Control and therapy of fish diseases, *Adv. Vet. Sci. Comp. Med.* 27:297–324.

Hajee, C.A.J., and Haagsma, N., 1995, Simultaneous determination of malachite green and its metabolite leucomalachite green in eel plasma using post-column oxidation, *J. Chromatogr. B.* 669:219–227.

Henderson, A.L., Schmitt, T.C., Heinze, T.M., and Cerniglia, C.E., 1997, Reduction of malachite green to leucomalachite green by intestinal bacteria, *Appl. Env. Microbiol.* 63:4099–4101.

Hines, G.A., and Watts, S.A., 1995, Non-steroidal chemical sex manipulation of tilapia, *J. World Aquacult. Soc.* 26:98–102.

IARC, 1977, IARC monographs on the evaluation of the carcinogenic risk of chemicals to man. Some miscellaneous pharmaceutical substances, Vol. 13, World Health Organization, Lyon, France.

IARC, 1980, IARC monographs on the evaluation of the carcinogenic risk of chemicals to humans: Some pharmaceutical drugs, Vol. 24, World Health Organization, Lyon, France.

Imray, F.P., and MacPhee, D.G., 1975, Induction of base-pair substitution and frameshift mutations in wild-type and repair-deficient strains of *Salmonella typhimurium* by the photodynamic action of methylene blue, *Mut. Res.* 27:299–306.

Kabata, Z., 1985, *Parasites and Diseases of Fish Cultured in the Tropics*, Taylor & Francis, London.

Kasuga, Y., Hishida, M., and Tanahashi, N., 1991, Simultaneous determination of malachite green and methylene blue in cultured fish by high-performance liquid chromatography, *J. Food Hyg. Soc. Japan* 32:137–141.

Kasuga, Y., Hishida, M., Tanahashi, N., and Arai, M., 1992, Studies on disappearance of malachite green in cultured rainbow trout, *J. Food Hyg. Soc. Japan* 33:539–542.

Kietzmann, M., Hapke, H.-J., Beeke, M., and Stehle, S., 1990, Problems in fish treatment with malachite green, *Dtsch. Tierarztl. Wschr.* 97:290–293.

Kim, S.G., Cho, J.Y., Chung, Y.S., Ahn, E.-T., Lee, K.-Y., and Han, Y.-B., 1998, Suppression of xenobiotic-metabolizing enzyme expression in rats by acriflavine, a protein kinase C inhibitor. Effects on epoxide hydrolase, glutathione *S*-transferases, and cytochromes P450, *Drug Metab. Disp.* 26:66–72.

Klein, E., Edelhauser, M., and Lippold, R., 1991, Bestimmung von malachitgrun-ruckstanden in speisefischen. 3. Mitteilung:gesamtmalachitgrun einschlieblich der leucoform, *Dtsch. Lebensm.-Rdsch.* 87:350–353.

Lanzing, W.J.R., 1965, Observations on malachite green in relation to its application to fish diseases. *Hydrobiol.* 25:426–441.

Law, F.C.P., Abedini, S., He, Y.T., and Greenlees, K., 1993, Absorption, distribution and elimination of ^{14}C-labeled malachite green by trout, Soc. Env. Toxicol. Chem. 14th Annual Meeting, November 14–18, Houston, TX.

Lim, J.T.E., Goh, C.L., Ng, S.K., and Wong, W.K., 1992, Changing trends in the epidemiology of contact dermatitis in Singapore, *Contact Dermatitis* 26:321–326.

MacPhee, D.G., and Imray, F.P., 1978, Frameshift mutagenesis by methylene blue plus visible light: implications for the Ames test, *Mut. Res.* 53:359–360.

Martin, R.L., 1968, Comparison of effects of concentrations of malachite green and acriflavine on fungi associated with diseased fish, *Prog. Fish-Cult.* 30:153–158.

Meinertz, J.R., Stehly, G.R., Gingerich, W.H., and Allen, J.L., 1995, Residues of [^{14}C]-malachite green in eggs and fry of rainbow trout, *Oncorhynchus mykiss* (Walbaum), after treatment of eggs, *J. Fish Dis.* 18:239–247.

Meyer, F.P., and Jorgenson, T.A., 1983, Teratological and other effects of malachite green on development of rainbow trout and rabbits, *Trans. Amer. Fish. Soc.* 112:818–824.

Meyer, F.P., and Schnick, R.A., 1989, A review of chemicals used for the control of fish diseases, *Rev. Aquat. Sci.* 1:693–710.

Munns, R.K., Holland, D.C., Roybal, J.E., Meyer, J.G., Hurlbut, J.A., and Long, A.R., 1992, Liquid chromatographic determination of methylene blue and its metabolites in milk, *J. AOAC Int.* 75:796–800.

Nakagawa, M., Murata, K., Shimokawa, T., Honda, T., Kojima, S., and Uchiyama, M., 1984, Determination of residual methylene blue and malachite green in muscle and liver of rainbow trout and eel, *Eisei Kagaku* 30:301–308.

Neidle, S., Pearl, L.H., Herzyk, P., and Berman, H.M., 1988, A molecular model for proflavine-DNA intercalation, *Nucleic Acids Res.* 16:8999–9016.

Okafor, N., and Taege, M., 1996, Toxicity of four commonly used chemotherapeutic compounds on hybrid fry of African catfish *Heteroclarias* (*Heterobranchus longifilis* x *Clarias gariepinus*), *J. Aquacult. Tropics* 11:113–118.

Paperna, I., 1984, Chemical control of *Amyloodinium ocellatum* (Brown 1931) (Dinoflagellida) infections: *in vitro* tests and treatment trials with infected fishes, *Aquacult.* 38:1–18.

Pfenning, A.P., 1998, Dyes. In: *Analytical Procedures for Drug Residues in Food of Animal Origin*. Chapt. 7. S.B. Turnipseed and A.R. Long, eds., Science Technology System, West Sacramento, CA. pp. 123–146.

Plakas, S.M., El Said, K.R., Stehly, G.R., and Roybal, J.E., 1995, Optimization of a liquid chromatographic method for determination of malachite green and its metabolites in fish tissues, *J. AOAC Int.* 78:1388–1394.

Plakas, S.M., El Said, K.R., Stehly, G.R., Gingerich, W.H., and Allen, J.L., 1996, Uptake, tissue distribution, and metabolism of malachite green in the channel catfish (*Ictalurus punctatus*), *Can. J. Fish. Aquat. Sci.* 53:1427–1433.

Plakas, S.M., El Said, K.R., Jester, E.L.E., Bencsath, F.A., and Hayton, W.L., 1997, Liquid chromatographic determination of acriflavine and proflavine residues in channel catfish muscle, *J. AOAC Int.* 80:486–490.

Plakas, S.M., El Said, K.R., Bencsath, F.A., Musser, S.M., and Hayton, W.L., 1998, Pharmacokinetics, tissue distribution and metabolism of acriflavine and proflavine in the channel catfish (*Ictalurus punctatus*), *Xenobiotica* 28:605–616.

Poe, W.E., and Wilson, R.P., 1983, Absorption of malachite green by channel catfish, *Prog. Fish-Cult.* 45:228–229.

Richards, R., 1977, Diseases of aquarium fish-4:Treatment, *Vet. Rec.* 101:166–167.

Rogers, A.M., and Back, K.C., 1982, Comparative mutagenicity of 4 DNA-intercalating agents in L5178Y mouse lymphoma cells, *Mut. Res.* 102:447–455.

Roybal, J.E., Munns, R.K., Holland, D.C., Hurlbut, J.A., and Long, A.R., 1992, Application of electrochemical and UV/visible detection to the LC separation and determination of methylene blue and its demethylated metabolites from milk, In: *Analysis of Antibiotic Drug Residues in Food Products of Animal Origin*, V.K. Agarwal, ed., Plenum Press, New York.

Roybal, J.E., Pfenning, A.P., Munns, R.K., Holland, D.C., Hurlbut, J.A., and Long, A.R., 1995, Determination of malachite green and its metabolite, leucomalachite green, in catfish (*Ictalurus punctatus*) tissue by liquid chromatography with visible detection, *J. AOAC Int.* 78:453–457.

Roybal, J.E., Pfenning, A.P., Turnipseed, S.B., Hurlbut, J.A., Long, A.R., 1996, Dye residues in foods of animal origin, In: *Veterinary Drug Residues*, W.A. Moats and M.B. Medina, ed., Chapt. 18, American Chemical Society, Wash., DC.

Rushing, L.G., and Thompson, H.C., Jr., 1997, Simultaneous determination of malachite green, gentian violet and their leuco metabolites in catfish or trout tissue by high-performance liquid chromatography with visible detection, *J. Chromatogr. B* 688:325–330.

Rushing, L.G., and Hansen, E.B., Jr., 1997, Confirmation of malachite green, gentian violet and their leuco analogs in catfish and trout tissue by high-performance liquid chromatography utilizing electrochemistry with ultraviolet-visible diode array detection and fluorescence detection, *J. Chromatogr. B* 700: 223–231.

Sagar, K., Smyth, M.R., Wilson, J.G., and McLaughlin, K., 1994, High-performance liquid chromatographic determination of the triphenylmethane dye, malachite green, using amperometric detection at a carbon fibre microelectrode, *J. Chromatogr A* 659:329–336.

Samuelson, J.S., Stewart, P.B., Kraft, S.C., and Farr, R.S., 1965, The effectiveness of acriflavine and ineffectiveness of proflavine as immunosuppressant agents, *J. Immunol.* 95:314–322.

Santamaria, L., Arnaboldi, A., Daffara, P., and Bianchi, A., 1979, Photocarcinogenesis by methoxypsoralen, neutral red and proflavine, *Boll. Chim. Farm.* 118.356–362.

Schmahl, G., Ruider, S., Mehlhorn, H., Schmidt, H., and Ritter, G., 1992, Treatment of fish parasites. 9. Effects of a medicated food containing malachite green on *Ichthyophthirius multifiliis* Fouquet, 1876 (Hymenostomatida, Ciliophora) in ornamental fish, *Parasitol. Res.* 78:183–192.

Schmidtke, L.M., and Carson, J., 1995, Characteristics of *Flexibacter psychrophilus* isolated from Atlantic salmon in Australia, *Dis. Aquat. Org.* 21:157–161.

Seenappa, D., and Manohar, L., 1983, *In vitro* effects of chemicals and disinfectants on the spores of *Myxobolus vanivilasae* Seenappa and Manohar, 1980 (Myxosporida: Protozoa), *Mysore J. Agricult. Sci.* 17:170–176.

Speck, W.T., and Rosenkranz, H.S., 1980, Proflavin: an unusual mutagen, *Mut. Res.* 77:37–43.

Speit, G., and Vogel, W., 1979, The effect on sister-chromatid exchanges of drugs and dyes by intercalation and photoactivation, *Mut. Res.* 59:223–229.

Stuart, N.C., 1983, Treatment of fish disease, *Vet. Rec.* 112:173–177.

Swarbrick, A., Murby, E.J., and Hume, P., 1997, Post-column electrochemical oxidation of leuco malachite green for the analysis of rainbow trout flesh using HPLC with absorbance detection, *J. Liq. Chrom. Rel. Technol.* 20:2269–2280.

Tonogai, Y., Ogawa, S., Ito, Y., and Iwaida, M., 1982, Actual survey on TLm (median tolerance limit) values of environmental pollutants, especially on amines, nitriles, aromatic nitrogen compounds and artificial dyes, *J. Toxicol. Sci.* 7:193–203.

Turnipseed, S.B., Roybal, J.E., Hurlbut, J.A., and Long, A.R., 1995a, Gas chromatographic/mass spectrometric confirmation of leucomalachite green in catfish (*Ictalurus punctatus*) tissue, *J. AOAC Int.* 78:971–977.

Turnipseed, S.B., Roybal, J.E., Rupp, H.S., Hurlbut, J.A., and Long, A.R., 1995b, Particle beam liquid chromatography-mass spectrometry of triphenylmethane dyes: application to confirmation of malachite green in incurred catfish tissue, *J. Chromatogr. B* 670:55–62.

Turnipseed, S.B., Roybal, J.E., Plakas, S.M., Pfenning, A.P., Hurlbut, J.A., and Long, A.R., 1997, Determination of methylene blue in channel catfish (*Ictalurus punctatus*) tissue by liquid chromatography with visible detection, *J. AOAC Int.* 80:31–35.

Wahli, T., Schmitt, M., and Meier, W., 1993, Evaluation of alternatives to malachite green oxalate as a therapeutant for ichthyophthiriosis in rainbow trout *Oncorhynchus mykiss*, *J. Appl. Ichthyol.* 9:237–249.

Wedemeyer, G.A., and Yasutake, W.T., 1978, Prevention and treatment of nitrite toxicity in juvenile steelhead trout (*Salmo gairdneri*), J. Fish. Res. Board Can. 35:822–827.

Wellborn, T.L., Jr., 1971, Toxicity of some compounds to striped bass fingerlings, Prog. Fish-Cult. 33:32–36.

Werth, G., 1958, Disturbances of the heredity pattern and production of tumors by experimental tissue anoxia, *Arzneimittel-Forsch.* 8:735–744.

Werth, G., and Boiteux, A., 1968a, Toxicity of triphenylmethane dyes. Malachite green as an uncoupling agent of oxidative phosphorylation *in vivo* and *in vitro*, *Archiv Toxikol.* 23:82–103.

Werth, G., and Boiteux, A., 1968b, The biologic activity of malachite green. Part VI: The detoxification of malachite green in the organism by the formation of leucomalachite green, *Arzneimittel-Forsch.* 18:39–42.

Willford, W.A., 1967, Toxicity of 22 therapeutic compounds to six fishes. U.S. Fish and Wildlife Service, *Investigations in Fish Control* 18, 10 pp.

Yu, Z., 1996, Disposition of proflavine and acriflavine in rainbow trout, Ph.D. Dissertation, The Ohio State University, Columbus, OH.

Yu, Z., Hayton, W.L., and Chan, K.K., 1997, Characterization of proflavine metabolites in rainbow trout, *Drug Metab. Disp.* 25:431–436.

12

UPTAKE, METABOLISM, AND ELIMINATION OF NICLOSAMIDE BY FISH

Verdel K. Dawson, Theresa M. Schreier, Michael A. Boogaard, and William H. Gingerich

U.S. Geological Survey
Biological Resources Division
Upper Midwest Environmental Sciences Center
2630 Fanta Reed Road
La Crosse, Wisconsin 54603

1. ABSTRACT

The lampricide 2′,5-dichloro-4′-nitrosalicylanilide (niclosamide) formulated as the 2-aminoethanol salt (Bayluscide) has been used in combination with 3-trifluoromethyl-4-nitrophenol (TFM) to control sea lampreys (*Petromyzon marinus*) in streams tributary to the Great Lakes for over 30 years. In addition, a bottom release formulation of Bayluscide has been used extensively to assess populations of larval lampreys in lenitic areas. The nature and magnitude of niclosamide residues in rainbow trout (*Onchorhynchus mykiss*) tissues and rates of elimination have been evaluated under laboratory and field exposure conditions. The uptake of niclosamide from water by rainbow trout is rapid, and residues reach equilibrium after 3 days of exposure. Only about 9% of total body niclosamide residues are found in edible fillet tissues after 24 hours of exposure. Metabolism of niclosamide by hydrolysis, reduction, or conjugation with glucuronic acid has been reported in fish, rats, cestodes, nematodes, midge larvae, mouse and sheep liver homogenates, and aerobic and anaerobic pond sediments. Rainbow trout exposed to niclosamide recently have been reported to have residues of the glucuronide conjugate and the sulfate ester in muscle fillet tissue. Fish transferred to niclosamide-free water after 24 hours of exposure to 0.02 mg/L of the chemical eliminated residues to below the limit of detection within 2–3 days.

Xenobiotics in Fish, edited by *Smith et al.*
Kluwer Academic / Plenum Publishers, New York, 1999.

2. INTRODUCTION

2.1. Background

The sea lamprey (*Petromyzon marinus*), a native parasite of the North Atlantic, entered the Great Lakes through shipping canals and was established in all the Lakes by 1940 (Smith, 1971). Lamprey predation contributed to the reduction of native lake trout (*Salvelinus namaycush*) populations to extinction in Lake Michigan and to remnant populations in Lakes Superior (Figure 1) and Huron. These changes led to the collapse of the commercial and sport fisheries in the Great Lakes and set the stage for sea lamprey control. Applegate *et al.* (1958) reported that 3-trifluoromethyl-4-nitrophenol (TFM) was selectively toxic to sea lamprey larvae. Lampricide treatments significantly reduced sea lamprey numbers and in conjunction with stocking, resulted in the gradual recovery of lake trout (Figure 1; Selgeby *et al.*, 1995).

Howell *et al.* (1964) reported that the addition of small amounts of 2′,5-dichloro-4′-nitrosalicylanilide (niclosamide) reduced the concentration of TFM needed to kill sea lamprey larvae without significantly affecting its selectivity. Niclosamide (Bayer 2353) is sold as the 2-aminoethanol salt (Bayer 73) under the commercial name Bayluscide and is available as a 70% wettable powder formulation for stream treatments. The U.S. Fish and Wildlife Service is the current registrant for niclosamide in the United States and Canada. The mode of action of niclosamide has not been completely delineated, but the selective toxicity of the material is probably based on the inability of sea lamprey larvae to conjugate and eliminate niclosamide. The basis for toxicity may be similar to that proposed for TFM by Lech and Statham (1975). In addition to being highly toxic to sea lamprey, niclosamide is also toxic to certain nontarget organisms (Marking and Hogan, 1967). For this reason, niclosamide is only used in combination treatments (Howell *et al.*, 1964) with TFM which reduces the requirement for the more expensive TFM, and as a granular, bottom-release formulation for survey and control of larval lamprey populations in lenitic habitats. When used in combination, especially for treatment of streams or rivers with high discharge rates, the recommended ratio of TFM to Bayluscide applied to streams ranges from 98:2 to 99.5:0.5 (Associate Committee on Scientific Criteria for Environmental Quality, 1985).

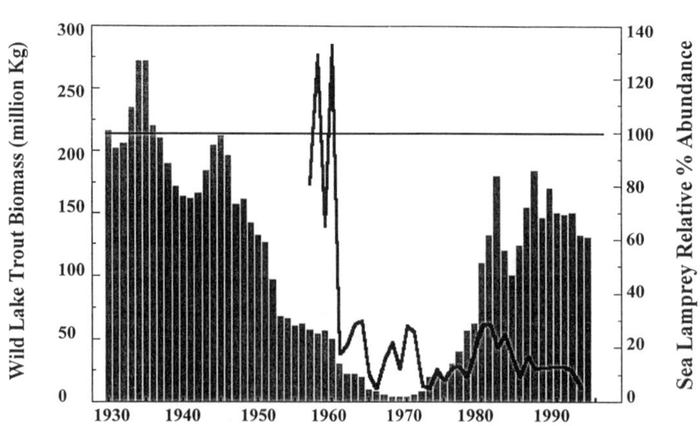

Figure 1. Yearly population estimates of lake trout (bars) and sea lamprey relative to average peak population before initiation of lampricide treatments (line) in Lake Superior. The U.S. Fish and Wildlife Service began lampricide treatments about 1960 (data courtesy of Mike Hanson, personal communication).

The use of lampricides to control sea lampreys in streams tributary to the Great Lakes is quite limited. Relatively few streams provide both the spawning substrate and burrowing habitat needed to support a population of sea lamprey larvae. Only 449 (<9%) of 5,339 streams tributary to the Great Lakes have ever been infested with sea lampreys. In a typical treatment year (1994), lampricides were applied to infested sections of 57 tributaries (1% of the total number of Great Lakes tributaries) and only 7 of those streams were treated with TFM and Bayluscide (Klar *et al.*, 1995).

The need to conduct a sea lamprey control treatment is determined through stream surveys for sea lamprey larvae. Combination treatments with TFM and niclosamide are recommended on streams or rivers with high discharge rates. Niclosamide is applied to the stream simultaneously but independently from the TFM. Both lampricides are metered into the stream for a period of time that insures the minimum lethal concentration (MLC) is maintained for 9 hours throughout the segment to be treated. This may require an application of up to 12 hours at the application site (Klar *et al.*, unpublished data). The application results in a concentration "block" of the lampricides that flows downstream. The concentrations of both lampricides in a treatment block are monitored periodically and rates of application are adjusted to maintain the desired concentrations. Typically, when used in combination with TFM, treatment concentrations of Bayluscide are less than 0.05 mg/L (Bregge, D., personal communication).

Marking and Olson (1975) reported that the concentration of TFM lethal to fingerling rainbow trout (*Oncorhynchus mykiss*) decreased as the pH of the treatment water decreased. The resulting correlation between pH and toxicity suggests that the bioavailable form of TFM is the lipid-soluble free phenol. This observation is supported by the findings of Hunn and Allen (1974) who reported that residues of TFM in channel catfish (*Ictalurus punctatus*) muscle tissue decreased with increasing pH. TFM has an ionizable proton with a pK_a of about 6.07 (Smith *et al.*, 1960) and when added to water will rapidly establish an equilibrium between the protonated and unprotonated forms.

Niclosamide also has an ionizable proton with a pK_a of about 6.25 (Williams and Heim, unpublished data), and therefore an ionization effect similar to that for TFM is expected. The equilibrium concentrations of ionized and unionized forms depend on the pH of the water, but are independent of the form of niclosamide added. Thus, at a particular pH, the toxicity of niclosamide will not depend on whether the free base or 2-aminoethanol salt was introduced; the bioavailability of niclosamide is the same.

2.2. Objectives

The purpose of this report is to review published literature on the uptake, metabolism, and elimination of niclosamide by fish and to describe results of current studies on niclosamide residues in fish.

3. PHYSICAL CHARACTERISTICS OF NICLOSAMIDE

Niclosamide (Figure 2) is a pale yellow crystal with molecular weight of 327.13 and a melting point of 225–230 °C. The ethanolamine salt of niclosamide (Bayluscide) is a yellow solid with a molecular weight of 388.1 and is available commercially as a 70% wettable powder formulation. A delayed-release formulation of Bayluscide is also available for treating larval sea lamprey populations in lenitic environments. As well as a lampricide, Bayluscide has been used as a molluscicide and an anthelmintic in veterinary medicine.

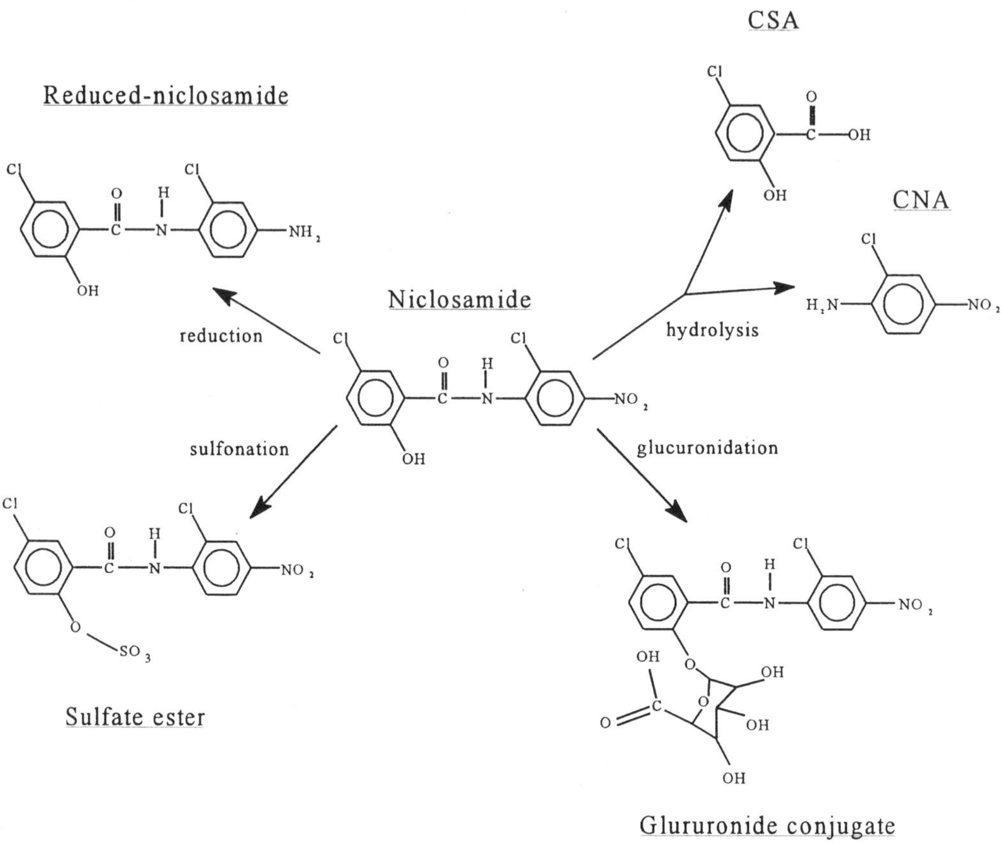

Figure 2. Chemical structure of niclosamide and several metabolites.

4. UPTAKE AND DISTRIBUTION OF NICLOSAMIDE IN FISH

Schultz and Harman (1978) exposed largemouth bass (*Micropterus salmoides*) to 0.05 mg/L of ^{14}C-niclosamide and measured residues in blood, brain, spleen, liver, kidney, bile, muscle, and head plus viscera at 2, 4, 8, 12, 24, 48, 72, and 144 hours of exposure. Residues were greatest (411 mg/kg) in the bile and least (0.67 mg/kg) in the muscle.

Hubert *et al.* (unpublished data) exposed rainbow trout to a sublethal concentration (0.005 mg/L) of ^{14}C-niclosamide for 35 days. Seven fish were sampled on exposure days 0, 3, 7, 14, 21, 28, 31, and 35 and dissected into samples of edible tissue, viscera, and residual carcass. Samples were oxidized and analyzed for radioactive residues by liquid scintillation. Radioactive residues accumulated rapidly in fillet tissue, viscera, and residual carcass during the first 3 days of exposure (Figure 3). After 3 days, residues tended to level off and remained relatively constant for the remainder of the exposure. During this time, the mean equilibrium bioconcentration factor for total niclosamide-equivalent residues for whole body was 215. The mean equilibrium bioconcentration factors were 49 for edible fillet tissue, 194 for residual carcass, and 916 for viscera (Hubert *et al.*, unpublished data). Approximately 80% of niclosamide residues concentrated in the viscera, primarily in the bile, whereas only about 9% was found in the edible fillets.

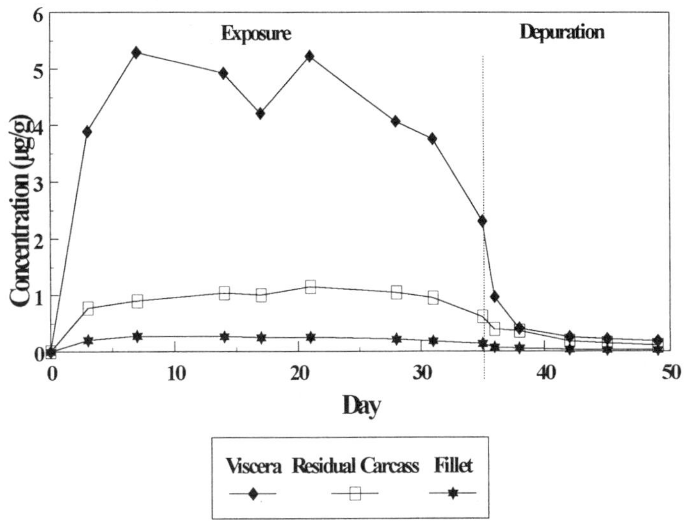

Figure 3. Concentration of ^{14}C-niclosamide-equivalents in viscera, edible tissue, and residual carcass of rainbow trout exposed to 0.005 mg/L of ^{14}C-niclosamide (Hubert at al., unpublished data).

Residues of niclosamide were measured in rainbow trout and channel catfish (200–400 g) that were exposed to 2 mg/L TFM and 0.02 mg/L of niclosamide during a stream treatment of the Ford River, Michigan in 1996. The fish were held in cages in the stream during the treatment and five fish of each species were taken for residue analysis at pretreatment and 6, 12, 18, 24, 48, 96, and 192 hours after the chemical block reached the exposure site. None of the fish exhibited signs of stress during the 12-hour treatment, although residue analysis indicated that they rapidly absorbed niclosamide. Residues were extracted from homogenized fillet tissue samples and quantified by high performance liquid chromatography (HPLC) using gradient elution and diode array detection. The maximum niclosamide concentration measured in edible fillet tissue was 0.040 ± 0.025 mg/kg for rainbow trout from the 12 hour sample and 0.047 ± 0.021 mg/kg for channel catfish from the 18 hour sample (Dawson *et al.*, unpublished data).

5. METABOLISM OF NICLOSAMIDE

Douch and Gahagan (1977) reported that a cestode (*Moniezia expansa*), a nematode (*Ascaris lumbricoides*), and enzymes from mouse and sheep liver homogenates can metabolize niclosamide by reduction to form 2′,5-dichloro-4′-aminosalicylanilide (reduced niclosamide), and by hydrolysis to form 5-chlorosalicylic acid (CSA) and 2-chloro-4-nitroaniline (CNA; Figure 2). Griffiths and Facchini (1979) described the formation of the glucuronide conjugate of niclosamide in rats and observed reduced niclosamide and its glucuronide conjugate in urine and 2′,5-dichloro-4′-salicylanilide in feces. Hydrolysis of niclosamide also reportedly occurs in midge larvae (*Chironomus tentans*; Kawatski and Zittel, 1977). Muir and Yarechewski (1982) found that ^{14}C-niclosamide in river and pond sediments degraded under both aerobic and anaerobic conditions to form reduced niclosamide (more than 50% of extractable residues) and CNA.

The authors also reported a considerable amount of ^{14}C-carbon dioxide was also trapped from respirometer flasks over a 93-day period suggesting a more complete degradation of niclosamide.

Statham and Lech (1975), in a study of niclosamide metabolism by rainbow trout, isolated a single, polar metabolite from bile. Analyses by analytical thin-layer chromatography (TLC), β-glucuronidase hydrolysis, acid hydrolysis, infrared spectroscopy, and mass spectrometry indicated that the material was the glucuronide conjugate of niclosamide.

A recent niclosamide metabolism study at the Upper Midwest Environmental Sciences Center (UMESC), La Crosse, WI, involved exposing rainbow trout (300 g) to 0.055 mg/L of ^{14}C-niclosamide for 24 hours. The exposed fish were dissected into subsamples of skin-on fillet, viscera, and residual carcass. The homogenized tissues were extracted and the resulting sample residues were analyzed by gradient HPLC using diode array detection and by TLC. Three distinct residues were isolated from the extracts of edible fillet tissue; parent niclosamide, the glucuronide conjugate of niclosamide, and the sulfate ester of niclosamide (Figure 2; Dawson *et al.*, unpublished data). Statham and Lech (1975) reported the observation of the glucuronide conjugate of niclosamide in gall bladder bile of rainbow trout. To the best of our knowledge, the presence of the glucuronide conjugate of niclosamide in fish muscle fillet tissue had not been reported prior to the study of Dawson *et al.* (unpublished data).

It is generally accepted that fish metabolize xenobiotics to create more polar compounds to facilitate urinary and biliary elimination. It is not known whether the concentrations of glucuronide conjugate we found in the muscle tissue resulted from formation in the cells of the tissue or whether it passed across the membrane from the plasma after formation in other tissue sites such as the liver or kidney. Dawson *et al.* (unpublished data) did not analyze plasma, however, Hunn and Allen (1975) reported that plasma concentrations of the glucuronide conjugate of TFM were about one third that of the concentration of TFM in rainbow trout following a 2-hour exposure to the lampricide. Based on the estimated plasma volume of rainbow trout fillet tissue reported by Gingerich *et al.* (1987), plasma concentrations of the glucuronide conjugate of niclosamide would have had to have been at least 17 mg/L to account for the glucuronide concentrations in fillet tissue unless some of the residues were within the cells of the tissues. In the case of the sulfate conjugate, the plasma concentration would have had to have been 170 mg/L if all of the conjugated residues in the fillet tissue samples were in the tissue interstitial plasma space and not within the cells of the tissues. It seems more reasonable to assume that the conjugated residues in the fillets were within the cells, but more study is needed to determine whether this is indeed true.

Kobayashi (1979) reported the formation of the sulfate conjugate of pentachlorophenol by goldfish (*Carassius auratus*). Layiwola and Linnecar (1981) reported the excretion of the sulfate conjugate of phenol by a number of fish species. James (1986) stated that xenobiotic substrates which contain hydroxy groups may be excreted as sulfate conjugates, but also reported that "the trout appears to be relatively defective in sulfate conjugation, as there was little or no evidence *in vivo* or in isolated perfused organ systems or hepatocytes for the presence of sulfate conjugates." The presence of the sulfate ester of niclosamide which had not been found previously in rainbow trout or other species of fish was confirmed by co-elution with a synthesized standard on HPLC and TLC, comparison of diode array spectra, and enzymatic digestion with sulfatase (Figure 4; Dawson *et al.*, unpublished data).

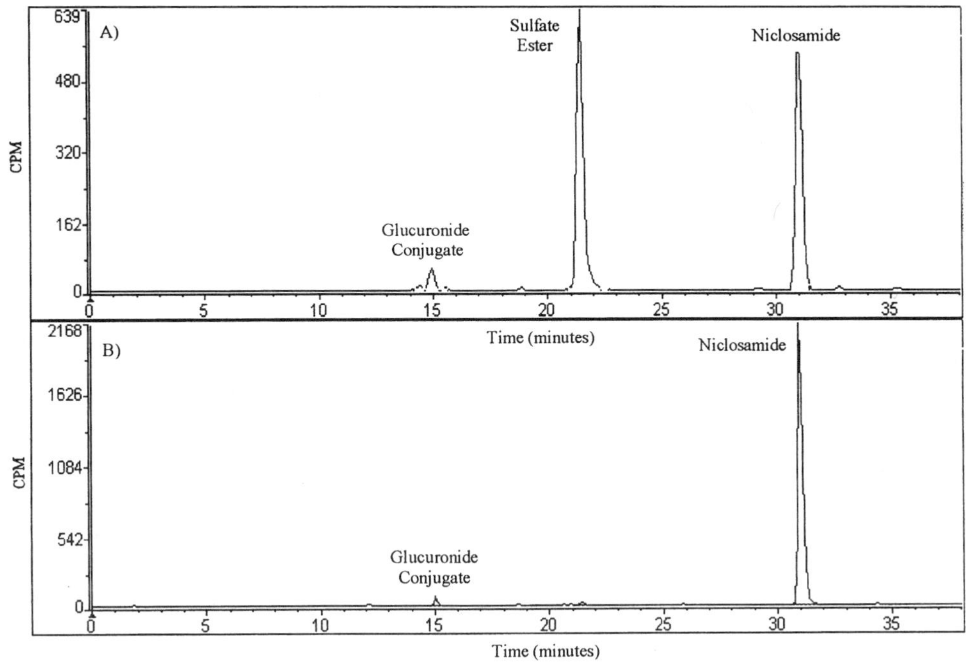

Figure 4. HPLC radiochromatograms of (A) methanol extract of fillet tissue from rainbow trout exposed to 0.055 mg/L of ^{14}C-niclosamide for 24 hours, and (B) radiochromatogram of the same extract after enzymatic digestion with sulfatase (Dawson *et al.*, unpublished data).

6. ELIMINATION OF NICLOSAMIDE

Statham and Lech (1975) showed that glucuronide conjugation and biliary excretion is an important pathway for the elimination of niclosamide by rainbow trout. They reported a bile to water ratio of 10,000 : 1 following a 24-hour exposure to niclosamide, indicating a high degree of biliary concentration. Relatively uniform elimination half-lives of 13.6, 12.8, 13.9, and 15.4 hours were reported for muscle, blood, heart, and liver, respectively.

Allen *et al.* (1979) catheterized rainbow trout and measured the excretion of residues in the urine after a 12-hour exposure to 0.05 mg/L of niclosamide in water or after intraperitoneal (ip) injection of 200 µg of the lampricide. The fish exposed by immersion began to excrete residues in the urine almost immediately, and continued to excrete residues for more than 60 hours after exposure. Rainbow trout injected with niclosamide excreted up to 25% of the dose in the urine and 20% was recovered in the bile at the end of the study; most residues were the glucuronide conjugate.

In a recent study at the UMESC, rainbow trout were exposed to 0.005 mg/L of ^{14}C-niclosamide for 35 days followed by a depuration period of 14 days. Seven fish were sampled on each of days 1, 3, 7, 10, and 14 of depuration and dissected into samples of edible tissue, viscera, and residual carcass. Then the samples were oxidized and analyzed for radioactive residues by liquid scintillation. Elimination of radioactivity from viscera and edible tissue was biphasic; both phases followed first-order kinetics. The half-lives for the two phases of residue elimination from fillet tissue were 2.9 and 121 hours (Hubert *et al.*, unpublished data).

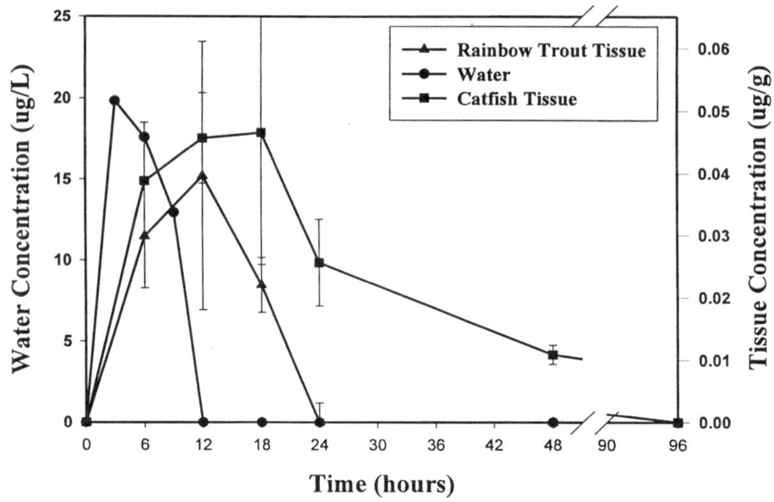

Figure 5. Mean concentrations of niclosamide in water samples taken during a lampricide treatment of the Ford River and in fillet tissue samples taken from rainbow trout and channel catfish exposed to the treatment (Dawson *et al.*, unpublished data).

Caged rainbow trout and channel catfish exposed to 2 mg/L of TFM and 0.02 mg/L of niclosamide for about 12 hours during the lampricide treatment of the Ford River, Michigan, were sampled (n = 5) for niclosamide residue analysis at pretreatment and 6, 12, 18, 24, 48, 96, and 192 hours after the chemical block reached the exposure site (Figure 5). Residues were extracted from homogenized fillet tissue samples and measured by HPLC using gradient elution and diode array detection. Residues decreased rapidly after the passage of the chemical block and were below the limit of detection within 24 hours after first exposure in rainbow trout and within 96 hours in channel catfish (Figure 5). Concentrations of niclosamide were higher and tended to persist longer in channel catfish than in rainbow trout. Parent niclosamide was the only detectible tissue residue which resulted from exposures to niclosamide in this study (Figure 6; Dawson *et al.*, unpublished data). The lack of measurable residues of the glucuronide conjugate or the sulfate ester was apparently due to the low concentration of niclosamide in the treated water. The low concentration of niclosamide used in the combined treatment with TFM was necessary to keep the lampricides sublethal for non-target fish.

7. CONCLUSIONS

The uptake of niclosamide from water by rainbow trout is rapid, and residues reach equilibrium after 3 days of exposure. Only about 9% of total body niclosamide residues in rainbow trout are found in edible fillet tissues. Rainbow trout can metabolize niclosamide to form the glucuronide conjugate and the sulfate ester of niclosamide. However, these metabolites were not found in fillets of rainbow trout or channel catfish exposed during a lampricide treatment, probably due to the low concentrations of niclosamide used in combination treatments with TFM. Rainbow trout use biliary and urinary routes to eliminate niclosamide residues. Rainbow trout transferred to

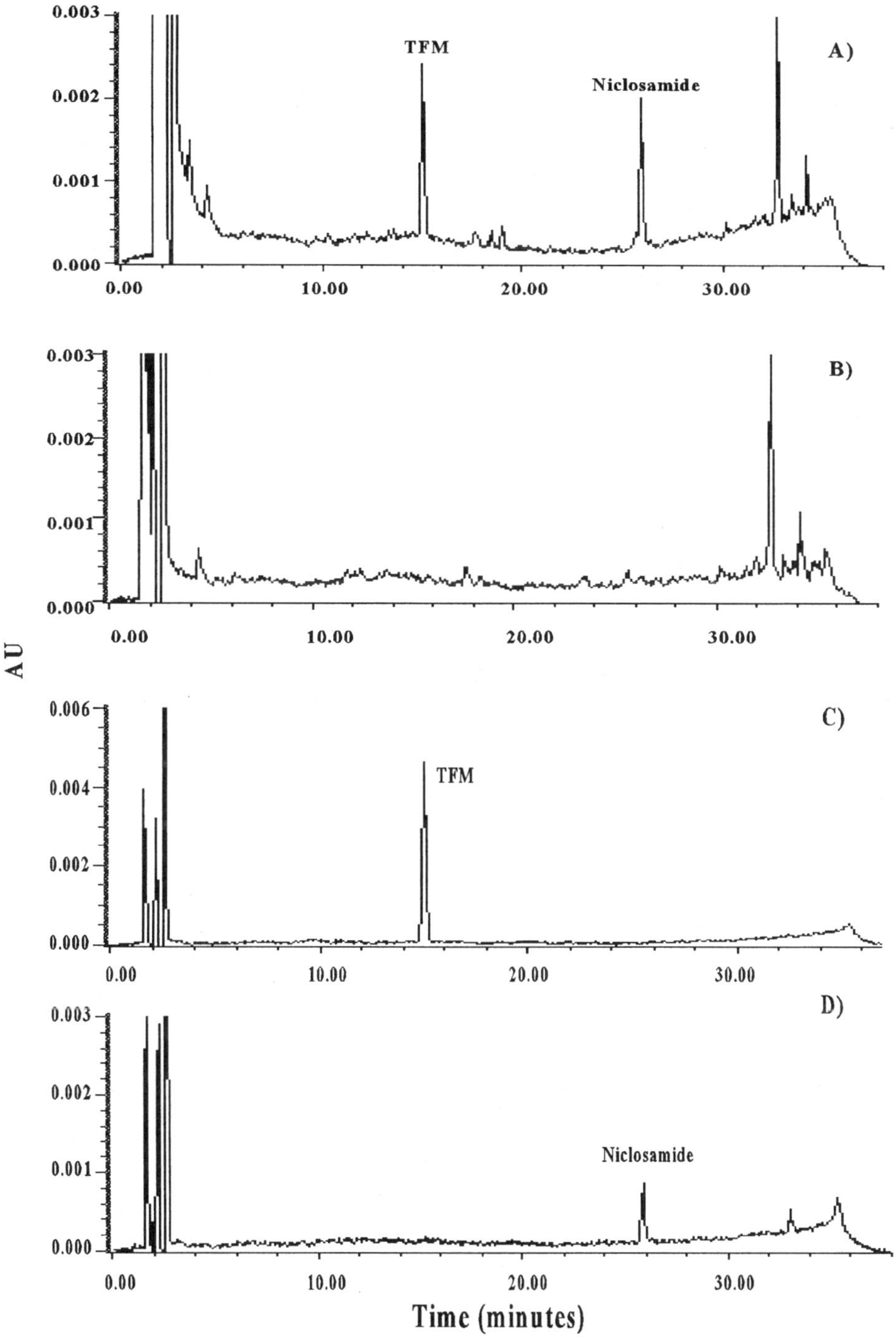

Figure 6. HPLC chromatograms of (A) acetone extract of fillet tissue from rainbow trout exposed to TFM and niclosamide for 12 hours during a lampricide treatment of the Ford River, (B) Control tissue, (C) TFM standard, and (D) niclosamide standard (Dawson *et al.*, unpublished data).

niclosamide-free water eliminate niclosamide residues to below the limit of detection within 2–3 days.

8. REFERENCES

Allen, J.L., Dawson, V.K., and Hunn, J.B., 1979, Excretion of the lampricide Bayer 73 by rainbow trout. In: *Aquatic Toxicology*. Marking, L.L., and R.A. Kimerle, eds. pp 52–61.

Applegate, V.C., Howell, J.H., and Smith, M.A., 1958, Use of mononitrophenols containing halogens as selective sea lamprey larvicides. *Science* 127:336–338.

Associate Committee on Scientific Criteria for Environmental Quality. 1985, TFM and Bayer 73: Lampricides in the Aquatic Environment. *Natl. Res. Counc. Can.*, No. 22488. 184 pp.

Douch, P.G.C., and Gahagan, H.M., 1977, The metabolism of niclosamide and related compounds by *Moniezia expansa*, *Ascaris lumbricoides* var *suum*, and mouse- and sheep-liver enzymes. *Xenobiotica*. 7:301–307.

Gingerich, W.H., Pityer, R.A., and Rach, J.J., 1987, Estimates of Plasma, Packed Cell and Total Blood Volume in Tissues of the Rainbow Trout (Salmo gairdneri). *Comp. Biochem. and Physiol.* 87A:251–256.

Griffiths, L.A., and Facchini, V., 1979, The major metabolites of niclosamide: indentification by mass spectrometry. *Recent Dev. Mass Spectrom. Biochem. and Med.* 2:121–126.

Howell, J.H., King, E.L., Jr., Smith, A.J., and Hanson, L.H., 1964, Synergism of 5, 2′-Dichloro-4′-nitrosalicylanilide and 3-trifluormethyl-4-nitrophenol in a selective lamprey larvicide. *Gt. Lakes Fish. Comm. Tech. Rep.* 8:1–21.

Hunn, J.B., and Allen, J.L., 1974, Movement of Drugs Across the Gills of Fishes. *Annu. Rev. Pharmacol.* 14:47–55.

Hunn, J.B., and Allen, J.L.,. 1975, Residue Dynamics of Quinaldine and TFM in Rainbow Trout. *Gen. Pharmacol.* 6:15–18.

James, M.O., 1986, Xenobiotic Conjugation in Fish and Other Aquatic Species. In: *Xenobiotic Conjugation Chemistry*. Paulson, G.D., J. Caldwell, D.H. Hutson, and J.J. Menn, eds. pp 29–47.

Kawatski, J.A., and Zittel, A.E., 1977, Accumulation, elimination, and biotransformation of the lampricide 2′,5-dichloro-4′-nitrosalicylanilide by *Chironomus tentans*. U.S. Fish Wild. Serv. *Investig. Fish Control.* 79:1–8.

Klar, G.T., Schleen, L.P., and Young, R.J., 1995, Integrated management of sea lampreys in the Great Lakes 1994, Annual report to the Great Lakes Fishery Commission. June 5–7, 1995, 75 pp.

Kobayashi, K., 1979, Metabolism of Pentachlorophenol in Fish. In: *Aquatic Toxicology*. Marking, L.L., and R.A. Kimerle, eds. pp 131–143.

Layiwola, P.J., and Linnecar, D.F.C., 1981, The Biotransformation of (Super(14)C)Phenol in Some Freshwater Fish. *Xenobiotica*, 11:167–171.

Lech, J.J., and Statham, C.N., 1975, Role of Glucuronide Formation in the Selective Toxicity of 3-Trifluoromethyl-4-nitrophenol (TFM) for the Sea Lamprey: Comparative Aspects of TFM Uptake and Conjugation in Sea Lamprey and Rainbow Trout. *Toxicol. Appl. Pharmacol.* 31:150–158.

Marking, L.L., and Hogan, J.W., 1975, Toxicity of Bayer 73 to fish. U.S. Fish Wildl. Serv. *Investig. Fish Control.* 19:1–13.

Marking, L.L., and Olson, L.E., 1975, Toxicity of the lampricide 3-Trifluoromethyl-4-nitrophenol (TFM) to nontarget fish in static tests. U.S. Fish Wildl. Serv. *Investig. Fish Control.* 60:1–27.

Muir, D.C.G., and Yarechewski, A.L., 1982, Degradation of niclosamide (2′,5-dichloro-4′-nitrosalicylanilide) in sediment and water systems. *J. Agric. Food Chem.* 30:1028–1032.

Schultz, D.P., and Harman, P.D., 1978, Uptake, distribution, and elimination of the lampricide 2′,5-dichloro-4′-nitro[^{14}C]salicylanilide (Bayer 2353) and its 2-aminoethanol salt (Bayer 73) by largemouth bass. *J. Agric. Food Chem.* 26:1226–1230.

Selgeby, J.H., Eshenroder, R.L., Krueger, C.C., Marsden, J.E., and Pycha, R.L., 1995, International Conference on Restoration of Lake Trout in the Laurentian Great Lakes. *J. Great Lakes Res.* Vol. 21, Supplement 1.

Smith, B.R., 1971, Sea lampreys in the Great Lakes of North America, in: *The Biology of Lampreys*. Vol. 1. Hardisty M.W., and I.C. Potter, eds., Academic Press, New York.

Smith, M.A., Applegate, V.C., and Johnson, B.G.H., 1960, Colorimetric determination of halogenated nitrophenols added to streams as sea lamprey larvicides. *Anal. Chem.* 32:1670–1675.

Statham, C.N., and Lech, J.L., 1975, Metabolism of 2′,5-dichloro-4′-nitrosalicylanilide (Bayer 73) in rainbow trout (*Salmo gairdneri*). *J. Fish. Res. Board Can.* 32:515–522.

13

UPTAKE, METABOLISM, AND ELIMINATION OF TFM BY FISH

Terrance D. Hubert, Jeffry A. Bernardy, Chue Vue, and William H. Gingerich

United States Geological Survey
Biological Resources Division
Upper Midwest Environmental Sciences Center
2630 Fanta Reed Road
La Crosse, Wisconsin 54603

1. ABSTRACT

The larval lampricide 3-trifluoromethyl-4-nitrophenol (TFM) has been used to control sea lamprey (*Petromyzon marinus*) in the Great Lakes basin for over 30 years. Studies have been conducted to (1) determine the nature and magnitude of TFM and TFM residues in fish fillet tissue under acute exposure conditions, and (2) determine the rates of uptake, elimination, and the nature and magnitude of TFM and TFM residues in fish tissues under subchronic exposure conditions. Data from these studies were submitted to the U.S. Environmental Protection Agency to support the reregistration of TFM. An historical perspective on the sea lamprey control program, data from previous research on the nature of TFM residues in fish, and the results of the studies conducted at UMESC are presented.

1. INTRODUCTION

1.1. History

The sea lamprey (*Petromyzon marinus)* became a problem in the Great Lakes after the Welland Canal was opened in 1829 (ACSCEQ, 1985). Adult sea lamprey are parasites which attach to species such as lake trout, whitefish, and chub and survive on their blood and body fluids. This results in either reduced size or death of the host. Sea lamprey were a major factor in the decline of populations of commercially and recreationally important fish species during the 1940s and 1950s. This, in turn, had a tremendous impact on the fishing industry of the Great Lakes. To address this problem, the Great Lakes

Xenobiotics in Fish, edited by *Smith et al.*
Kluwer Academic / Plenum Publishers, New York, 1999.

Fishery Commission was formed in an agreement between the United States and Canada in 1955. The charge of the Commission, which continues today, is to develop research programs and to recommend measures that will permit maximum sustained productivity of fish populations in the Great Lakes. The Commission also has the responsibility for oversight of the Sea Lamprey Control Program.

The Commission initiated a program to identify chemicals that would be selectively toxic to sea lamprey. Approximately 6,000 chemicals, primarily organic, were surveyed and, of this total, 15 halogenated mononitrophenols were found to have some degree of selective toxicity to the lamprey. Of these, 3-trifluoromethyl-4-nitrophenol (TFM) was found to be the most effective (Applegate *et al.*, 1958). Sea lamprey control treatments with TFM were initiated in Lake Superior in 1958 (ACSCEQ, 1985). The program was expanded to include Lakes Michigan and Huron in 1960. Lake Ontario was added in 1971, and the first treatments in Lake Erie began in 1986.

The target of the control program is the larval stage of the lamprey, known as the ammocoete. Sea lamprey spawn in tributaries of the Great Lakes that have suitable habitat (ACSCEQ, 1985). After the eggs hatch, the larvae burrow into the sediment where they live for up to seven years, feeding on organic matter in the stream. After an initial TFM treatment streams must be retreated every 3 to 5 years due to recolonization.

Alternative control measures are being explored. These include physical barriers to streams where lamprey spawn, such as electrical weirs and high-current raceways. Also under development is a program in which male lampreys are captured, injected with the chemosterilant bisazire, and released into spawning areas. Sterile male lampreys interfere with the spawning process. Although these programs have some impact on sea lamprey populations, by far the most effective means of sea lamprey control continues to be the chemical control program.

TFM is the most effective chemical for controlling sea lamprey, but it is also the most expensive. In instances where large amounts of TFM would be required, such as treatment of a river with a high discharge rate, it is often used in combination with Bayluscide (niclosamide; Bayer 73) to reduce the amount of costly TFM required. When used in combination, the TFM : Bayer 73 ratio ranges from 98 : 2 to 99.5 : 0.5 (ACSCEQ, 1985).

In 1988, Congress required that all chemicals registered with the Environmental Protection Agency (EPA) before 1984 go through a reregistration process. This has necessitated the reregistration of TFM and Bayer 73. The Upper Midwest Environmental Sciences Center (UMESC) has been charged with the responsibility of providing data required by the EPA for reregistration. Both niclosamide and TFM are required to be registered with Health Canada, and the UMESC has coordinated this effort.

1.2. TFM Physical Characteristics

The structure of TFM is shown in Figure 1. TFM is a phenol containing a nitro group in the para position and a trifluoromethyl group in the meta position. The mole-

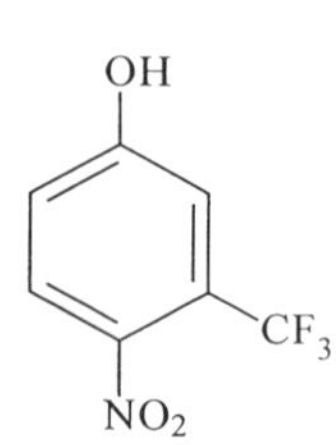

Figure 1. Structure of TFM (3-trifluoromethyl-4-nitrophenol).

cular weight is 207.11 g/mole. The compound has a melting point of 75–76 °C and a pK_a of 6.07 (ACSCEQ, 1985). It is soluble in alcohol, acetone, ether, benzene, toluene, and dimethylformamide. The sodium salt is infinitely soluble in water, and is the form used in the sea lamprey control program.

2. ACCUMULATION AND METABOLISM OF PHENOLS

Lech *et al.* (1978) studied the uptake and metabolism of pentachlorophenol in rainbow trout. Animals were exposed to a 0.025 µg/mL concentration of the chemical in water for 24 hours. Though not expressly quantitated, the uptake was described as rapid. Analysis of tissues following the exposure period showed the concentration of parent pentachlorophenol to be 16 µg/g in the liver, 6.5 µg/g in the blood, 6.0 µg/g in the fat, and 1.0 µg/g in the muscle. Elimination of the chemical was also rapid, with reported half-lives of 9.8 hr for liver, 6.2 hr for blood, 23 hr for fat, and 6.9 hr for muscle.

Niimi and Cho (1983) studied the accumulation of pentachlorophenol in salmonids. Fish were fed diets containing from 40–3,000 µg/kg pentachlorophenol for up to 110 days. At the 40 µg/kg dose rate, animals maintained whole body levels of 2 µg/kg parent compound over the study period, while at the 3,000 µg/kg dose rate the concentration of parent increased to 40 µg/kg for the first 40 days, then decreased to 20 µg/kg at the end of the study. The biological half-life of the parent was estimated to be approximately 7 days.

Rogers *et al.* (1988) studied the bioconcentration of chlorophenols in juvenile chinook salmon. Fish were subject to exposure conditions that simulated winter conditions in the upper Fraser River. Fingerling chinook were exposed to waterborne tetrachlorophenol and pentachlorophenol at a concentration of 2 µg/g for a period of 62 days at a temperature of 0.7 °C. In whole fish, concentrations of tetrachlorophenol and pentachlorophenol were 224 ng/g and 431 ng/g, repsectively, at the end of the exposure period.

Two major metabolic pathways exist for phenolic compounds in fish species, sulfation and glucuronidation (Figure 2). Kobayashi and coworkers (1977, 1978, 1979) reported that goldfish exposed to pentachlorophenol excreted the xenobiotic in the form of the sulfate conjugate. Examination of the bile, however, revealed the presence of the glucuronide conjugate. Lech *et al.* (1978) reported high concentrations pentachlorophenol glucuronide in the bile of rainbow trout in their research. Layiwola and Linnegar (1981) demonstrated that both glucuronide and sulfate conjugates were excreted by bream, perch, roach, and rudd when exposed to ^{14}C-phenol, though only the sulfate conjugate was excreted by goldfish, guppy, minnow, and perch.

Stehly and Hayton (1989) demonstrated that biliary metabolites of pentachlorophenol alone were poor indicators of the major routes of elimination of this chemical. The amounts of metabolites in bile and exposure water were species-dependent, and all of the metabolites excreted into the water were sulfate conjugates,

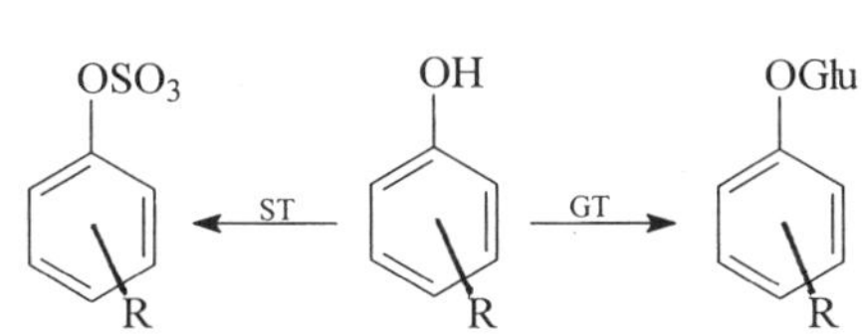

Figure 2. Metabolism of phenols by fish. R = H, alkyl, or halogen; GT = glucuronyltransferase; ST = sulfotransferase; Glu = glucuronic acid.

while those in the bile were glucuronide conjugates. An excellent review by Clarke *et al.* (1991) showed that the synthesis of glucuronides by microsomal UDP-glucuronyltransferase is a signifcant pathway of inactivation and excretion of both endogenous and xenobiotic compounds in fish. Formation of *O*-glucuronides, *N*-glucuronides, *S*-glucuronides, and *C*-glucuronides have been reported in fish. Finally, the *in vivo* metabolism of pentachlorophenol and aniline in arctic charr larvae was studied by Cravedi *et al.* (1995). Fish exposed to ^{14}C-pentachlorophenol in a static bath for 48 hours metabolized the material into sulfate and glucuronide conjugates.

3. UPTAKE AND DISTRIBUTION OF TFM

Schultz and coworkers studied the uptake, metabolism, and elimination of TFM in largemouth bass (Schultz *et al.*, 1979). Animals were exposed to a 1.0 μg/g solution of ^{14}C-TFM for up to 24 hours. The greatest concentrations of TFM-related residues after 8 hours were in brain and muscle tissue, and in blood, liver, kidney, and head/viscera after 12 hours. Concentrations of radioactive residues increased throughout the exposure period in the bile.

A field trial to characterize the uptake of TFM, and elimination of TFM and TFM-glucuronide, used rainbow trout (*Oncorhynchus mykiss*) and channel catfish (*Ictalurus punctatis*) as the test organisms (Hubert *et al.*, unpublished report). Rainbow trout were chosen because they inhabit the streams where sea lamprey control treatments occur. Channel catfish were chosen to provide data on a species that is a bottom-feeder (EPA, 1982). This study was conducted during a regularly scheduled sea lamprey control treatment of the Ford River in the upper peninsula of Michigan in June of 1996. The U.S. Fish and Wildlife Service (US FWS) Sea Lamprey Control groups based in Marquette and Ludington, Michigan were responsible for conducting the treatment. The target treatment concentration was set at 1.5 times the minimum lethal concentration for sea lamprey, or 4.1 mg/L, based on the water quality parameters of the Ford River. The objective of the lamprey control operation was to maintain the minimum lethal TFM concentration for a period of 12 hours. Such treatment results in what is termed a "chemical block" of lampricide which flows downstream with the current.

Rainbow trout and channel catfish were approximately 1 year old and weighed approximately 300 g and 275 g, respectively. The site selected for the exposure of rainbow trout and channel catfish was the upper reach of the Ford River. The site selected for placement of the fish that served as untreated controls was located upstream from the TFM application point. At the exposure site, which was downstream from the TFM application point, rainbow trout and channel catfish were placed into wire-mesh cages. Fish and water samples were collected before the treatment began and at 6, 12, 18, 24, 48, 96, and 192 hours after the chemical block reached the downstream site. At each sample interval, a single cage containing fish exposed to the chemical was removed from the downsream site, and unexposed rainbow trout and channel catfish were removed from the cage at the upstream site. Frozen samples of skin-on fillet for rainbow trout and skinless fillet for channel catfish were removed from the fish and frozen. Samples were transported on dry ice to the UMESC where they were transferred to frozen storage at or below −20 °C.

Analysis of water samples collected at the downstream site during treatment showed the maximum concentration of TFM to be 2,700 ng/mL at the 6 hour sampling interval. This was well below the target concentration of 4,100 ng/mL. Samples collected and analyzed by the US FWS indicated a TFM concentration of approximately

2,800 ng/mL, a value consistent with that measured in our laboratory. TFM concentration decreased to approximately 1,900 ng/mL by the 12 hour interval; TFM was not detectable in water samples after 96 hours (limit of detection = 0.81 ng/mL). No TFM was detected in the control or pretreatment water samples.

Rainbow trout and channel catfish fillet samples were homogenized with dry ice. The thawed samples were extracted with methanol and water and the analytes of interest were concentrated from the extracts by solid-phase extraction and analyzed by HPLC. Maximum residues of TFM and TFM-glucuronide in rainbow trout fillet tissue were measured in fish collected at the 12-hour sampling interval. The average concentration of TFM in those samples was 300 ± 130 ng/g, while the concentration of TFM-glucuronide was 400 ± 180 ng/g. In channel catfish, the highest concentration of TFM was found in the fillet samples collected at 6 hours and the highest concentration of TFM-glucuronide was found in samples collected at 12 hours. The mean concentration of TFM at the 6-hour interval was 1,400 ± 240 ng/g, while that of TFM-glucuronide was 900 ± 360 ng/g at the 12-hour sampling interval. In general, concentrations of TFM and TFM-glucuronide were higher in channel catfish than in rainbow trout, but the residues persisted longer in rainbow trout.

In another study rainbow trout selected for a laboratory bioaccumulation study were approximately 2 g at the initiation of the test (Hubert *et al.*, unpublished report). An aqueous solution of ^{14}C-TFM was administered to the test tank at a flow rate that would give a constant concentration of approximately 62 μg/L, which is 0.01 the reported 96-hour LC_{50} value (Marking and Olson, 1975). The exposure period was 28 days, followed by a 15-day depuration period.

Samples of test and control fish were collected at 13 intervals during the exposure phase and 8 intervals during the depuration phase. Each fish was weighed and dissected into samples of edible tissue (skin, bone, fillet, tail, and fins), viscera, and head. Determination of uptake and elimination rate constants and bioconcentration factors were based on total TFM-equivalent radioactivity. Therefore, the values determined for these variables likely represent their greatest magnitude.

Radioactive residues accumulated rapidly in rainbow trout during the first 24 hours of the exposure. Radioactive residues in the tissues plateaued by 72 hours and then remained relatively constant over the next 600 hours (25 days). The estimated uptake rate for TFM into whole body, based on the initial uptake curve was 0.0098 L/hour. Concentration of radioactive residues were 0.26 μg/g in whole body, 0.05 μg/g in edible tissue, 0.10 μg/g in head, and 1.6 μg/g in viscera after the first 24 hours of the exposure phase. Viscera contained the highest percentage of radioactive residues during the uptake period at approximately 80%, followed by edible tissue at approximately 11%, and head at approximately 9% (Figure 3).

The apparent equilibrium period ran from days 3 to 28 (72 to 672 hours) of the exposure. During this time, the mean apparent equilibrium bioconcentration factor for whole body, based on total radioactive residues in tissue and the concentration of parent TFM in water, was 8.33. The mean apparent equilibrium bioconcentration factors were 0.83 for the edible tissue, 1.00 for the head, and 46.4 for the viscera, based on total radioactivity in each component. Wet weight, whole body TFM-equivalent concentration of radioactivity during the equilibrium period averaged 0.44 μg/g. The highest tissue concentration of parent-equivalent radioactivity during the equilibrium period was 3 μg/g in the viscera collected on several different days during the exposure period. TFM-equivalent residues in all tissues ranged from 0.027 to 3 μg/g. During the equilibrium period, viscera contained an average of 89% of the radioactive residues; the head, 4%; and the edible tissue 7%.

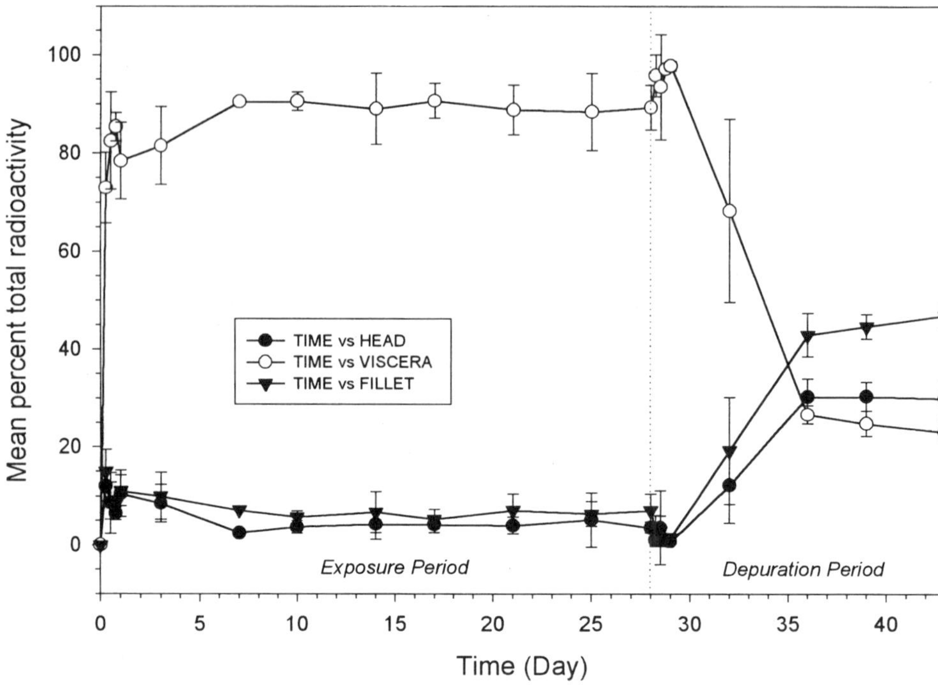

Figure 3. Mean percent of total radioactivity in edible tissue, head, and viscera of rainbow trout during exposure to 60 μg/L ^{14}C-TFM and during a depuration period (Hubert *et al.*, unpublished report). The apparent equilibrium period was from day 3 to day 28 of the exposure. Values represent the mean of seven replicate samples at each sample interval.

In a study designed to determine the nature of TFM residues, rainbow trout were exposed to 2.1 mg/L ^{14}C-TFM for 24 hours (Hubert *et al.*, unpublished report). Test animals used were approximately 300 g and 1.5 years old. Fish were exposed in an aerated static water bath. Samples of skin-on fillet, viscera, and residual carcass were frozen and then homogenized with dry ice and subsamples were analyzed to determine the quantity of radioactive residues in each tissue. Mean tissue concentrations of radioactivity (TFM-equivalents) after the 24 hour exposure period were approximately 57.0 μg/g in the viscera, 0.740 μg/g in the residual carcass (head, bones, and fins) and 0.472 μg/g in the fillet. Mean percent distribution of TFM was 91% (1.95 μg) in the viscera, 5% (0.094 μg) in the residual carcass and 4% (0.092 μg) in the skin-on fillet.

4. METABOLISM OF TFM

Lech and coworkers have done extensive work on the metabolism of TFM in rainbow trout (Lech and Costrini, 1972; Lech *et al.*, 1973; Lech, 1973, and Lech and Statham, 1975). In the earliest study, rainbow trout of an unspecified age were exposed to 0.5 ppm ^{14}C-TFM in an aerated static water bath at 12°C, and pH 7.2. Bile collected from each fish after euthanization was pooled. Dosing water from the exposure tank was acidified to pH 4.0, pumped through XAD-2 resin and the XAD-2 resin was washed with methanol to collect TFM related materials. Bile was diluted with water and subjected to the same purification procedure as the tank water. Samples were analyzed by thin-layer chromatography on silica gel. The major metabolite was TFM-glucuronide, which was isolated from both bile and tank water. The identity of the conjugate was confirmed by preparation of the standard material from liver extracts employing UDPGT, and by hydrolysis experiments using β-glucuronidase with and without saccharo-1,4-lactone, a known inhibitor of the enzyme. The identity was further confirmed analytically by thin-layer chromatography, uv-visible and infrared spectroscopy, gas chromatography, and mass spectrometry.

In the same study, Lech demonstrated the reduction of the nitro group of TFM *in vitro* by nitroreductase to produce the aminophenol. The reaction was enhanced by adding FAD and employing an NADPH regenerating system. Experiments with liver and kidney extracts produced N-acetylated-reduced-TFM in a reaction that required the addition of acetyl-CoA. However, neither of these metabolites have been demonstrated to be produced *in vivo* in rainbow trout (Figure 4).

Additional research by Lech and Statham established glucuronidation of TFM as a major contributor to the selective toxicity of TFM in the sea lamprey (Lech and Statham, 1975). When sea lamprey ammocoetes and rainbow trout were exposed to identical water concentrations of the lampricide, the ammocoetes exhibited a higher rate of accumulation of TFM and an increased steady-state concentration compared to the rainbow trout. In vitro assays using glucuronyl transferase showed that the rate of formation of the glucuronide in sea lamprey was lower than rainbow trout. Studies with living organisms revealed that sea lamprey had a much higher ratio of free to conjugated TFM than rainbow trout. Lech conducted experiments in which trout and sea lamprey were pretreated with salicylamide, a known inhibitor of glucuronyl transferase, before being exposed to TFM. Pretreatment with salicylamide greatly reduced the LC_{50} for trout, but did not alter the LC_{50} for the lamprey.

Schultz and coworkers made limited attempts to characterize ^{14}C-TFM residues in largemouth bass (Schultz *et al.*, 1979). The work confirmed the presence of TFM-

Figure 4. Metabolism of TFM by rainbow trout (Lech and Costrini, 1972). Reduced TFM, N-acetylated-reduced-TFM, and sulfated TFM have not been identified in the tissues of living animals.

glucuronide in the bile. Hexane extracts of samples of muscle and head/viscera contained ^{14}C-TFM and unidentified radioactive components. The same tissue samples were subsequently extracted with methanol, and the methanol extracts contained small amounts of ^{14}C-TFM but large amounts of unidentified, polar radioactive components.

Kane *et al.* (1994) studied the glucuronidation kinetics of TFM in sea lamprey, channel catfish, rainbow trout, and bluegill in vitro with hepatic microsomes. The maximal velocity was much greater in bluegills, channel catfish, and rainbow trout than in sea lamprey. Furthermore, the binding affinities varied significantly in the order of bluegill > rainbow trout > channel catfish > sea lamprey. The efficiency of glucuronidation in all species was influenced more by the binding affinities than by the maximal velocity of the reaction. The ratio of maximal velocity/binding affinity was shown to be lower for species that were previously reported to be more sensitive to TFM.

As part of the study to determine the nature of TFM residues in fish, which was described in Section 3, efforts were made to characterize residues in the edible fillet tissue, the primary route of human exposure (EPA, 1982). Samples were extracted with methanol and water, which removed 99% of the total terminal radioactive residue (TRR). Analysis of the methanol extracts by high-performance liquid chromatography (HPLC) and thin-layer chromatography (TLC) revealed the presence of two radioactive peaks which accounted for approximately 81% of the TRR. The radioactive peak corresponding to TFM accounted for approximately 44% of the TRR. The second radioactive peak accounted for approximately 36% of the TRR and was suspected to be the glucuronide conjugate of TFM. In a pilot study (Hubert and Bernardy, unpublished report), TFM-glucuronide was isolated from the bile of rainbow trout treated with 1.0 mg/L ^{14}C-TFM. The UV spectrum of the metabolite isolated from fillet tissue was similar to the UV spectrum of TFM-glucuronide isolated from the bile. Hydrolysis experiments on the fillet tissue metabolites were conducted using β-glucuronidase. As described by Lech (1973), saccharo-1,4-lactone, a specific inhibitor of β-glucuronidase, inhibited the hydrolysis of the metabolite. These experiments established the identity of the second peak of radioactivity as the glucuronide conjugate of TFM.

As part of the bioaccumulation study described in Section 3, the nature of TFM residues in edible tissue and viscera samples collected at one sampling interval during the exposure phase of this study was examined. Methanol extraction removed an average of 98% of the radioactive residues from edible tissue. Relative to the original concentration of TFM-equivalent radioactive residues in the tissue sample, the concentration of TFM-equivalent radioactive residues extracted into this fraction was 0.041 μg/g. An additional 3% were extracted into the water, and approximately 4% remained in the pellet. Chromatographic analysis of the methanol extract indicated that approximately 90% of the residues were the glucuronide conjugate of TFM, and 10% was parent TFM (Table 1).

Methanol extracts of viscera contained 97% of the TRR in viscera. Relative to the original concentration of TFM-equivalent radioactive residues in the tissue sample, the concentration of TFM-equivalent radioactive residues in this fraction was 2.8 μg/g. An additional 2.2% was extracted into the water, and 4.6% remained in the pellet. Chromatographic analysis of the methanol extract showed that 36% of the residue was the glucuronide conjugate of TFM and 57% was parent TFM. The remaining 7% of the radioactivity was represented by three unknowns, which were chromatographically more polar than TFM.

The current theory on the metabolism of xenobiotics in fish holds that biotransformation into more polar metabolites facilitates elimination in the urine or bile. Conjugated metabolites, if present in muscle tissue, would be expected to be present in low

Table 1. Distribution of TFM-equivalent radioactivity in extracts of edible tissue and viscera of rainbow trout exposed to 60 μg/L ^{14}C-TFM. Tissues were sampled at 504 hours (21 days) of the exposure phase (Hubert *et al.*, unpublished report).

Tissue	Residue	Concentration (μg/g)	Percent
Edible Tissue	TFM	0.006 ± 0.0057	10 ± 13
	TFM-Glu	0.036 ± 0.0030	90 ± 13
Viscera	TFM	1.4 ± 0.05	57 ± 5.2
	TFM-Glu	0.9 ± 0.17	36 ± 4.1

concentrations relative to the parent compound. The observation that TFM-glucuronide is present in the fillet tissue in concentrations approaching the concentration of parent TFM for the acute exposure, and exceeding the concentration of parent TFM for the chronic, low level exposure is not consistent with what might be expected.

5. ELIMINATION OF TFM

Schultz *et al.* (1979), exposed largemouth bass to 1.0 mg/L ^{14}C-TFM for 12 hours after which the fish were transferred to lampricide-free flowing water for up to 72 hours. The concentrations of radioactive residues in blood, brain, bile, liver, kidney, muscle, and head/viscera decreased with time during the depuration. No TFM residues were detected in the muscle tissue 12 hours after the fish were transferred to fresh water.

A field study with rainbow trout and channel catfish showed that TFM residue concentrations decrease rapidly, becoming less than 100 ng/g within 6 to 12 hours after the TFM chemical block has passed the exposure site. In the bioaccumulation study, it was found that the elimination of TFM residues from the whole body of rainbow trout was monoexponential, with a rate constant of 0.0387 L/h ($T_{1/2}$ = 17.9 h). Elimination of TFM residues from the viscera was also monoexponential, with a rate constant of 0.0371 L/h. During the depuration phase, concentrations of radioactive residues in the viscera, head, and edible tissue decreased from 2.1 μg/g, 0.01 μg/g, and 0.02 μg/g, respectively, to less than the limit of detection (viscera LOD = 0.00230 μg/g; head LOD = 0.00120 μg/g; edible tissue LOD = 0.000519 μg/g).

6. CONCLUSIONS

In summary, the major metabolite of TFM in fish exposed to TFM is TFM-glucuronide. The presence of TFM-glucuronide in the muscle tissue of exposed fish at concentrations greater than the parent material in both acute, high concentration exposures or chronic, low concentration exposures is not consistent with current theory regarding the disposition of glucuronide conjugates in fish, and needs further investigation. While no reduced-TFM, N-acetylated-reduced-TFM, or sulfated-TFM were detected in the fillet tissue of rainbow trout, the compounds may be present in the viscera of exposed animals. They do not represent major elimination products of TFM. Uptake of TFM is rapid, with the viscera being the main depot. Residues in the fillet tissue are low when compared to the viscera. Whole body elimination of TFM and its metabolites to concentrations below the limit of analytical detection is also rapid, generally being complete within hours of transfer of the animal to fresh water.

7. ACKNOWLEDGMENTS

We would like to thank Dr. Guy Stehly and Dr. Ronald Amel for their helpful comments on the preparation of this manuscript. We would like to extend our gratitude to the numerous staff who were involved in the technical aspects of these studies. We also thank the Great Lakes Fishery Commission who provided the funding to conduct the studies described in this manuscript.

8. REFERENCES

Applegate, V.C., Howell, J.H., and Smith, M.A., 1958, Use of mononitrophenols containing halogens as selective sea lamprey larvicides. Science 127(3294):336–338.

ACSCEQ. 1985, TFM and Bayer 73: Lampricides in the aquatic environment. Nat. Res. Counc. Can. Assoc. Comm. Sci. Crit. Environ. Qual. Publ. Ottawa, Canada. No. 22488.

Clarke, D.J., George, S.G., and Burchell, B., 1991, Glucuronidation in fish. Aquat. Toxicol. 20:35–56.

Cravedi, J.P., Gillet, C., and Monod, G., 1995, *In vivo* metabolism of pentachlorophenol and aniline in arctic charr (*Salvelinus alpinus L.*) Larvae. Bull. Environ. Contam. and Tox. 54:711–716.

EPA. 1982, Pesticide Assessment Guidelines, Subdivision O: Residue Chemistry. U.S. Environ. Prot. Agency. PB83-153981. p. 25.

Hubert, T.D., and Bernardy, J.A., Unpublished report to EPA.

Hubert, T.D., Bernardy, J.A., Vue, C., and Gingerich, W.H., Unpublished report to EPA.

Hubert, T.D., Bernardy, J.A., Boogaard, M.A., Vue, C., and Gingerich, W.H. Unpublished report to EPA.

Kane, A.S., Kahng, M.W., and Reimschuessel, R., 1994, UDP- glucuronyltransferase kinetics for 3-trifluoromethyl-4-nitrophenol (TFM) in fish. Trans. Am. Fish. Soc. 123:217–222.

Kobayashi, K., Kimura, S., and Shimizu, E., 1977, Studies on the metabolism of chlorophenols in fish—IX: Isolation and Identification of pentachlorophenyl-β-glucuronide accumulated in the bile of goldfish. Bull. JPN Soc. Sci. Fish. 43:601–607.

Kobayashi, K., 1978, Metabolism of pentachlorophenol in fishes. In "Pentachlorophenol". R. Rao, ed. Plenum Press: 89–105.

Kobayashi, K., and Nakamura, N., 1979, Isolation and identification of a conjugated-PCP excreted in the urine of goldfish. Bull. JPN Soc. Sci. Fish. 45:1001–1003.

Layiwola, P.J., and Linnegar, D.F.C., 1981, The biotransformation of ^{14}C-phenol in some freshwater fish. Xenobiotica. 11:167–171.

Lech, J.J., and Costrini, N.V., 1972, *In Vitro* and *in vivo* metabolism of 3-trifluoromethyl-4-nitrophenol (TFM) in rainbow trout. Comp. Gen. Pharm. 3:160–166.

Lech, J.J., 1973, Isolation and identification of 3-trifluoromethyl-4-nitrophenyl glucuronide from the bile of rainbow trout exposed to 3-trifluoromethyl-4-nitrophenol. Toxicol. Appl. Pharmacol. 24:114–124.

Lech, J.J., Pepple, S., and Anderson, M., 1973, Effects of novobiocin on the acute toxicity, metabolism, and biliary excretion of 3-trifluoromethyl-4-nitrophenol in rainbow trout. Toxicol. Appl. Pharmacol. 25:542–552.

Lech, J.J., and Statham, C.N., 1975, Role of glucuronide formation in the selective toxicity of 3-trifluoromethyl-4-nitrophenol (TFM) for the sea lamprey: comparative aspects of TFM uptake and conjugation in sea lamprey and rainbow trout. Toxicol. Appl. Pharmacol. 31:150–158.

Marking, L.L., and Olson, L.E., 1975, Toxicity of the lampricide 3- trifluoromethyl-4-nitrophenol (TFM) to nontarget fish in static tests. U. S. Fish and Wildl. Ser. Investig. Fish Control. No. 61.

Niimi, A.J., and Cho, C.Y., 1983, Laboratory and field analysis of pentachlorophenol (PCP) accumulation by salmonids. Water Res. 17:1791–1795.

Rogers, I.H., Servizi, J.A., and Levings, C.D., 1988, Bioconcentration of chlorophenols by juvenile chinook salmon (*Onchorhynchus tshawytscha*) overwintering in the upper Fraser River: Field and Laboratory tests. Water Poll. Res. J. Can. 23:100–113.

Schultz, D.P., Harmon, P.D., and Luhning, C.W., Uptake, metabolism, and elimination of the lampricide 3-trifluoromethyl-4-nitrophenol by largemouth bass (*Micropterus salmoides*). J. Agric. Food Chem. 27:328–331.

Stehly, G.R., and Hayton, W.L., 1989, Metabolism of pentachlorophenol by fish. Xenobiotica. 19:75–81.

14

METABOLISM, ELIMINATION, AND PHARMACOKINETICS OF THE FISH ANESTHETIC BENZOCAINE

Jeffery R. Meinertz, Guy R. Stehly, and William H. Gingerich

Upper Midwest Environmental Sciences Center
U.S. Geological Survey
Biological Resources Division
2630 Fanta Reed Road
La Crosse, WI 54603

1. ABSTRACT

Benzocaine (ethyl para-aminobenzoate), a mammalian local anesthetic, is an effective general anesthetic agent for fish when administered by bath exposure. Its proposed use in aquaculture is to sedate fish and reduce mortality during transport and/or spawning operations in hatcheries or during stocking operations. Benzocaine is metabolized into at least three compounds by acetylation and hydrolysis. Benzocaine and its acetylated metabolite are rapidly eliminated across the gills while the more polar de-ethylated and de-ethylated-acetylated metabolites are excreted at slower rates primarily in the urine. Benzocaine and the metabolite, acetylbenzocaine, are rapidly lost from the edible fillet tissue after prolonged exposures to solutions of benzocaine. During an exposure, benzocaine is rapidly distributed into the highly perfused tissues and rapidly eliminated. Elimination of benzocaine from plasma is characterized by a very rapid initial phase followed by a relatively slow terminal phase. During the initial elimination phase, the majority of benzocaine is lost from the fish while only a small portion of the benzocaine is lost during the terminal elimination phase. Benzocaine pharmacokinetic parameters of uptake, clearance, and metabolic clearance are influenced by temperature, i.e., the pharmacokinetic parameters increase with increasing water temperature. The research presented in this manuscript indicates that benzocaine has the characteristics appropriate for a fish anesthetic with a short withdrawal time.

Xenobiotics in Fish, edited by *Smith et al.*
Kluwer Academic / Plenum Publishers, New York, 1999.

2. INTRODUCTION

Currently, the only approved anesthetic for use on fish in the U.S. is MS-222 (methanesulfonate salt of ethyl meta-aminobenzoate; Finquel®; Figure 1). MS-222 is used extensively in U.S. aquaculture as a general fish anesthetic to reduce mortality and injury of fish during transport, spawning and stocking operations. Despite its wide use, MS-222 has a limiting restriction associated with its use. According to the use label for MS-222, salmonids, ictalurids, escocids, and percids exposed to MS-222 can not be made available for human consumption for 21 days after exposure (Federal Joint Subcommittee on Aquaculture, 1994). The 21-day withdrawal period is undesirable in many aquaculture situations, therefore the approval of an anesthetic with a shorter withdrawal time is preferred.

Benzocaine (ethyl para-aminobenzoate), a structurally similar compound to MS-222 (Figure 1), has been used effectively as a general anesthetic for fish when administered by bath (Dawson and Gilderhus, 1979; Gilderhus and Marking, 1987; Allen, 1988; Gilderhus, 1989; Gilderhus *et al.*, 1991). In addition, benzocaine is approved for use as a local anesthetic in human medicine. As a general anesthetic for fish, benzocaine's desirable characteristics of rapid induction and recovery times are similar to those of MS-222. Because of benzocaine's desirable characteristics, it has the potential to be approved by the US Food and Drug Administration (FDA) as a fish anesthetic.

Early research concerning the metabolism and elimination of benzocaine by fish was limited to a study by Allen (1988) in which benzocaine residues (benzocaine and metabolites with a free aromatic amino group) in muscle of largemouth bass (*Micropterus salmoides*) and rainbow trout exposed to benzocaine declined to the assay detection limit (0.5–1.0 ppm) after 4 to 8 h in benzocaine free water. Additional research related to benzocaine metabolism and elimination included studies with the anesthetic MS-222 (Walker and Schoettger, 1967; Hunn *et al.*, 1968; Stenger and Maren, 1974). MS-222 was reported to be eliminated primarily across the gills and secondarily by the kidney. Fish were thought to biotransform MS-222 by at least two pathways, namely acetylation of the primary amine group and hydrolysis of the ethyl ester group. The parent form and the hydrophobic meta acetylated product were eliminated predominantly across the gills, whereas both hydrophobic and hydrophillic residues were isolated in urine.

Because metabolism, elimination, and pharmacokinetic data of benzocaine in fish is limited, we report in this manuscript additional information useful in evaluating benzocaine as an anesthetic with a short withdrawal time. The research presented in this manuscript is an overview of recently completed studies conducted at the Upper Midwest Environmental Sciences Center, La Crosse, WI, designed to described the metabolism, elimination, and pharmacokinetics of benzocaine by fish.

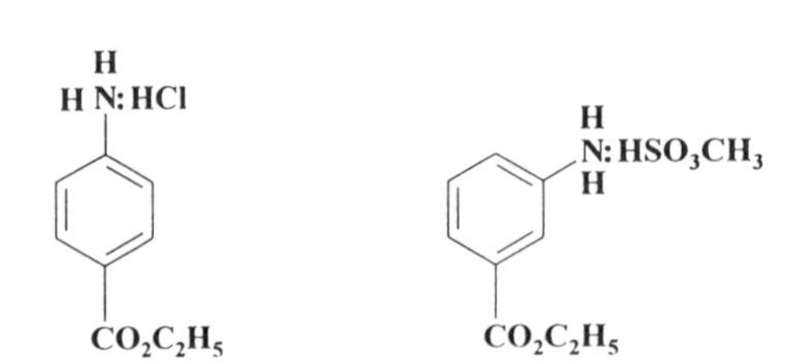

Benzocaine Hydrochloride

MS-222

Figure 1. Chemical structure of benzocaine (ethyl p-aminobenzoate; formula weight, 165.2) and MS-222 (ethyl m-aminobenzoic acid methanesulfonate salt; formula weight, 261.3).

Table 1. Profile of the elimination routes of benzocaine residues after a single arterial injection of ^{14}C-benzocaine hydrochloride (20.8 μg/kg) into rainbow trout.

			Percent of Dose Recovered			
Fish identification number	Gender	Weight (g)	Branchial elimination by 3 h	Renal elimination by 24	Biliary elimination by 24 h	Total
1	female	628	56.1	9.1	sample lost	65.2
2	female	653	53.7	10.6	2.5	66.7
3	male	676	56.3	12.5	2.5	71.3
4	female	704	63.5	6.4	1.6	71.5
5	male	595	66.3	6.5	1.2	74.1
		mean ± sd	59.2 ± 5.4	9.0 ± 2.6	2.0 ± 0.7	69.7 ± 3.7

3. BENZOCAINE METABOLISM AND ROUTES OF ELIMINATION

To characterize the metabolism and elimination of benzocaine (Meinertz *et al.*, 1991), rainbow trout were exposed to ^{14}C-benzocaine with a bolus injection through a dorsal aortic cannula (Beyenbach and Kirschner, 1975) and branchial, renal, and biliary pathways monitored for residue elimination. Branchial elimination is the primary excretory route for benzocaine residues. Within 1 min after administration, benzocaine residues were detected in the effluent water from the exposure chamber and by 3 h, 59.2% of the dose was detected in the effluent water (Table 1). Renal elimination of benzocaine residues is a minor elimination route. Only 2.7% of the dose was collected from a urinary bladder catheter (Wood and Randall, 1973) by 3 h after the injection and 9.0% by 24 h. Biliary elimination of benzocaine residues is the least dynamic of the three elimination routes. Only 2.0% of the dose was found in the bile 24 h after the injection.

Benzocaine and acetylbenzocaine were the only residues in water and the profile of these two residues changed through time (Table 2). At 3 min after injection, benzocaine represented 87.3% of the radioactive residues and acetylbenzocaine represented 12.7%. By 60 min after injection benzocaine represented 32.7% of the residues and acetylbenzocaine represented 67.4%.

Gills are an optimal site for the elimination of these compounds since benzocaine and acetylbenzocaine have attributes that are conducive (small molecular size, high lipid solubility, and neutrality) to branchial diffusion (Hunn and Allen, 1974; Stenger and Maren, 1974; Hayton and Barron, 1990). Transport of benzocaine residues across the skin may have contributed to the residues in the water (Ferreira *et al.*, 1984), but loss

Table 2. Profile of the branchial elimination of benzocaine residues after a single arterial injection of ^{14}C-benzocaine hydrochloride (20.8 μg/kg) into rainbow trout. Data are presented as the mean ± sd of the percent of dose recovered from individual fish.

Time after injection (min)	n	*p*-Aminobenzoic acid	Acetyl-*p*-aminobenzoic acid	Benzocaine	Acetyl-benzocaine
3	4	not detected	not detected	87.3 ± 4.5	12.7 ± 4.5
15	5	not detected	not detected	65.6 ± 14.7	34.4 ± 14.7
60	5	not detected	not detected	32.7 ± 12.7	67.4 ± 12.7

Table 3. Profile of the renal elimination of benzocaine residues after a single arterial injection of ^{14}C-benzocaine hydrochloride (20.8 μg/kg) into rainbow trout. Data are presented as the mean ± sd of the percent of dose recovered from individual fish.

Time after injection (h)	n	*p*-Aminobenzoic acid	Acetyl-*p*-aminobenzoic acid	Benzocaine	Acetyl-benzocaine
1	5	7.6 ± 3.3	59.7 ± 11.3	19.5 ± 7.0	8.0 ± 2.4
8	5	2.6 ± 2.6	90.3 ± 2.2	0.1 ± 1.3	0.1 ± 0.2
24	5	1.0 ± 1.4	96.6 ± 2.0	not detected	not detected

across skin is less likely than branchial elimination because of the relatively small vascular space in the dermal layers of salmonids (Hawkes, 1983) and the relatively small surface area in comparison to that of the gills.

The residue profile in urine at 1 h after the exposure consisted of modest quantities of the less polar residues, benzocaine and acetylbenzocaine, and greater quantities of the more polar residues, *p*-aminobenzoic acid and acetyl-*p*-aminobenzoic acid. By 20 h after the exposure, only *p*-aminobenzoic acid and acetyl-*p*-aminobenzoic acid were detected in the urine with acetyl-*p*-aminobenzoic acid existing as the primary residue (Table 3).

Based on the sampling scheme devised for this study, 70% of the benzocaine dose was recovered by 24 h after administration. This value is probably a low estimation of the benzocaine residues that are lost from a fish within 24 h of exposure. It is likely that additional benzocaine residues were eliminated in the water after 3 h so that by 24 h, only trace amounts of benzocaine residues remained in the fish (Hayton *et al.*, 1996).

The elimination and metabolic profile of benzocaine are similar between rainbow trout and channel catfish (Hayton *et al.*, 1996). Branchial elimination of unchanged benzocaine appears to be the primary route of elimination for both species while biliary elimination of benzocaine residues constitutes a minor pathway. Channel catfish and rainbow trout both form acetylbenzocaine, PABA, and acetylPABA, with the acetylated conjugates existing as the primary metabolites. Because metabolic profiles in each species are dominated by acetylated products, a primary metabolic pathway was proposed reflecting acetylation of benzocaine's amine group followed by hydrolysis of the ester bond (Figure 2). A secondary metabolic route was also proposed showing hydrolysis of benzocaine's ester bond followed by acetylation of the amine group.

4. DEPLETION OF BENZOCAINE RESIDUES FROM FILLET TISSUE

Legal use of therapeutic aquaculture drugs and anesthetics in the USA depends on approval by the FDA. An attribute of a drug that must be characterized before approval is granted is its rate of depletion from edible tissues. Characterizing residue depletion allows the FDA to establish withdrawal times for exposed fish and ensure that residue concentrations are below permitted concentrations before fish are made available for human consumption. To characterize the depletion of benzocaine and acetylbenzocaine from fish fillet tissue, channel catfish were exposed to a 70 mg/L benzocaine solution for 5 min then a 35 mg/L solution for 30 min; fish were sampled immediately after the 30 min exposure, and at 1, 24, and 48 h after the exposure. Benzocaine residues were

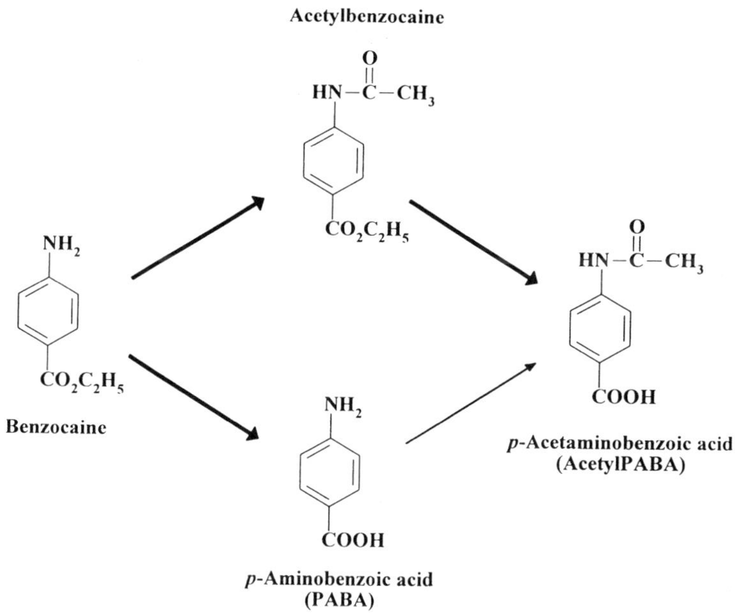

Figure 2. Proposed metabolic pathways for benzocaine in rainbow trout.

rapidly depleted from channel catfish fillet tissue. Benzocaine concentrations in the fillet tissue immediately after exposure were near the concentration of benzocaine in the 30 min exposure bath and declined nearly 10-fold within 1 hour (Table 4). After 24 h of exposure, benzocaine concentrations in the fillet tissue were near or below the method quantitation limit (22 ng/mL) and by 48 h, benzocaine could not be detected in the fillet tissue. The acetylbenzocaine concentration was about 800 ng/g immediately after the exposure and at 1 h after the exposure. Twenty-four hours after the exposure terminated, acetylbenzocaine was not detected in the fillet tissue.

In a study by Hayton *et al.* (1996), total benzocaine residues in channel catfish fillets were near the concentration of the 30 min exposure bath immediately after the end of the exposure. However, the concentration of benzocaine in the white muscle was notably lower than the benzocaine concentration we reported in the fillet tissue from our study.

Table 4. Concentrations of benzocaine and acetylbenzocaine in skinless fillets from channel catfish exposed to a 70 mg/L benzocaine solution for 5 min then to a 35 mg/L benzocaine solution for 30 min. Analyte concentrations were quantified using high performance liquid chromatography with absorbance detection.

Time after exposure (h)	n	Mean benzocaine (ng/g)	Mean acetylbenzocaine (ng/g)
0	3	40,832	782
1	3	6,230	803
24	2	17	not detected
48	3	not detected	not detected

The explanation for this difference may be related to processing time. In the study by Hayton *et al.* (1996), fish were briefly placed in chipped ice and water after the exposure to rinse the exposure solution from body surfaces and further anesthetize the fish. Although the duration of immersion in the benzocaine-free water was not specified, placement of the fish in benzocaine-free water for further anesthetization must have taken an appreciable amount of time. Based on the information we present in this manuscript, a considerable amount of benzocaine may have been rapidly lost from the fish during the immersion time. Despite the difference in benzocaine concentrations immediately after the exposure, benzocaine concentrations decreased rapidly through 4 h after the exposure in each study. In the study by Hayton *et al.* (1996), the depletion of total benzocaine residues from the fillet tissue after 25 h was much slower and was characterized by a half-life of about 8 days.

5. BENZOCAINE PHARMACOKINETICS

The plasma pharmacokinetics of benzocaine were characterized in rainbow trout after arterial and waterborne administration. Initial pharmacokinetic studies with benzocaine performed with bolus injections of benzocaine into the dorsal aorta revealed the rapid distribution and elimination characteristics of benzocaine (Meinertz *et al.*, 1996). The steepness and highly variable plasma benzocaine concentrations in the initial phase led to difficulties estimating distributive phase pharmacokinetic parameters for benzocaine. Despite these challenges, the pharmacokinetic parameters, volume of distribution at steady state (V_{ss}), mean residence time (MRT), total body clearance (Cl_b), and terminal phase half-life ($t_{1/2(\gamma)}$), were estimated. The objective of a subsequent study was to compare the pharmacokinetic parameters between groups of fish acclimated to different water temperatures. Because of the distributive phase was rapid after arterial dosing, it was unlikely that comparable benzocaine pharmacokinetic data would be produced by bolus dosing fish in the dorsal aorta. Therefore, an exposure method was developed to expose fish to benzocaine by bath and provide pharmacokinetic information that could be compared between the temperature groups (Stehly *et al.*, 1998).

5.1. Benzocaine Pharmacokinetics Developed from an Arterial Exposure

In this study, the pharmacokinetics of benzocaine in fish not under the symptomatic influence of the anesthetic (no loss of equilibrium) were compared with those of fish under the influence of the anesthetic (loss of equilibrium followed by equilibrium recovery). Selection of the single bolus doses for this study were based on the results from preliminary studies in which fish were dosed at levels less than 6 mg/kg and greater than 9 mg/kg. Most fish dosed at levels greater than 9 mg/kg lost and never regained equilibrium while fish dosed at levels less than 6 mg/kg did not maintain sufficient plasma concentrations of benzocaine for an adequate period of time. Optimum dose levels were found at 9 mg/kg where most fish displayed the desired symptomatic responses and at 6 mg/kg where most fish did not display symptomatic responses and maintained plasma benzocaine concentrations above quantitation limit concentrations throughout the sampling period.

Typically in single arterial bolus dosing studies, plasma pharmacokinetic parameters are calculated for each dosed animal, then the mean with a standard deviation are reported for each parameter. However, because of excessive variability in data from individual fish, one depuration curve was generated for each dose level from the mean benzocaine concentrations at each sample time. Original estimates for the *A* coefficient were excessively high as a result of the initial steep decline in benzocaine concentrations, resulting in unrealistically low estimates for V_{ss}, therefore, an alternative method was used to calculate the coefficient. The *A* coefficients were calculated with the equation, mean dose ÷ (42.4 mL/kg × 0.754), where the mean dose is the mean of all doses administered, 42.4 mL/kg is the estimated blood volume of rainbow trout (Olson, 1992), and 0.754 is the average plasma volume of the exposed fish at the time of dosing. The calculated *A* coefficients (187,130 ng/mL for the 6 mg/kg dose group and 283,529 ng/mL for the 9 mg/kg dose group) were used to estimate the other model parameters for each dose. The data from both dose levels were best represented by a three compartment model (Figure 3). The pharmacokinetic parameters developed from these models are presented in Table 5.

Differences in the pharmacokinetic parameters of fish given an anesthetizing dose of benzocaine and those given a nonanesthetizing dose were insignificant. The magnitude of V_{ss} for both dose levels indicated that the distribution of benzocaine outside the plasma is minimal while the *MRT* for both dose levels indicated that plasma clearance is extremely rapid. The slight differences in V_{ss} and Cl_b between the dose levels were thought to be the result of the variable initial phase data. Despite the rapid elimination of benzocaine from plasma within the first 20 min after dosing, quantifiable benzocaine

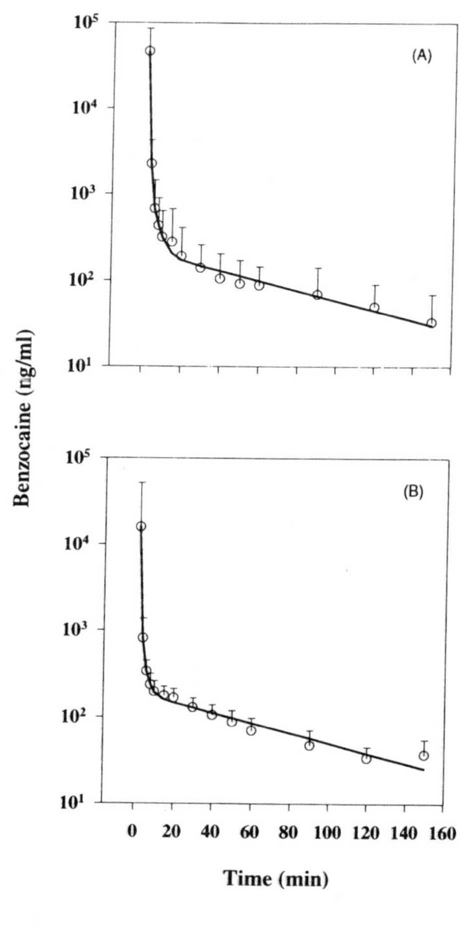

Figure 3. Depuration of benzocaine from the plasma of rainbow trout after arterial bolus administration of benzocaine at nominal concentrations of (A) 6 mg/kg ($n = 8$) and (B) 9 mg/kg ($n = 10$). Data points are the mean benzocaine concentrations determined by high performance liquid chromatography with absorbance detection and the vertical bars represent the sample standard deviations. The lines represent the fit of data to a three compartment model.

Table 5. Pharmacokinetic parameters estimated for benzocaine in rainbow trout following intraarterial bolus administration at nominal concentrations of 6 mg/kg ($n = 8$) and 9 mg/kg ($n = 10$).

	Dose (mg/kg)	
Parameter	6	9
A (ng/mL)	188,659	283,139
B (ng/mL)	164	184
C (ng/mL)	156	94
α (min^{-1})	0.769	1.46
β (min^{-1})	0.082	0.044
γ (min^{-1})	0.008	0.006
MRT (min)	8.9	8.7
Cl_b (mL/min/kg)	12.5	18.0
V_{ss} (mL/kg)	112	156
$t_{1/2(\gamma)}$ (min)	89.4	108.5

concentrations persisted throughout the sampling period. The $t_{1/2(\gamma)}$ for benzocaine was 89 min for fish dosed at 6 mg/kg and 109 min for fish dosed at 9 mg/kg.

5.2. Benzocaine Pharmacokinetics Developed from a Bath Exposure

Normally, vascular administration is preferred to unambiguously determine pharmacokinetics of a compound (no confounding effects of absorption), even if it is not the normal route of exposure. However, vascular administration of benzocaine was problematic. A study exposing rainbow trout to benzocaine by water bath was subsequently developed to overcome the limitations of vascular administration (i.e., the data variability associated with the rapid distributive phase). For these studies, the affects of temperature on benzocaine pharmacokinetics was evaluated with rainbow trout acclimated at 6, 12, or 18 °C for more than one month.

The unique model design for this study was based on the assumption that the water bath and highly perfused tissues were one compartment and that modestly perfused tissues were a second compartment. This assumption was supported by data from preliminary studies where the interaction of benzocaine between the exposure bath water and highly perfused tissues was characterized and where the loss of benzocaine from the recirculating exposure system was characterized. Data from these studies indicated that benzocaine in the exposure bath was near equilibrium with benzocaine in the plasma within 2 min after immersing fish in the bath (plasma benzocaine concentration about 2.4 times water benzocaine concentration) and that any benzocaine lost from the exposure system during an experiment was either absorbed or metabolized by the fish.

After fish were introduced to an exposure bath, the concentration of benzocaine in the water declined in a biphasic manner. Loss was characterized by an initial rapid uptake phase (0 to 45 min) representing distribution and a terminal linear phase (45 to 240 min) representing metabolism (Figure 4). The mean coefficients (A and B) and exponents (α and β) for a biexponential equation determined by curve stripping are presented in Table 6.

The linear terminal phase of benzocaine depletion from the exposure baths (45 to 240 min) suggested that benzocaine accumulation in the tissues was complete and that benzocaine was in equilibrium with all tissues. The following factors supported the

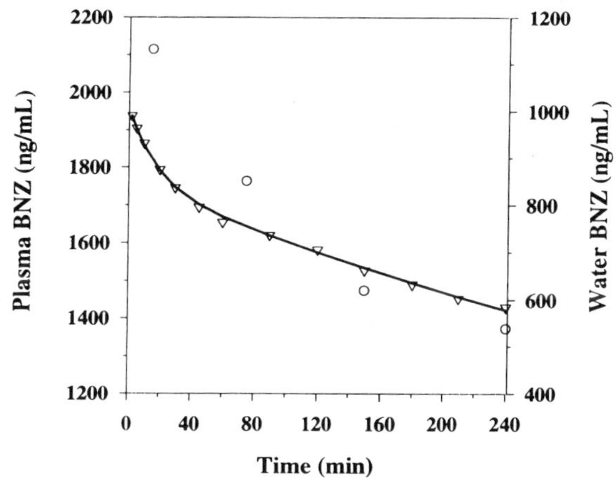

Figure 4. Representation of benzocaine in exposure water and rainbow trout plasma after immersing a rainbow trout in a recirculating 1 mg/mL benzocaine solution at 6 °C for 4 h. Data points are the benzocaine concentrations in water (●) and plasma (▲) determined by high performance liquid chromatography with absorbance detection. The line represents the fit of data to a two compartment model.

Table 6. The pharmacokinetic parameters estimated for benzocaine after immersing a rainbow trout in a recirculating 1 mg/mL benzocaine solution. Fish were acclimated to the treatment temperature, 6 (n = 7), 12 (n = 10), and 18 °C (n = 6), for at least one month before exposure to benzocaine. Data are presented as the mean and standard deviation in parentheses. Significant differences within a column ($p \leq 0.05$) are denoted by different superscripts.

Temp.	*A* (ng/mL)	*B* (ng/mL)	α (min^{-1})	β (min^{-1})	Uptake clearance (mL/min/kg)	Metabolic clearance (mL/min/kg)	V_{ss} (mL/kg)
6 °C	142.1	848.5	0.0412	0.00111	581[a]	15.2[a]	2,369
	(40.5)	(32.8)	(0.0102)	(0.00035)	(179)	(4.1)	(678)
12 °C	119.8	848.1	0.0412	0.00114	674	17.5	2,708
	(31.3)	(30.6)	(0.0050)	(0.00034)	(201)	(3.1)	(544)
18 °C	90.4	839.6	0.0650	0.00129	1154[b]	22.3[b]	3,260
	(24.8)	(45.2)	(0.0198)	(0.00023)	(447)	(4.2)	(1,182)

concept that benzocaine is equilibrated with the fish tissues by 45 min: the tissue accumulation characteristics of benzocaine, the extraction ratio of benzocaine, and the half-life of benzocaine.

Benzocaine is not particularly lipophilic (log octanol : water partition coefficient of benzocaine is 1.98; Downes *et al.*, 1994), therefore it would be limited in it's ability to accumulate in tissues with low perfusion and high tissue to blood partition coefficients (e.g., adipose tissue). Also, benzocaine is readily metabolized by fish reducing the capacity of the fish to accumulate it (Meinertz *et al.*, 1991; Hayton *et al.*, 1996).

Simply defined, the extraction ratio is the ratio of the elimination rate to the presentation rate (Benet and Zia-Amirhosseini, 1995). In this definition, the elimination rate can be affected by any process causing loss from the system (e.g., metabolism or uptake by tissues). Since clearance is directly related to extraction ratios, the following equation is typically used to calculate extraction ratios:

$$\text{extraction ratio of organ} = \text{organ clearance}/\text{organ blood flow}. \quad (1)$$

To estimate the extraction ratios for our study, the following equation was used:

$$\text{extraction ratio} = \text{metabolic clearance}/(\text{cardiac output} \times 2.24), \quad (2)$$

where the metabolic clearance was derived from the benzocaine concentration versus time curve for the exposure bath, cardiac output values of 19.8 mL/min/kg at 6 °C, 38.7 mL/min/kg at 12 °C, and 62.5 mL/min/kg at 18 °C were derived from Barron *et al.* (1987), and the constant, 2.24, was the mean ratio of plasma benzocaine concentration : water benzocaine concentration during the exposure.

Typically, drugs with an extraction ratio of more than 70% are considered high extraction ratio drugs, whereas drugs with a ratio less than 30% are considered low extraction ratio drugs (Benet and Zia-Amirhosseini, 1995). The extraction ratio for benzocaine ranged from 16% at 18 °C to 34% at 6 °C classifying benzocaine as a low to moderate extraction ratio drug. A high extraction ratio (>70%) could suggest that benzocaine is lost from the water by processes other than metabolism (e.g., uptake by tissue), therefore, the low to moderate extraction ratio classification suggests that benzocaine uptake by tissue is minimal during the metabolic phase of the curve.

The half-life of a drug is usually defined as the time it takes for the plasma concentration or the amount of drug in the body to reduce by 50%, however when a drug is being administered by infusion, half-life also represents the time it takes for drug concentrations to reach 50% of the expected steady state concentration (Benet and Zia-Amirhosseini, 1995). Conventionally, 3.3 half-lifes are considered the time required to achieve 90% of steady-state under constant infusion. With the information provided from the arterial exposure study and assuming the exposure of fish to benzocaine in a bath is equivalent to constant infusion, more than 3.3 half-lifes of benzocaine occur during 45 min of exposure.

The uptake clearances increased with temperature and were significantly different ($p \leq 0.05$) in fish exposed at 6 °C and 18 °C (Table 6). The increase in uptake clearance is probably a direct result of increased ventilation at the higher temperatures. Despite the increases with temperature, uptake clearances were relatively high with respect to ventilation volumes for similar sized trout (177 mL/min/kg at 12 °C; Nichols *et al.*, 1991) and supports the contention that benzocaine uptake is limited by branchial blood flow (Hayton *et al.*, 1996). Since there were no other routes for benzocaine loss from the system, the elevated uptake clearances were probably a result of increased ventilation caused by benzocaine exposure.

The metabolic clearances increased with an increase in temperature and were significantly different ($p \leq 0.05$) between groups of fish exposed at 6 °C and 18 °C. The increase in metabolic clearances probably resulted from an increased enzyme activity rate related to the thermodynamics effects of temperature.

The mean benzocaine concentrations in plasma 240 min after initiating the exposures were 1538 ng/mL at 6 °C, 1463 ng/mL at 12 °C, and 1224 ng/mL at 18 °C; all values were significantly different ($p \leq 0.05$). At 2.5 min after the fish were transferred to benzocaine free water, mean plasma concentrations of benzocaine were 242 ng/mL at 6 °C, 98 ng/mL at 12 °C, and 76 ng/mL at 18 °C; all values were significantly different ($p \leq 0.05$). Even though differences between benzocaine concentrations at single time points are not measurements describing benzocaine pharmacokinetics, the differences support the notion that some underlying pharmacokinetic changes occur as a result of temperature.

The terminal phase elimination rate constants for benzocaine and resulting plasma half-lifes for individual fish were highly variable. Half-lifes were 60.8 ± 30.3 (mean ± sd) min at 6 °C, 35.9 ± 13.0 min at 12 °C, and 42.4 ± 21.0 min at 18 °C. Despite the variability, all half-lifes indicated that benzocaine is rapidly eliminated from the plasma after a prolonged bath exposure.

6. CONCLUSION

Benzocaine is metabolized by acetylation and hydrolysis to acetylbenzocaine, *p*-aminobenzoic acid, and acetyl-*p*-aminobenzoic acid. Benzocaine residues are eliminated by branchial, renal, and biliary pathways. Branchial elimination is the major route of loss of benzocaine while renal and biliary elimination account for less important excretory routes. Benzocaine residues eliminated by the gill are benzocaine and acetylbenzocaine. The benzocaine residues eliminated in urine include benzocaine, acetylbenzocaine, PABA, and acetylPABA.

Benzocaine is rapidly lost from the fillets and plasma of fish. The rapid distribution and elimination characteristics of benzocaine from plasma makes initial phase pharmacokinetic parameters difficult to estimate after a single bolus arterial administration. A three-compartment pharmacokinetic model best describes benzocaine concentrations in plasma regardless of whether the doses result in overt signs of anesthesia. The V_{ss} and Cl_b of benzocaine increase in a dose dependant manner, however, since the *MRT* is similar at both dosages, the differences in the V_{ss} and Cl_b is probably due to the data's variability. The model parameter, V_{ss} indicates that distribution of benzocaine outside the plasma is minimal for each dose. Relative to benzocaine's initial rapid distribution and elimination from plasma, terminal phase benzocaine elimination is slow.

Benzocaine is rapidly absorbed by fish immersed in a solution of benzocaine. The uptake clearance and metabolic clearance of benzocaine significantly increase with increasing water temperatures. The V_{ss} of benzocaine has a trend for increasing with temperature but the increases are not significant. The initial phase of benzocaine elimination from plasma indicates temperature related decreases while terminal phase benzocaine elimination does not indicate temperature related differences.

7. REFERENCES

Federal Joint Subcommittee on Aquaculture, 1994, Guide to Drug, Vaccine, and Pesticide use in Aquaculture, Texas Agriculture Extension Service, Texas A&M University System, College Station, Texas, publication number B-5085, 68 pages.

Dawson, V.K., and Gilderhus, P.A., 1979, Ethyl p-aminobenzoate, (benzocaine): efficacy as an anesthetic for five species of freshwater fish, *U.S. Fish Wildl Serv Invest Fish Control*. 87:1–5.

Gilderhus, P.A., and Marking, L.L., 1987, Comparative efficacy of 16 anesthetic chemicals on rainbow trout, *N Am J Fish Manage*. 7:288–292.

Gilderhus, P.A., 1989, Efficacy of benzocaine as an anesthetic for salmonid fishes, *N Am J Fish Manage*. 9:150–153.

Gilderhus, P.A., Lemm, C.A., and Woods, L.C., 1991, Benzocaine as an anesthetic for striped bass, *Prog Fish-Cult*. 53:105–107.

Allen, J.L., 1988, Residues of benzocaine in rainbow trout, largemouth bass, and fish meal, *Prog Fish-Cult*. 50:59–60.

Stenger, V.G., and Maren, T.H., 1974, The pharmacology of MS-222 (ethyl *m*-aminobenzoate) in *Squalus acanthias*, *Comp Gen Pharmacol*. 5:23–35.

Hunn, J.B., Schoettger, R.A., and Willford, W.A., 1968, Turnover and urinary excretion of free and acetylated MS-222 by rainbow trout, *Salmo gairdneri*, *J Fish Res Board Can*. 25:25–31.

Walker, C.R., and Schoettger, R.A., 1967, Residues of MS-222 in four salmonids following anaesthesia, *U.S. Fish Wildl Serv Invest Fish Control*. 15:3–11.

Meinertz, J.R., Gingerich, W.H., and Allen, J.L., 1991, Metabolism and elimination of benzocaine by rainbow trout, *Oncorhynchus mykiss*, *Xenobiotica*. 21:525–533.

Beyenbach, K.W., and Kirschner, L.B., 1975, Kidney and urinary bladder function of the rainbow trout in Mg and Na excretion, *Am J Physiol*. 229:389–393.

Wood, C.M., and Randall, D.J., 1973, The influence of swimming activity on sodium balance in the rainbow trout (*Salmo gairdneri*), *J Comp Physiol.* 82:207–233.

Hunn, J.B., and Allen, J.L., 1974, Movement of drugs across the gills of fishes, *Annu Rev Pharmacol.* 14:47–55.

Hayton, W.L., and Barron, M.G., 1990, Rate-limiting barriers to xenobiotic uptake by the gill, *Environ Toxicol Chem.* 9:151–157.

Ferreira, J.T., Schoonbee, H.J., and Smit, G.L., 1984, The uptake of the anaesthetic benzocaine hydrochloride by the gills and the skin of three freshwater fish species, *J Fish Biol.* 25:35–41.

Hawkes, J.W, 1983, Skin and scales, in: *Microscopic Anatomy of Salmonids: An Atlas*, W.T. Yasutake, and J.H. Wales, ed., Washington, D.C., U.S. Fish and Wildlife Service, Resource publication 150, pages 14–23.

Hayton, W.L., Szoke, A., Kemmenoe, B.H., and Vick, A.M., 1996, Disposition of benzocaine in channel catfish, *Aquat Toxicol.* 36:99–113.

Meinertz, J.R., Stehly, G.R., and Gingerich, W.H., 1996, Pharmacokinetics of benzocaine in rainbow trout (*Oncorhynchus mykiss*) after intraarterial dosing, *Aquaculture.* 148:39–48.

Stehly, G.R., Meinertz, J.R., and Gingerich, W.H., 1998, Effect of temperature on the pharmacokinetics of benzocaine in rainbow trout (*Oncorhynchus mykiss*) after bath exposure, *J Vet Pharmacol and Ther.* 21:121–127.

Olson, K.R., 1992, Blood and extracellular fluid volume regulation: role of the renin-angiotensin system, kallikrein-kinin system, and atrial natriuretic peptides, in: *Fish Physiology*, W.S. Hoar, D.J. Randall, and A.P. Farrell, ed., XII, Part B, Academic Press, New York, pages 135–254.

Downes, H., Kienle, E.A., and Pederson, C., 1994, Metamorphosis and steady state anesthetic concentrations of tricaine, benzocaine, and ethanol, *Comp Biochem and Physiol.* 107C:95–103.

Benet, L.Z., and Zia-Amirhosseini, P., 1995, Basic principles of pharmacokinetics, *Toxicol Pathol.* 23:115–123.

Barron, M.G., Tarr, B.D., and Hayton, W.L., 1987, Temperature-dependence of di-2-ethylhexyl phthalate (DEHP) pharmacokinetics in rainbow trout, *Toxicol Appl Pharmacol.* 88:305–312.

Nichols, J.W., McKim, J.M., Lein, G.J., Hoffman, A.D., and Bertelsen, S.L., 1991, Physiologically based toxicokinetic modeling of three waterborne chloroethanes in rainbow trout, *Oncorhynchus mykiss*, *Toxicol and Appl Pharmacol.* 110:374–389.

15

METABOLISM AND PHARMACOKINETICS OF SULFAMONOMETHOXINE IN EDIBLE FISH SPECIES

Ryuji Ueno

Faculty of Bioresources
Mie University
Tsu, Japan

1. ABSTRACT

Sulfamonomethoxine has been efficacious in the treatment of various bacterial diseases of fish. The present paper reviews recent studies on the metabolism and pharmacokinetics of sulfamonomethoxine in edible fish species such as yellowtail, rainbow trout, and eel. N^4-Acetylated metabolites were the major metabolites whereas glucuronic acid conjugates were the minor metabolites. Moreover, a double conjugated metabolite, the glucuronic acid conjugate of N^4-acetylsulfamonomethoxine was found in three species. Serum concentrations after intravascular administration could be fitted to be a two-compartment model and those after oral administration to a one-compartment model in all three species. The absorption parameters of sulfamonomethoxine in eel were substantially different from those in rainbow trout and yellowtail. In eel, the drug was assimilated slowly and cleared very slowly. On the other hand, in yellowtail, sulfamonomethoxine was well distributed to the tissues and was cleared quickly. As an application of the pharmacokinetic studies, the Japanese dosage regimens of sulfamonomethoxine in three fish species were evaluated.

2. INTRODUCTION

In recent years, various drugs have been used for the control of fish diseases as fish production has increased in Japan. As a result, drugs residues remaining in the tissues of fish raise questions of quality and safety of fish as food. In mammals, the efficacy, safety and residues of drugs have usually been estimated by pharmacokinetic analyses using compartmental models. That is, by using pharmacokinetics various parameters of drug disposition may be measured. These parameters are related to the duration and intensity

Xenobiotics in Fish, edited by *Smith et al.*
Kluwer Academic / Plenum Publishers, New York, 1999.

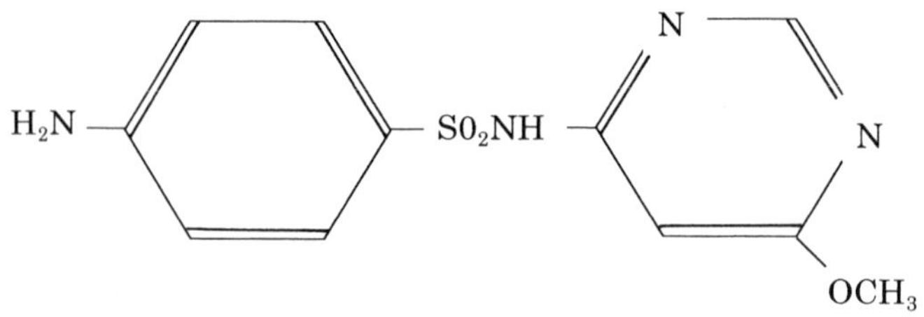

Figure 1. Chemical structure of sulfamonomethoxine.

of drug action and are used to obtain a meaningful measure of the drug's usefulness, possible toxicity, residue, etc. In aquatic animals, a few researchers have investigated the pharmacokinetics of sulfonamides such as sulfadimethoxine and sulfadimidine (Barron *et al.*, 1988; Grondel *et al.*, 1986; Kleinow *et al.*, 1987; Kleinow and Lech, 1988; Kleinow *et al.*, 1992; Michel *et al.*, 1990; Squibb *et al.*, 1988; van Ginneken *et al.*, 1991). However, few papers have been concerned with the pharmacokinetics of sulfamonomethoxine, which is a long-acting sulfonamide in fish.

Sulfamonomethoxine [4-amino-N-(2-methoxypyrimidinyl) benzenesulfonamide] is the most common sulfonamide among the chemotherapeutic drugs and is regulated in conformity with the Pharmaceutical Law in Japan (Figure 1). The usage of sulfamonomethoxine has been registered or approved for 4 saltwater fish species (yellowtail, red sea bream, silver salmon and horse mackerel) and 5 freshwater fish species (carp, eel, rainbow trout, ayu fish and tilapia). Sulfamonomethoxine has been used effectively against cultured fish diseases such as vibriosis, furunculosis, and red fin disease because of its broad-spectrum activity against gram-positive cocci and gram-negative bacilli.

The present paper reviews recent studies on the metabolism and pharmacokinetics of sulfamonomethoxine in edible fish species such as yellowtail, rainbow trout and eel, which are the most common cultured fish in Japan (Ueno *et al.*, 1994; Ueno, 1998; Uno *et al.*, 1993; Uno *et al.*, 1997).

3. TISSUE LEVEL OF SULFAMONOMETHOXINE IN FISH AFTER ORAL ADMINISTRATION

3.1. Pharmacokinetic Analysis

The most common method of pharmacokinetic evaluation is to assume that the drug concentration vs. time can be described by one of several compartmental models, and to fit the data to an equation consistent with the assumed model using a non-linear least-squares regression. In our experiments, pharmacokinetic analysis was applied assuming a one- or two-compartmental model with first-order absorption using the non-linear-squares program MULTI (Yamaoka *et al.*, 1981). The selection of the models was judged by Akaike's information criterion (Yamaoka *et al.*, 1978).

The serum sulfamonomethoxine concentration of eel, rainbow trout and yellowtail was analyzed with the program for best fit to a one-compartment model (Ueno *et al.*, 1994; Uno *et al.*, 1993). The following equations were obtained:

$$C_t = 141[e^{-0.005t} - e^{-0.064t}] \text{ for eels (Ueno, 1998)} \quad (1)$$

Table 1. Pharmacokinetic parameters for sulfamonomethoxine following oral administration.

Parameters	Eel	Rainbow trout	Yellowtail
Dose (mg/kg)	400	300	400
Weight (g)	160	240	357
Water temp. (°C)	27	15	22
Ka (h^{-1})	0.064	0.10	1.59
Ke (h^{-1})	0.005	0.02	0.15
$Ta_{1/2}$ (h)	10.8	6.9	0.44
$T_{1/2}$ (h)	139	32.9	4.53
AUC (μg·h/ml)	28,400	3,080	932
MRT (h)	232	70.6	7.94

Ka, Absorption rate constant; Ke, Elimination rate constant; $Ta_{1/2}$, Absorption half-life; $T_{1/2}$, Elimination half-life; AUC, Area under serum concentration-time curve; MRT, Mean residence time.

$$C_t = 75.9[e^{-0.021t} - e^{-0.10t}] \text{ for trout (Uno } et\ al., 1993) \tag{2}$$

$$C_t = 173[e^{-0.153t} - e^{-1.59t}] \text{ for yellowtail (Ueno } et\ al., 1994) \tag{3}$$

Where C_t and t are the serum concentration and the time, respectively.

Only limited information is available on the pharmacokinetics of sulfamonomethoxine in mammals. That is, Carli *et al.* (1993) reported that, in the horse, the drug was best described by a one-compartment model after oral administration.

In channel catfish (Michel *et al.*, 1988) and rainbow trout (Uno *et al.*, 1993) the pharmacokinetics of sulfadimethoxine after oral administration was described by a one-compartment model.

Table 1 shows the pharmacokinetics parameters for sulfamonomethoxine following oral administration. The drug in yellowtail (absorption half-life, $Ta_{1/2}$: 0.44 h) was assimilated more rapidly from the digestive tract than it was in eel ($Ta_{1/2}$: 10.8 h) or rainbow trout ($Ta_{1/2}$: 6.9 h). On the other hand, the excretion of sulfamonomethoxine was much slower in eel (elimination half-life, $T_{1/2}$: 139 h) than it was in rainbow trout ($T_{1/2}$: 32.9 h) and yellowtail ($T_{1/2}$: 4.53 h). In comparison, the $Ta_{1/2}$ and $T_{1/2}$ of sulfadimethoxine in rainbow trout were 5.0 and 24.5 h, respectively (Uno *et al.*, 1993). The area under the curve (AUC) for sulfamonomethoxine could not be compared directly because these values are affected by the drug dose. Therefore, the values of AUC per dose were recalculated; the values of AUC per dose were significantly different among the fish (71.0 for eel, 10.2 for rainbow trout, and 2.3 for yellowtail). The AUC for sulfadimethoxine in rainbow trout was reported to be 10.6. Hara *et al.* (1967) reported less absorption of sulfonamides by rainbow trout than by mammals.

3.2. Tissue Distribution

Sulfamonomethoxine was detected in the non-alimentary tissues of all fish species tested within 1 hour after administration of an oral dose (Ueno *et al.*, 1994; Ueno, 1998; Uno *et al.*, 1993). This suggests that a substantial amount of sulfamonomethoxine assimilation occurred in the stomach, where sulfamonomethoxine (pK_a of 5.9), would exist primarily in an un-ionized form. In fact, rapid gastric assimilation of several drugs has been shown to occur in fish (Droy *et al.*, 1990; Ueno *et al.*, 1988a; Ueno *et al.*, 1988b; Ueno *et al.*, 1995; Uno *et al.*, 1992a; Uno *et al.*, 1992b; Uno *et al.*, 1993).

In eels (Ueno, 1998) and rainbow trout (Uno *et al.*, 1993) the tissue levels of sulfamonomethoxine peaked between 9 hours and 4 days, but peaked within 6 hours of administration in yellowtail (Ueno *et al.*, 1994). At the highest peak tissue concentration, the concentrations of sulfamonomethoxine were observed in the order of kidney ≥ liver > serum > muscle for eel and serum > kidney ≥ liver > muscle for rainbow trout and yellowtail. Similar results were obtained for tissue levels of sulfamonomethoxine in yellowtail by Matsushima *et al.* (1971) and rainbow trout by Hara *et al.* (1967). A considerable amount of sulfamonomethoxine was observed in the bile of all fish (Hara and Inoue, 1967).

In eel, the peak levels of N^4-acetylsulfamonomethoxine were observed at 7 days in the serum and liver, and at 11 days in the muscle (Ueno, 1998). In rainbow trout (Uno *et al.*, 1993), the peak levels of N^4-acetylsulfamonomethoxine were at 1 day in the serum, 3 days in the muscle, 2 days in the liver, and 5 days in the kidney. On the other hand, the peak levels of N^4-acetylsulfamonomethoxine in yellowtail (Ueno *et al.*, 1994) were at 6 hours in the serum, muscle, liver and kidney. The N^4-acetylsulfamonomethoxine level was much higher in the internal organs, especially the gall bladder (i.e., bile) of three fish species. The bile levels remained above 1,000 µg/ml for up to 10 days in rainbow trout and 40 days in eel. Kleinow *et al.* (1992) also reported a high concentration of N^4-acetylsulfadimethoxine in the bile of rainbow trout and suggested the possibility of enterohepatic recirculation of the parent drug.

3.3. Metabolites of Sulfamonomethoxine

Figure 2 shows the metabolic pathway of sulfamonomethoxine in fish. Sulfamonomethoxine, its N^4-acetylated metabolite and their glucuronic acid conjugates were detected and large amounts of N^4-acetylsulfamonomethoxine were observed in bile at 20–32 days post dosing in all tissues among fish. N^4-Acetylsulfamonomethoxine was the main excretory metabolite in rainbow trout and yellowtail. At the time of maximum concentration post dosing (Cmax), the percentage of N^4-acetylsulfamonomethoxine present as metabolites was high in liver of rainbow trout (77.5%) and in kidney of yellowtail (72.3%). However, a high concentration of glucuronic acid conjugate of N^4-acetylsulfamonomethoxine was observed in liver of eel (Ueno, 1998) and in the bile of rainbow trout (Uno *et al.*, 1993) and yellowtail (Ueno *et al.*, 1994).

It has been reported that N^4-acetylated metabolites and glucuronic acid conjugates

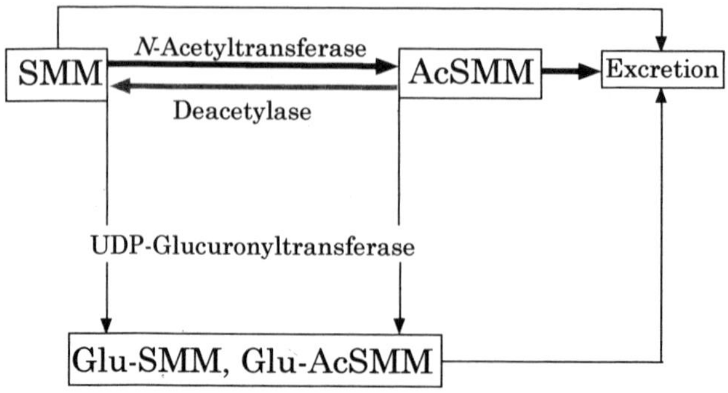

Figure 2. Metabolic pathway of sulfamonomethoxine in fish. SMM, Sulfamonomethoxine; AcSMM, N^4-Acetylsulfamonomethoxine; Glu-SMM, Glucuronic acid conjugate of sulfamonomethoxine; Glu-AcSMM, Glucuronic acid conjugate of N^4-acetylsulfamonomethoxine.

are two major types of excretory metabolites of sulfonamides in mammals (Vree *et al.*, 1990a; Vree *et al.*, 1990b). In fish, Grondel *et al.* (1986) and van Ginneken *et al.* (1991) reported the kinetics of N^4-acetylated metabolites of sulfadimidine in the plasma of carp and rainbow trout by HPLC analysis. Ishida (1986) reported the existence of N^4-acetylated and glucuronidated metabolites of five sulfonamides in bile and urine of rainbow trout by thin-layer chromatography. Our results also show that sulfamonomethoxine undergoes acetylation and glucuronidation in fish. N^4-Acetylated metabolites were the major metabolites whereas glucuronic acid conjugates were the minor metabolites. Moreover, we found a double conjugated metabolite, the glucuronic acid conjugate of N^4-acetylsulfamonomethoxine, in fish as well as mammals.

The minimum inhibitory concentrations (MICs) of sulfamonomethoxine for pathogenic bacteria such as *Aeromonas salmonicida, Aeromonas hydrophila* and *Vibrio anguillarum* are between 1.6 to 3.2 μg/ml (Tanaka, 1977). After the dose (two times higher than normal) of sulfamonomethoxine in our experiment, the eel (400 mg/kg) and rainbow trout (300 mg/kg) would maintain the MIC for more than 40 days and 7 days, respectively. In yellowtail (400 mg/kg) however, sulfamonomethoxine concentrations would fall below the MIC within 2 days. If the objective of treating yellowtail with sulfamonomethoxine is to maintain the MIC, then the administered dose will need to be much greater than the recommended dose. This may be accomplished by increasing the dose of the drug or the frequency of dose administration.

The Pharmaceutical Affairs Law in Japan stipulates certain withdrawal periods for therapeutics used to treat fish diseases; the withdrawal period for sulfamonomethoxine is 30 days for eel and rainbow trout, and is 15 days for yellowtail. Our results show that the current withdrawal periods are reasonable for the elimination of the parent drugs in rainbow trout and yellowtail but not in eel. However, a significant amount of N^4-acetylated metabolites was still detected in bile of the three species even beyond the withdrawal period established for each fish.

3.4. Acetylation

The acetylation of sulfamonomethoxine can be calculated from the following equation (Vree *et al.*, 1980; Nouws *et al.*, 1983):

$$\text{Acetylation (\%)} = \text{AUC}_{(\text{AcSMM})} / \text{AUC}_{(\text{SMM})} + \text{AUC}_{(\text{AcSMM})} \times 100 \tag{4}$$

Where (AcSMM) and (SMM) are concentrations of N^4-acetylsulfamonomethoxine and sulfamonomethoxine, respectively.

The acetylation for eel, rainbow trout and yellowtail following oral administration was 3.8, 49 and 72%, respectively. Hara and Inoue (1969) and Matsushima *et al.* (1971) reported that the acetylation of sulfamonomethoxine in the sera of rainbow trout and yellowtail was calculated to be 0–8% and 15.2% respectively, by using the modified method of Bratton and Marshall (1939). It has been known that acetylation is a widely distributed reaction in mammals and is catalyzed by the enzyme *N*-acetyltransferase. These differences in acetylation may be due to differences in *N*-acetyltransferase activity among the fish. In mammals, 'slow' and 'fast' acetylator phenotypes are related to the toxicity of isoniazid, which is normally cleared from the body by *N*-acetylation (Gibson and Skett, 1986). It would be interesting if divergent acetylator phenotypes were also present in fish.

Acetylated metabolites of sulfamonomethoxine lose their antibacterial action and tend to be less water-soluble than the parent drugs, potentially leading to crystalluria, a contributing factor to renal damage. Deacetylation of N^4-acetylated metabolites of sulfonamides has been reported in mammals (Nouws *et al.*, 1983; Shimoda *et al.*, 1997; Souich *et al.*, 1978; Vree *et al.*, 1981; Vree *et al.*, 1983). In a preliminary experiment, we have also observed the deacetylation of sulfamonomethoxine and sulfadimethoxine in the liver of rainbow trout (unpublished data). Therefore, activation of the N^4-acetylated metabolite, by a deacetylation process, likely converts the metabolite back to the parent drug. Further investigations are needed to elucidate these points from the standpoint of food safety.

4. SERUM LEVEL OF SULFAMONOMETHOXINE IN FISH AFTER INTRAVASCULAR ADMINISTRATION

4.1. Pharmacokinetic Analysis

The serum level of sulfamonomethoxine in fish was fitted by MULTI using a two-compartment model (Yamaoka *et al.*, 1981). The following equations were obtained:

$$C_t = 987[e^{-2.74t}] + 425[e^{-0.008t}] \text{ for eels (Ueno, 1998)} \tag{5}$$

$$C_t = 221[e^{-1.61t}] + 124[e^{-0.02t}] \text{ for trout (Uno } et\ al., \text{ 1997)} \tag{6}$$

$$C_t = 210[e^{-1.32t}] + 161[e^{-0.12t}] \text{ for yellowtail (Uno } et\ al., \text{ 1997)} \tag{7}$$

Carli *et al.* (1993) reported that the pharmacokinetics of sulfamonomethoxine after intravascular administration to horses could be evaluated by a two-compartment model. The pharmacokinetics of sulfadimethoxine in lobster (Barron *et al.*, 1988) and channel catfish (Michel *et al.*, 1990) after intrapericardial and intravenous administration were also described by two-compartment models. In carp (van Ginneken *et al.*, 1991) the pharmacokinetics of sulfadimidine after intravenous administration was analyzed by a two-compartment model.

Table 2 shows the pharmacokinetic parameters for sulfamonomethoxine following intravascular administration. Differences in serum $T_{1/2}$ for the distribution in eel, rainbow trout and yellowtail ($T_{1/2\alpha}$: 0.25, 0.43 and 0.53 h, respectively) after intravascular administration show that there were slight differences in the clearance of sulfamonomethoxine from the blood. However, significant differences in $T_{1/2}$ for the elimination ($T_{1/2\beta}$) and body clearance (Cl_B) were observed in eel, rainbow trout and yellowtail ($T_{1/2\beta}$: 86.6, 34.7 and 5.78 h and Cl_B: 3.38, 18.5, 66.7 ml/kg/h, respectively). Thus, eel had the slowest terminal elimination rate among the three species. In comparison, horse, after given a dose of sulfamonomethoxine exhibited $T_{1/2a}$, $T_{1/2\beta}$ and Cl_B values of 0.21 h, 3.6 h and 72 ml/kg/h, respectively (Carli *et al.*, 1993). Sulfonamides are weak organic acids of moderate liposolubility and they disperse in most body tissues with an apparent steady-state distribution (Vss) usually in the range 0.3 to 0.8 l/kg (Davitiyananda and Rasmussen, 1974; Rasmussen *et al.*, 1979). Similar Vss values were found for sulfamonomethoxine (eel, 0.45 l/kg; rainbow trout, 0.83 l/kg; yellowtail, 0.56 l/kg). The Vss values were greater for fish than those reported for pigs (0.341 l/kg) (Shimoda *et al.*, 1997). Keilnow *et al.* (1992) reported a Vss of 0.5 l/kg in rainbow trout that had been administered sulfadimethoxine.

Table 2. Pharmacokinetic parameters for sulfamonomethoxine following intravascular administration.

Parameters	Eel	Rainbow trout	Yellowtail
Dose (mg/kg)	200	100	100
Weight (g)	156	171	638
Water temp. (°C)	27	15	21
α (h^{-1})	2.74	1.61	1.32
β (h^{-1})	0.008	0.02	0.12
Ke (h^{-1})	0.025	0.061	0.25
K_{12} (h^{-1})	1.91	0.98	0.55
K_{21} (h^{-1})	0.82	0.59	0.64
$T_{1/2\alpha}$ (h)	0.25	0.43	0.53
$T_{1/2\beta}$ (h)	86.6	34.7	5.78
AUC (μg·h/ml)	59,100	5,400	1,500
MRT (h)	129	34.3	7.45
Cl_B (ml/kg/h)	3.38	18.5	66.7
Vss (l/kg)	0.45	0.83	0.56
Vc (l/kg)	0.14	0.29	0.27

α, β, Values related to the slopes of distribution and terminal phases, respectively, of biexponential drug disposition curve; Ke, Elimination rate constant; K_{12}, K_{21}, Distribution rate constant from between the central and peripheral compartment; $T_{1/2\alpha}$, $T_{1/2\beta}$, Distribution half-life and elimination half-life of drug; AUC, Area under serum concentration-time curve, Cl_B, Total body clearance; MRT, Mean residence time; Vss, Apparent steady state distribution; Vc, Apparent volume of central compartment.

4.2. Bioavailability, Mean Absorption time and Drug Absorption Rate

Bioavailability (F), expressed as the extent of bioavailability, may be calculated from the following equation:

$$F\ (\%) = [(\text{AUC p.o.} * \text{dose i.v.} \div (\text{AUC i.v.} * \text{dose p.o.})] * 100 \qquad (8)$$

where p.o. and i.v. represent the oral and intravascular administration, respectively.

The bioavailabilities of sulfamonomethoxine for eel, rainbow trout and yellowtail were 24, 19 and 16%, respectively. In general, the low bioavailability of the drugs is caused by incomplete absorption or by biotransformation in the intestinal wall or liver. However, Kleinow *et al.* (1992) reported that oral bioavailability of sulfadimethoxine in rainbow trout varied with the chemical form (salt versus free drug). Hara *et al.* (1967) also showed that the absorption of sulfamonomethoxine (free form) was low in rainbow trout. Therefore, low bioavailabilities in our experiments is due to the lower solubility of the free drug in water. However, catfish exhibited a similar bioavailability pattern with two chemical forms of sulfadimethoxine (Squibb *et al.*, 1988). For comparison, the bioavailability of sulfamonomethoxine was reported to be 84% in horse (Carli *et al.*, 1993).

The mean absorption time (MAT) or rate of bioavailability was calculated by the following equation (Yamaoka and Tanigawa, 1983):

$$\text{MAT} = \text{MRT p.o.} - \text{MRT i.v.} \qquad (9)$$

where MRT represents the mean residence time.

The values of MAT for eel, rainbow trout and yellowtail were 103, 36 and 0.5 h, respectively. Wagner and Nelson (1964) reported that drug absorption rate could be calculated from the serum level versus time data using the equation shown in Figure 3 when the behavior of the drug is expressed by a one-compartment model.

This equation relates the cumulative amount of drug absorbed after a certain time to the amount of drug ultimately absorbed, rather than to the dose administered. The value of Ke following intravascular administration was used for the calculation of Wagner and Nelson's equation. The time required for sulfamonomethoxine absorption (TDA, over 90% of the maximum absorption) was 600 h for eel, 120 h for rainbow trout and 12 h for yellowtail.

4.3. Protein Binding of Sulfamonomethoxine

The protein binding of a drug can influence the concentration of active drug in the blood and its distribution into the tissues. The percentages of sulfamonomethoxine that bound to serum protein *in vivo* were 6.4% for rainbow trout and 5.8% for yellowtail. Protein binding was constant over the concentration range tested at each sampling time point. In contrast, Shimoda *et al.* (1983) reported that the protein binding of sulfamonomethoxine was 60–65% in pigs. Kleinow and Lech (1988) and Michel *et al.* (1990) reported that the protein binding of sulfadimethoxine was 17.3% for rainbow trout and 18.2% for catfish. Baggot *et al.* (1976) reported that the protein binding of sulfadimethoxine was 80% in dogs. Therefore, most of the drug might be available for rapid transport to the tissues with little binding in the plasma of fish compared with mammals.

Figure 4 and Table 3 shows the absorption of sulfamonomethoxine in fish. The

$$\text{Fraction absorbed} = \frac{At}{A_\infty} = \frac{Ct + Ke\int_0^t Cdt}{Ke\int_0^\infty Cdt}$$

Figure 3. Equation of Wagner and Nelson's. At, Amount of drug absorbed to time t; A_∞, Amount of drug ultimately absorbed; Ct, Concentration at time t; Ke, Elimination rate constant.

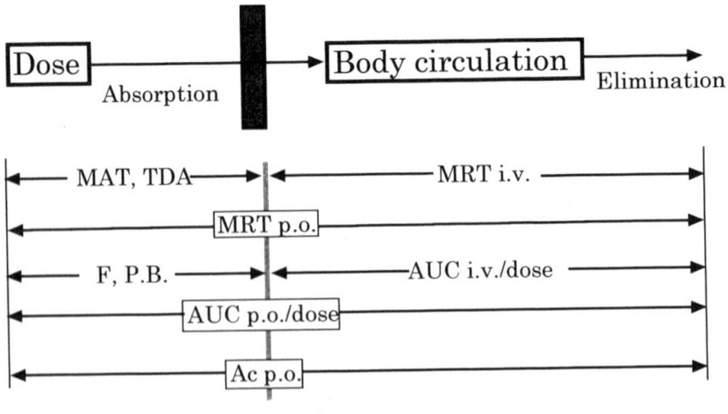

Figure 4. Absorption and elimination parameters of sulfamonomethoxine in fish; MAT, Mean absorption time. The left and right halves of the figure relate the absorption and elimination of the drug after administration, respectively. TDA, Time required for drug absorption; MRT, Mean residence time; F, Bioavailability; P.B., Protein binding; AUC, Area under the concentration-time curve; Ac, Acetylation; p.o., Oral administration; i.v., Intravascular administration.

Table 3. Absorption kinetics of sulfamonomethoxine in fish.

Parameters	Eel	Rainbow trout	Yellowtail
MRT i.v. (h)	129	34.3	7.45
MRT p.o. (h)	232	70.6	7.94
MAT (h)	103	36.3	0.45
TDA (h)	600	120	12
AUC i.v./Dose	276	54	15
AUC p.o./Dose	71.0	10.3	2.33
F (%)	24	19	16
P.B. (%)	—*	6.4	5.8
Ac p.o. (%)	3.8	49	72

*Not calculated.

MRT, Mean residence time; MAT, Mean absorption time; TDA, Time required for drug absorption, calculated by Wagner-Nelson's method; AUC, Area under serum concentration-time curve; F, Bioavailability; P.B., Protein binding; Ac, Acetylation; p.o., oral administration; i.v., intravascular administration.

absorption parameters of sulfamonomethoxine in eel were substantially different from those in rainbow trout and yellowtail. In eel, the drug is assimilated slowly and clears very slowly. This is in direct contrast to the data for sulfamonomethoxine in yellowtail, in which the drug is well distributed to the tissues and clears quickly. However, sulfamonomethoxine in eel is significantly more available than it is in the other fish, that is, the value of AUC/dose in eel was 7 times and 30 times higher than in rainbow trout and yellowtail, respectively.

5. EVALUATION OF DOSAGE REGIMENS

As an application of the pharmacokinetic studies the Japanese dosage regimens of sulfamonomethoxine in three fish species were evaluated. The serum level during multiple oral dosing of a constant dose (Cn) and average steady state serum concentration (Css) can be estimated according to the formula shown in Figure 5.

Figure 6 shows serum levels of sulfamonomethoxine after multiple oral administration of a constant dose given once a day in eel (200 mg/kg), rainbow trout (150 mg/kg) and yellowtail (200 mg/kg). In eel and rainbow trout, the serum levels of the drug were maintained above the MIC as explained in the previous section. That is, the MIC was rapidly exceeded and remained very high in the serum with repeated dosing. It seems to be an excessive dosing. Moreover, the high values of Css (540 μg/ml for eel, 62 μg/ml for rainbow trout) were calculated. However, yellowtail, fell below the MIC within 1 day (Css: 20 μg/ml). That is, yellowtail might be underdosed. As a result, it was suggested that the current dosage regimens of sulfamonomethoxine were not adequate for these three fish species, especially yellowtail. It has been well-known that an excess dosing cause not only the economical problem of fish farm but environmental pollution, and that an underdosing does the outbreak of drug-resistant bacteria. Further investigations are needed to compare these theoretical calculations with results obtained under actual conditions of the dosage regimens used on fish farms.

In conclusion, differences in the pharmacokinetics of sulfamonomethoxine can be attributed to species-dependent physiological differences in renal function, muscle constitution, enzyme activity, vascularization and so on (van Ginneken *et al.*, 1991). Moreover, Nielsen and Rasmussen (1977) suggested that drug pharmacokinetic profiles vary

$$Cn = \frac{F*Dose*Ka}{Vd(Ka-Ke)} \left[\frac{1-e^{(-n*Ke*t)}}{1-e^{(-Ke*t)}} e^{(-Ke*t)} - \frac{1-e^{(-n*Ka*t)}}{1-e^{(-Ka*t)}} e^{(-Ka*t)} \right]$$

$$Css = \frac{F*Dose}{Vd*Ke*\tau}$$

Figure 5. Equation for multiple dosing; Cn, Serum concentration during multiple oral dosing of a constant; Css, Average steady state serum concentration; Ka, First-order absorption rate constant; F, Bioavailability; Vd, Apparent volume of distribution; Ke, First-order elimination rate constant; τ, Dosage interval; t, Dosage time.

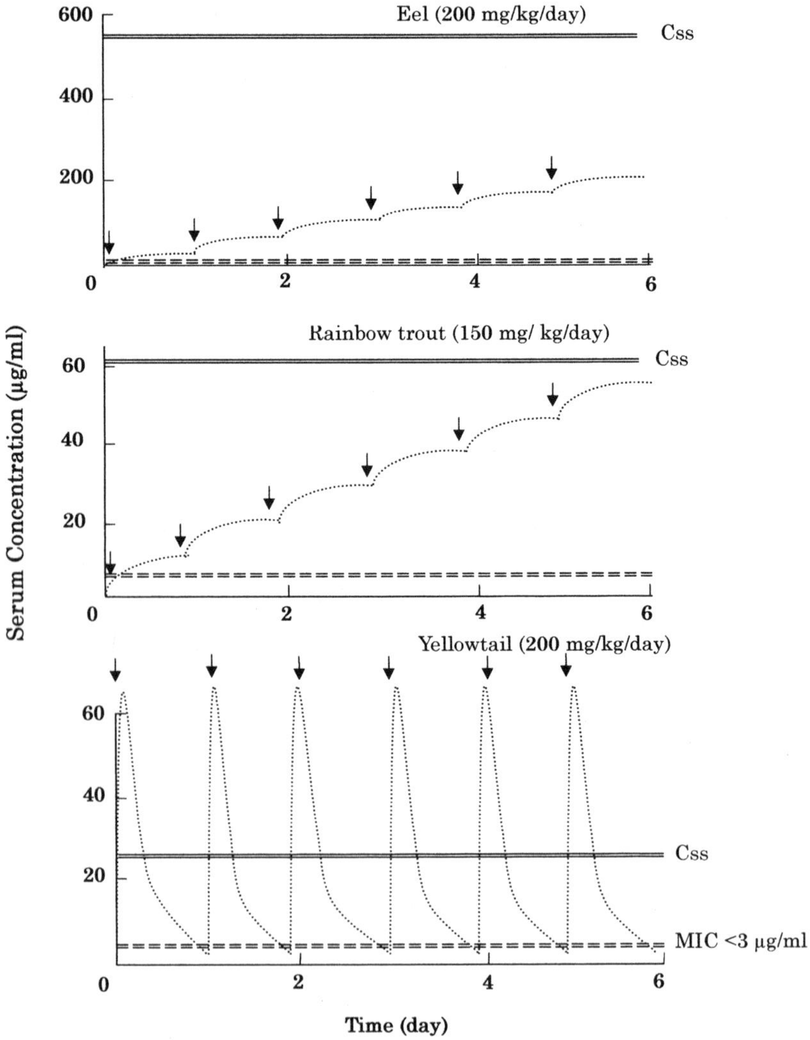

Figure 6. Predicted concentration of sulfamonomethoxine during multiple oral dosing in fish; Css, Average steady state serum concentration; MIC, Minimum inhibitory concentration.

from one animal species to another and it is not possible to predict elimination in one animal using results obtained from another. Therefore, pharmacokinetics of drugs used in fish might still be limited to the species studied.

6. REFERENCES

Baggot, J.D., Ludden, T.M., and Powers, T.E., 1976, The bioavailability, disposition kinetics and dosage of sulphadimethoxine in dogs, *Can. J. Comp. Med.*, 40:310.

Barron, M.G., Gedutis, C., and James, O.M., 1988, Pharmacokinetics of sulphadimethoxine in the lobster, *Hamarus americanus*, following intrapericardial administration, *Xenobiotica*, 18:269.

Bratton, A.C., and Marshall, E.K., 1939, A new coupling component of sulfanilamide determination, *J. Biol. Chem.*, 128:537.

Carli, S., Sonzogni, O., Villa, R., Bignazzi, R., and Montesissa, C., 1993, Pharmacokinetic profile of sulphamonomethoxine-trimethoprim in horses after intravenous, intramuscular and oral administration, *Res. Vet. Sci.*, 54:184.

Davitiyananda, D., and Rasmussen, F., 1974, Half-lives of sulfadoxine and trimethoprim after single intravenous infusion in cows, *Acta Vet. Scand.*, 15:356.

Droy, B.F., Goodrich, M.S., Lech, J.J., and Kleinow, K.M., 1990, Bioavailability, disposition and pharmacokinetics of ormetoprim in rainbow trout (*Salmo gairdneri*), *Xenobiotica*, 20:147.

Gibson, G.G., and Skett, P., 1986, *Introduction to Drug Metabolism*, Chapman and Hall, New York.

Grondel, J.L., Nouws, J.F.M., and Haenen, O.L.M., 1986, Fish and antibiotics: Pharmacokinetics of sulphadimidine in carp (*Cyprinus carpio*), *Vet. Immunol. Immunopathol.*, 12:281.

Hara, T., Inoue, S., and Saito, M., 1967, Concentration of sulfonamides in the tissue of rainbow trout after oral dosage of the sodium salts, *Bull. Jpn. Soc. Sci. Fish.*, 33:624.

Hara, T., and Inoue, S., 1967, Maintenance dose of tissue levels of sulfamonomethoxine in rainbow trout after oral administration, *Fish Path.* 1:35.

Hara, T., and Inoue, S., 1969, Blood levels of sulfamonomethoxine in rainbow trout after oral administration, *Fish Path.*, 3:9.

Ishida, N., 1989, Metabolites of five sulfa drugs in the bile and urine of rainbow trout, *Nippon Suisan Gakkaishi*, 55:2163.

Kleinow, K.M., Beilfuss, W.L., and Lech, J.J., 1987, Pharmacokinetics of sulfadimethoxine in free swimming trout. *Toxicologist*, 7:20.

Kleinow, K.M., and Lech, J.J., 1988, A review of the pharmacokinetics and metabolism of sulfadimethoxine in the rainbow trout (*Salmo gairdneri*), *Vet. Human Toxicol.*, 30:26.

Kleinow, K.M., Beilfuss, W.L., Jorboe, H.H., Droy, B.F., and Lech, J.J., 1992, Pharmacokinetics, bioavailability, distribution, and metabolism of sulfadimethoxine in the rainbow trout (*Oncorhynchus mykiss*), *Can. J. Fish. Aquat. Sci.*, 49:1070.

Matsushima, M., Matsubara, S., Ai, K., and Inoue, S., 1971, Toxicity of sulfa drugs in yellowtail, *Fish Path.*, 5:112.

Michel, C.M.F., Squibb, K.S., and O'Connor, J.M., 1990, Pharmacokinetics of sulphadimethoxine in channel catfish (*Ictalurus punctatus*), *Xenobiotica*, 20:1299.

Nielsen, P., and Rasmussen, F., 1977, Half-life, apparent volume of distribution and protein-binding for some sulfonamides in cows, *Res. Vet. Sci.*, 22:205.

Nouws, J.F.M., Vree, T.B., Baakman, M., and Tijhuis, M., 1983, Effect of age on the acetylation and deacetylation reactions of sulphadimidine and N^4-acetylsulphadimidine in calves, *J. Vet. Pharmacol. Ther.*, 6:13.

Rasmussen, F., Gelsa, H., and Nielsen, P., 1979, Pharmacokinetics of sulfadoxine and trimethoprim in horses. *J. Vet. Pharmac. Ther.*, 2:245–255 (1979).

Shimoda, M., Tsuboi, T., Kokue, E., and Hayama, T., 1983, Dose-dependent pharmacokinetics of intravenous sulfamonomethoxine in pigs, *Jpn. J. Pharmacol.*, 33:3903.

Shimoda, M., Okamoto, K., Sikazwe, G., Fujii, C., and Son, D.S., 1997, Deacetylation as a determinant of sulphonamide pharmacokinetics in pigs, *Vet. Q.*, 19:186.

Souich, P.D., McLean, A.J., Lalka, D., Jenkins, E.1., Haegele, K.D., and McNay, J.L., 1978, Sulfadiazine handling in the rabbit. 1. Pseudosaturation of *N*-acetyltransferase, *J. Pharmacol. Exp. Ther.*, 7:221.

Squibb, K.S., Michel, C.M.F., Zelikoff, J.T., and O'Connor, J.M., 1988, Sulfadimethoxine pharmacokinetics and metabolism in the channel catfish (*Ictalurus punctatus*), *Vet. Human Toxicol.*, 30:31.

Tanaka, J., 1977, *Expatiation on Fish Drugs*, Soft Science Inc., Tokyo.

Ueno, R., Horiguchi, Y., and Kubota, S.S., 1988a, Levels of oxolinic acid in cultured yellowtail after oral administration, *Nippon Suisan Gakkaishi*, 54:479.

Ueno, R., Okumura, M., Horiguchi, Y., and Kubota, S.S., 1988b, Levels of oxolinic acid in cultured rainbow trout and amago salmon after oral administration, *Nippon Suisan Gakkaishi*, 54:485.

Ueno, R., Uno, K., and Aoki, T., 1994, Pharmacokinetics of sulphamonomethoxine in cultured yellowtail after oral administration, *Food Research International*. 27:33.

Ueno, R., Uno, K., and Aoki, T., 1995, Pharmacokinetics and bioavailability of oxytetracycline in cultured yellowtail *Seriola quinqueradiata*, *Dis. in Asian Aquacul.*, 2:523.

Ueno, R., 1998, Pharmacokinetics and bioavailability of sulphamonomethoxine in cultured eel, *Fish Path.*, 33:000.

Uno, K., Aoki, T., and Ueno, R., 1992a, Pharmacokinetic study of oxytetracycline in cultured rainbow trout, amago salmon, and yellowtail, *Nippon Suisan Gakkaishi*, 58:1151.

Uno, K., Aoki, T., and Ueno, R., 1992b, Pharmacokinetics of nalidixic acid in cultured rainbow trout and amago salmon, *Aquaculture*, 102:297.

Uno, K., Aoki, T., and Ueno, R., 1993, Pharmacokinetics of sulphamonomethoxine and sulphadimethoxine following oral administration to cultured rainbow trout (*Oncorhynchus mykiss*). *Aquaculture*, 115:209.

Uno, K., Aoki, T., and Ueno, R., 1993, Pharmacokinetics of sodium nifurstyrenate in cultured yellowtail after oral administration, *Aquaculture*, 116:331.

Uno, K., Aoki, T., Ueno, R., and Maeda, I., 1997, Pharmacokinetics and metabolism of sulphamonomethoxine in rainbow trout (*Oncorhynchus mykiss*) and yellowtail (*Seriola quinqueradiata*) following bolus intravascular administration, *Aquaculture*, 153:1.

van Ginneken, V.J.T., Nouws, J.F.M., Grondel, J.L., Driessens, F., and Degen, M., 1991, Pharmacokinetics of sulphadimidine in carp (*Cyprinus carpio* L.) and rainbow trout (*Salmo gairdneri* Richardson) acclimated at two different temperature levels, *Vet. Q.*, 13:88.

Vree, T.B., O'Reilly, W.J., Hekster, Y.A., Damsma, J.E., and vander Kleijin, E., 1980, Determination of the acetylator phenotype and pharmacokinetics of some sulfonamides in man, *Clin. Pharmacokinet.*, 5:274.

Vree, T.B., Hekster, Y.A., Damsma, J.E., Tijhuis, M., and Friesen, W.T., Pharmacokinetics and mechanism of renal excretion of short acting sulfonamides and N^4-acetylsulfonamide derivatives in man, *Eur. J. Clin. Pharmacol.*, 20:283.

Vree, T.B., Reekers-Kitting, J.J., Hekster, C.A., and Nouws, J.F.M., 1983, Acetylation and deacetylation of sulfonamides in dogs, *J. Vet. Pharmacol. Ther.* 6:153.

Vree, T.B., BenekenKolmer, E.W.J., Martea, M., Bosch, R., Hekster, Y.A., and Shimoda, M., 1990a, Pharmacokinetics, N^4-glucuronidation and N^4-acetylation of sulfadimethoxine in man, *Pharm. Weekbl. Sci.*, 12:51.

Vree, T.B., BenekenKolmer, E.W.J., Martea, M., Bosch, R., and Shimoda, M., 1990b, High-performance liquid chromatography of sulphadimethoxine and its N^4-glucuronide, N^4-acetyl and N^4-acetyl- N^1—glucuronide metabolites in human plasma and urine, *J. Chromatogr.*, 526:119.

Wagner, J.G., and Nelson, E., 1964, Kinetic analysis of blood levels and urinary excretion in the absorptive phase after single doses of drug, *J. Pharm. Sci.*, 53:1392.

Yamaoka, K., Nakagawa, T., and Uno, T., 1978, Application of Akaike's information criterion (AIC) in the evaluation of linear pharmacokinetic equations, J. *Pharmacokinet. Biopharm.*, 6:165.

Yamaoka, K., Tanigawara, Y., Nakagawa, T., and Uno, T., 1981, A pharmacokinetic analysis program (MULT) for microcomputer, *J. Pharmacobio-Dyn.*, 4:879.

Yamaoka, K., and Tanigawara, Y., 1983, Statistical moments, In: *Pharmacokinetics Using Personal Computer*, Nankoh-doh Press, Tokyo.

INDEX